ACB 9956

D0208095

Philosophy of **Science**

The Historical Background

Edited by
Joseph J. Kockelmans

Transaction Publishers
New Brunswick (U.S.A.) and London (U.K.)

Second paperback edition copyright © 1999 by Transaction Publishers, New Brunswick, New Jersey. Originally published in 1968 by The Free Press.

This book is printed on acid-free paper that meets the American National Standard for Permanence of Paper for Printed Library Materials.

Library of Congress Catalog Number: 99-11434
ISBN: 0-7658-0602-9
Printed in the United States of America

Library of Congress Cataloging-in-Publication Data

Philosophy of science : the historical background / Joseph J. Kockelmans, editor.
    p. cm. — (Science and technology studies)
    Originally published: New York : Free Press, 1968.
    Includes bibliographical references and index.
    ISBN 0-7658-0602-9 (paper)
    1. Science—Philosophy. I. Kockelmans, Joseph J., 1923–   II. Series.
Q175.P51294   1999
501—dc21
                                      99-11434
                                         CIP

CONTENTS

PREFACE

Some excellent anthologies on *Philosophy of Science* have been published in recent years. They present essays and book selections on a variety of subjects generally considered to be basic for an understanding of contemporary philosophy of science. There are also some studies available in which philosophy of science is approached historically, but these are relatively few in number and do not include longer selections from original sources. To the best of my knowledge, no attempt has been made until now to present an anthology of original texts compiled for the exclusive purpose of providing immediate contact between the reader and philosophy of science's unique historical background. Such has been my goal for this anthology because I am convinced that without a thorough knowledge of the historical development of the discipline it is virtually impossible to understand genuinely and to be able to make an objective evaluation of the various trends in contemporary philosophy of science. Admittedly a thorough knowledge of the history of philosophy of science will not be acquired solely by reading an anthology, this or any other. Some knowledge of science and its history as well as a substantial familiarity with major philosophical views of the past should be present. Nevertheless it is my conviction that what the reader is about to encounter here constitutes a mighty inroad into a complicated territory, partly because it presents original material not readily accessible through ordinary channels, and partly because it orders and integrates the on-going development of thinking in the vital area of philosophy of science over years of productive efforts by the greatest minds of the past three centuries.

I had another goal in mind for this anthology, too. Those who have taught a course on philosophy of science at the undergraduate level may have experienced certain difficulties stemming from the fact that many students who are interested in the subject may, however, lack either a profound insight into philosophy or a clear idea of science and its development. In this situation I believe this anthology could render a service to all concerned in that the selections do not presuppose more than a general insight into classical physics and chemistry. Those who desire additional background material or explanations of basic philosophical issues may find them easily in a good history of philosophy or through class discussion.

I have limited myself to the natural sciences in this anthology not only because of a wealth of material versus space restrictions, but also because philosophical reflections on the psychological and social sciences did not begin to appear until much later than most of the studies which appeared in the time period covered here. Even within the bounds chosen for the anthology many outstanding and appropriate issues and studies were not included; in some instances length was the prohibitive factor, in others availability; in certain cases it was impossible to isolate a section or chapter of a book without rendering it unintelligible. At all times it was a delicate matter to make final decisions about the authors to be represented, and particularly to choose one piece of an author's work over another. In retrospect it is my belief that the anthology gives an objective idea of the origin and development of what is called philosophy of science. A more specified justification of the choices made will appear in the Introduction.

In introducing the selections I have tried to keep the remarks at a minimum so that the space could be devoted to the original material itself. I wished only to familiarize the reader in the sense of a "briefing" with the author about to be encountered, and to mention the insights which the reader would find presupposed in the body of the selection. Whoever wants additional information will hopefully find a source in the selective bibliography which is listed for each author. Here again the introductory character of the anthology dictated a reasonable minimum.

All the selections in this anthology have been reprinted exactly and without changes except for correction of a few typographical errors and obvious mistakes. As a consequence there are variations in spelling throughout the selections and the footnote policy is not always uniform. Where it seemed to be necessary to delete part of the original text I have so indicated in the usual way (. . .).

Grateful acknowledgment is made to the publishers, authors, and translators represented here for their generosity in granting permission to reprint selections from copyright material and for their kind cooperation throughout the work. The author expresses his sincere appreciation to Mr. H. M. McConnell, Assistant Director of The Free Press, for endorsing the project, and for his consistent cooperation from beginning to completion.

Department of Philosophy   *Joseph J. Kockelmans, Ph.D.*
University of Pittsburgh

Introduction

Writing on the history of the philosophy of science is a project which from the outset involves a number of difficult problems, even when the study is limited to philosophy of the natural sciences, as is the case here.

First there is the crucial question of what is to be understood by the expression "philosophy of science." The answer to this question has direct bearing on the localizing of the historical origin of philosophy of science as "earlier" or "later." To give an example, if one were to adopt the viewpoint that philosophy of science was a specialized part of analytic or phenomenological philosophy, then its historical origin could not be placed prior to the origin of analytic philosophy or phenomenology themselves. If on the other hand it is seen as a logical analysis which clarifies the basic ideas and methods of the natural sciences, then its genesis must have been sometime during the nineteenth century. However, if one were to define philosophy of science as a philosophical reflection upon the principles, methods, and results of the natural sciences, then in all likelihood it would have had its beginnings in the mostly implicit self-reflections of our oldest scientists, as well as in the philosophical works of a Bacon, Descartes, Hume, Leibniz, or others of their stature. But finally, if it were to be defined as *any* philosophical reflection upon science, then perhaps its genesis could be traced to the post-Aristotelian philosophy, with its numerous treatises on the "division of the sciences."

One conclusion seems to be evident here, namely, that it would be completely unacceptable for an historian to decide a priori what philosophy of science is supposed to be and how it is to be conceived in greater detail. If the historian is permitted such a conception, then it ought to represent the results of his historical investigations, rather than his arbitrarily predecided starting point.

On the other hand it is also evident that the historian cannot begin his work without a certain notion of what philosophy of science is supposed to be. But in this regard it seems to me to be vitally important that in making his provisional decision he tries to avoid unnecessary limitations based on arbitrary preconceptions. In compiling this anthology I followed guidelines which do not, I believe, imply an undesirable one-sidedness; my rationale was along the following lines. First of all, generally speaking it is not correct to say that the history of the philosophy of science must coincide with the history of

1

the sciences themselves, or with the history of philosophy as a whole. For one reason, there cannot be any doubt that philosophy was able to, and in fact did, develop for many centuries and in many particular modes without any direct interest in or influence from science. Secondly, a denial of this thesis would immediately confront us with the extremely complicated question concerning the origin of the natural sciences. The view that the natural sciences just fell from a blue sky in the seventeenth and eighteenth centuries appears to be historically incorrect. Although indeed Kepler, Galileo, Newton, Huygens, Boyle, and many others substantially contributed to the development and systematization of the natural sciences, their genuine origin can nonetheless be traced to the late Middle Ages, if not to the time between Archimedes and Ptolemy. Whatever else one thinks about the genesis of natural science, in my opinion it is wise to maintain the thesis that there could be no genuine philosophy of science in the proper sense as long as the sciences in question were not yet constituted *and sufficiently developed.*

Secondly, one may state without danger of one-sidedness that philosophy of science consists in an *explicit* reflection upon, or analysis of, science, and that this reflection or analysis must be genuinely philosophical. Here the expression "genuinely philosophical" should be understood in the broad sense intended by those commonly considered to be "the great" philosophers of the past.

Adopting these views, namely, that philosophy of science in the strict sense of the term presupposes a science which has already reached a certain form of perfection, and also that it must consist of an *explicit* philosophic reflection upon these sciences, leads immediately to the conclusion that the historical origin of philosophy of science is to be found not earlier than the latter part of the eighteenth century. This is not to say that it is unimportant to make a thorough study of the implicit self-reflections upon their own work, principles, methods, and results of scientists such as Kepler, Galileo, Newton, and others. Nor does it follow from this view that a study of the philosophies of Bacon, Descartes, Leibniz, Hume, and others is not of utmost importance for a genuine understanding of the history of the philosophy of science. For as the reader will note, the first who attempted a systematic philosophy of science were not only quite familiar with the ideas of the scientists and philosophers just mentioned but in many cases explicitly took their starting point from their works.

There is still another point to be carefully considered by the historian who wishes to avoid onesidedness. He must realize that the natural sciences constitute a complex realm that is connected in manifold ways with other realms of man's experience. For this reason a philosopher can reflect upon science from many different points of view, e.g., logical, methodological, epistemological, ontological, "social," or the generally cultural aspects of science. These various perspectives from which philosophers of the past have looked at the natural sciences have frequently been determined largely by the general outlook they had regarding philosophy's task and meaning. Certainly which of the perspectives and which of the general philosophical views is to be considered the most important is an issue which can be decided only at the end of an historical investigation and not claimed in an a priori way beforehand.

More specifically, for an empiricist philosopher, philosophy of science will consist substantially in a radical analysis of fundamental concepts and basic methods in science. For Kant, on the other hand, philosophy of science is to be concerned with the a priori conditions which make an enterprise like natural science possible at all. For Hegel, finally, philosophy of science consists in a radical attempt to understand the final meaning of science within the general perspective of an absolute knowledge of the Absolute in regard to its own dialectical development from Idea, via Nature, to objective Spirit. In my opinion the historian as such has no right to decide in advance that some of these views were meaningless and therefore, *for that reason*, could be passed over in an account of the historical development of philosophy of science. If he wishes to set limits and omit certain views, then he should do so on other, legitimate grounds. Among the latter, practical considerations are very often the most pressing. And this brings me to a further theme.

Because of the staggering amount of material available as contrasted with the limited number of pages at my disposal, I have attempted to delineate the subject matter of the anthology as reasonably as possible by staying within the bounds of philosophy of science as such, thereby excluding all information about the historical development of philosophy proper, and also of the sciences. Much could be said for the view that any good history of the philosophy of science should include a thoroughgoing survey of the basic principles of the philosophical trends and views which are presupposed in the various conceptions of the philosophy of science: it is, in fact, a view which I wholeheartedly endorse. Nevertheless, I did not follow it because a number of excellent books on the history of philosophy are available; in these the reader can easily find whatever information he needs concerning the many philosophical "systems" developed over the past four centuries. In addition, basic texts of the great philosophers are available in inexpensive editions, so that anyone who wishes to concentrate on a particular author or philosophical school has ready access to the material needed. And finally regarding the use of this anthology in a philosophy course, I think that in a sound course on the history of the philosophy of science, the instructor will find many opportunities to acquaint the students with the basic principles of the underlying philosophies involved. Analogous considerations pertain also to the history of the natural sciences, many excellent introductory works having been published in the last two decades.

It seems to me that a similar restriction is possible for logical and methodological investigations, too, without a real risk of nearsightedness. It is an undeniable fact that the history of the philosophy of science overlaps to a great extent the history of logic and general methodology; it would not therefore be difficult to "fill" a few anthologies with important texts on logic and general methodology. One might even hope for such a book in the future, so that in these areas, also, basic texts would be more easily accessible to everyone. In my schema I have limited my choices to texts which were written in immediate contact with the natural sciences. Where I have chosen certain texts with a clearly logical or methodological content, it was because the authors in question developed their views precisely in connection with the natural sciences and their underlying philosophy. This is why texts from Mill, Peirce, Broad, and others appear here, but texts from authors such as Bolzano, Boole, De Morgan, Frege, Zermelo, Hilbert, and many others do not.

I should expect that many readers will be amazed at the absence of texts from Hegel, Marx, Comte, Lotze, Nietzsche, and Bergson, among others. I must confess that it was very difficult for me to make a decision but I eventually decided against including their work on the grounds that these authors tried not so much to contribute something substantial to the philosophy of science as to found their philosophical views on science, or as was generally the case, to interpret the highly important phenomenon "science" from the general philosophical outlook they had adopted on other counts. If these reasons are still not acceptable to the reader, then I must point again to the space limitations as well as the fact that the views of the authors not represented here will be found in anthologies dealing with the history of nineteenth-century philosophy.

For the rest may I say that in most cases my choice was determined by the idea that a history of the philosophy of science ought to shed light on all the important trends, schools, and movements within this realm of philosophy, even though some of these trends have been shown to be relatively unfruitful.

Another problem to be solved along the way was the demarcation of the line between "past" and "present." I am convinced that the explicit philosophical reflection on the theory of relativity and quantum mechanics constitutes the essential part of our contemporary philosophy of science, and so I decided to exclude all papers which have as their central theme one of the issues immediately connected with these theories. I felt I could do so without fear of incompleteness in that most of the relevant "older" papers on these topics can be found in anthologies dealing substantially with the contemporary scene. However, I did take papers of authors who also dealt with these theories, but have chosen essays which do not explicitly deal with, nor presuppose, the theories. My strongest reason for adopting this viewpoint is that evolution and development is still proceeding in these areas, and it is not the task of the historian to report on events which have not yet reached their relative completion.

This was my reason, also, for not including selections written by phenomenologists and authors influenced by the philosophy of Wittgenstein. As far as phenomenology is concerned, most of the important essays and studies on philosophy of science were written after 1927, the date I chose as the upper limit for this survey of the history of the philosophy of science. This is obviously true for Heidegger, Merleau-Ponty, and Suzanne Bachelard, and it is even true for thinkers such as Husserl and Roman Ingarden. To be sure, some interesting papers were published before 1927, but I was not able to choose a few selections capable in themselves of giving a real idea of phenomenology's contribution to contemporary philosophy of science. For this reason, it seemed better to leave the realm completely untouched, the same applying to studies written under the influence of Wittgenstein's later philosophy.

In conclusion, I readily agree with any reader that other choices could have been made in many instances. Nevertheless I believe that the anthology as it stands now gives a rather objective idea of some important events in the history of philosophy of science.

SELECTIVE BIBLIOGRAPHY

To the best of my knowledge there is no book in the English language which systematically treats the history of philosophy of science. There are, however, some important essays dealing with certain aspects and periods in philosophy of science's history; some of them are included in the general bibliography that follows. For further bibliographic information about the authors whose works are represented here, refer to sources listed after the introduction to each selection.

Alexander, P., "The Philosophy of Science: 1850–1910" in D. J. O'Connor, ed., *A Critical History of Western Philosophy*. New York, 1964. Pp. 402–425.

Bavink, B., *The Natural Sciences*. New York, 1932.

Burtt, E. A., *The Metaphysical Foundations of Modern Science*. New York, 1932.

Dampier, W., *A History of Science and Its Relations to Philosophy and Religion*. Cambridge, 1948.

Dingle, H., *The Scientific Adventure. Essays in the History and Philosophy of Science*. New York, 1953.

Enriques, F., *Problems of Science*. Chicago, 1914.

Frank, Ph., *Modern Science and Its Philosophy*. New York, 1949.

Jörgenson, J., *The Development of Logical Empiricism. International Encyclopedia of Unified Science*, Vol. II, no. 9. Chicago, 1951.

Madden, E. H., ed., *Theories of Scientific Method: The Renaissance Through the Nineteenth Century*. Seattle, 1960.

———, ed., *The Structure of Scientific Thought*. Boston, 1960.

Nagel, E., *The Structure of Science*. New York, 1961.

In addition to the selections which are reprinted in this anthology, some other selections, important from an historical point of view, can be found in the following anthologies:

Danto, A. and Morgenbesser, S., eds., *Philosophy of Science*. New York, 1960.

Feigl, H. and Brodbeck, M., eds., *Readings in the Philosophy of Science*. New York, 1953.

Wiener, P. P., ed., *Readings in Philosophy of Science*. New York, 1953.

Part I
THE BEGINNING:
1786-1850

Chapter 1
IMMANUEL KANT
(1724–1804)

INTRODUCTION

*I*MMANUEL KANT was born in Königsberg on April 22, 1724. He entered the Collegium Fridericianum for his secondary education in 1732, then eight years later registered as a theology student at the University of Königsberg. In a short time, however, the preparatory sciences (natural science and philosophy) had so appealed to him that he decided to become a philosopher. He finished his course work in 1748, but owing to financial difficulties delayed taking his degree, and worked instead as a private tutor for several families in his neighborhood. Finally in 1755 he presented his thesis *De Igne* and was graduated; later that year his dissertation, entitled *Primorum cognitionis metaphysicae principium nova dilucidatio*, qualified him for *Privatdocent*. Kant taught as *magister legens* at the University of Königsberg for fifteen years before becoming a full professor of philosophy in 1770. After a span of some twenty-six years of regular teaching, he ceased lecturing in 1796 and devoted his attention completely to his writing. He died on February 12, 1804, after a life of study, teaching, and writing; he had not married, traveled, or been involved in political or other activities, preferring to the exclusion of all else the life of a scholar.

Kant was educated by Martin Knutzen in the philosophical tradition of Leibniz and Wolff; thus his earlier publications were written within this general perspective. Between 1762 and 1770 English empiricism influenced his thinking to an extent quite evident in his major writings of that period. In his inaugural address, *De mundi sensibilis atque intelligibilis forma et principiis* (1770), a new, highly personal element began to manifest itself. In this lecture it became clear that Kant thought he had found a means of transcending Leibniz' dogmatism as well as Hume's skepticism. However, these ideas were destined to ferment for twelve years as Kant was gradually formulating his critical solution, described at length for the first time in his famous *Critique of Pure Reason* (1781). During the fifteen years which followed, Kant revised this work and proceeded to apply it to the various realms of man's experience, the physical sciences, ethics, aesthetics, history, religion, and politics. During the last eight years of his life, he worked on a restatement of his philosophy; the notes which were meant as material for a revision of certain major parts of his philosophical "system," predominant among them his investigations concerning the metaphysical principles of the natural sciences, were published posthumously by Adickes in 1920 under the title *Kant's Opus Postumum*.

In carefully analyzing and comparing the dogmatic rationalism defended in the Leibnizo-Wolffian tradition and skeptic empiricism as proposed by Hume, Kant gradually came to the conclusion between 1770 and 1781 that metaphysics, understood as the speculative study of suprasensible realities, that is, as demonstrated doctrine about God, human freedom, and immortality, was impossible. In Kant's view the main concern of the philosopher had to be the question of whether such a metaphysics was possible at all. "My purpose is to convince those who find it worth their while to occupy themselves with metaphysics: that it is absolutely necessary to suspend their work for the present, to regard everything that has happened hitherto as not having happened, and before all else first to raise the question: 'whether such a thing as metaphysics is possible at all'." (*Prolegomena*, Preface).

A thoroughgoing study of this question led Kant to the position that classical metaphysics is impossible because it transcends the limits of man's possible knowledge. This however does not mean that *all* metaphysics is impossible. By showing the impossibility of the rationalist metaphysics Kant attempted also to lay firm foundations for another type of metaphysics, hoping to transcend Hume's skepticism. The new metaphysics was to be a careful and critical investigation of those rational principles and conditions which man must necessarily employ and observe within experience, rather than outside it, whether this experience be our experience of physical things, or our moral, political, or religious experience.

In the *Critique of Pure Reason* Kant comes to the conclusion that our knowledge is to be restricted to the knowledge of phenomena. In restricting the cognitive use of the categories of our understanding to phenomena, it follows for Kant that we cannot know the things as they are in themselves independent from their appearance to man; we cannot know noumena in the sense of knowing their essential characteristics; neither are we entitled to assert dogmatically *that* they are. With respect to physical reality this position entails that all forms of philosophy of nature in the traditional sense of the term be excluded. The question concerning the essence of material things becomes here a meaningless one. However there are empirical sciences of nature whose subject matter does not consist in nature as it *is in itself,* but in nature insofar as it, as phenomenon, is given in experience. Thus it follows that it is and remains meaningful to ask about the a priori conditions which are to be fulfilled by a datum in order that it may be a valid subject of investigation for a particular empirical science. But in this case, too, the question concerning the very *essence* of that datum, taken as it allegedly is in itself, remains completely out of consideration. In Kant's view it was the task of his *Metaphysical First Principles of Natural Science* to trace those a priori conditions which must be fulfilled by inorganic material things, taken exclusively insofar as they manifest themselves in our experience as phenomena, to be possible objects of our empirical investigations in the natural sciences.

Some ten years after the publication of this book Kant began to wonder whether or not he could further develop the argument presented in the *Metaphysical First Principles of Natural Science,* namely that, starting from the categories of our understanding and one single empirical concept, matter, the basic laws of physical science could be deduced. He thought it possible at that time to construct a priori not merely the general outline of the science of nature but many of its details as well. By making extensive use of the concepts of force and ether, he attempted to elaborate the construction he had in mind. It is quite probable that in so doing he became increasingly aware of the necessity of reconstructing his whole theory of knowledge along lines which would make the "transition from metaphysics to physics" plausible. In view of the fact that an understanding of the relationship between Kant's books *Metaphysical Foundations* and *Opus Postumum* is vital for an understanding of his position in regard to philosophy of science in general, some remarks on the subject are relevant here.

Kant was convinced in 1786 that, in regard to corporeal nature, the *Metaphysical First Principles of Natural Science* which he had just published materialized at least in principle what he referred to in the *Critique of Pure Reason* as "Metaphysics of Nature" (*Ib.,* pp. 14 and 37). There is no doubt that at that time he intended his book title to suggest more than simply a treatment of elementary principles of a metaphysics of the natural sciences; in fact the book was meant to deal with the *metaphysical* principles of natural science in an exhaustive way (*Ib.*, p. 13). There is no indication in the book that the application of the principles to empirical laws of nature and the multiplicity given in experience might somehow involve complications.

In the *Metaphysics of Morals* (1796) it becomes evident that in this case, as well as in a metaphysics of natural science, there is a gulf between the a priori principles brought to light in the two branches of metaphysics, and the concrete data of experience which constitute the content of our moral consciousness and the immediate subject matter of natural science. In this event, according to Kant, there is need for a new science whose main task it will be to bridge this gulf (E. Adickes, *Kant's Opus Postumum,* pp. 156–158).

From 1796 until 1803, one year before his death, Kant was almost exclusively engaged in preparing material for a work that was to be entitled: *On the Transition from the "Metaphysical First Principles of Natural Science" to Physics.* His unfinished "manuscripts" have been published as *Opus Postumum.* The work includes different collections of notes and short essays, some of which are of an exclusively epistemological and metaphysical character; most of the collections, however, deal with problems connected with a new approach to philosophy of science. The latter collections are pertinent here; undoubtedly the others were written with some of the problems in the last part of the *Critique of Pure Reason* in mind.

The sketches and notes immediately concerned with the projected study "On the Transition" deal for the most part with a large-scale ether theory which is connected with a separate treatise entitled "On the Elementary System of Motor Forces." In the latter Kant takes up in succession the most characteristic properties of matter, such as ponderability, states of aggregation, drop shape, capillarity, cohesion, friction, metallic luster, and so on. However, neither in the title of the work, nor in the manner in which Kant introduces the new discipline within the realm of related sciences, is there any mention of, or reference to, an ether theory. Kant describes the new science merely as one which must build a bridge to span the gulf between the *Metaphysical First Principles of Natural Science* and physics by enlarging the domain of the transcendental and aprioric considerations concerning the concept of matter (Adickes, p. 155).

As might be expected in a collection of notes pertaining to the first draft of a systematic work, there is much repetition. Also, whereas some ideas are dealt with quite adequately, many others are only mentioned in passing. And it is not always simple to grasp the exact meaning of Kant's statements, nor to decide which of his many conflicting views is his final one. Study of the chronology of the notes by Adickes and others has solved some of the interpretation difficulties, but in many instances it is not possible to determine how Kant would have developed his thoughts—which ideas he would finally have maintained, discarding others, and how, ultimately, he would have reconciled various conflicting points of view.

The *Metaphysical First Principles of Natural Science* deals with the concept of matter as that which is subject to motion in space, and with its laws insofar as all of this can be determined with the help of pure concepts and a priori principles. The meaning and goal of these investigations was to give to physics a rigorously scientific, that is, systematic character. For physics, as the study of matter's moving forces and their laws, is a merely empirical science, and as such unable to transcend the character of an aggregate of observations which are arbitrarily pieced together. Or as Kant later expresses it, physics is concerned with the laws of the moving forces of matter insofar as they are given in experience, and as such its methods, observation and experiment, can yield only empirical concepts out of which a genuine system can never be framed (*zimmern*). That is why in its historical development physics has come to fragmentary results only, although it is true that these were important and of far-reaching consequences. Kant maintains, however, that notwithstanding its grandiose success physics is still lacking a stable goal, with the result that its investigations are not guided by one idea, a whole which is founded from within and which clearly delimits itself (*Op. Post.*, Vol. XXI, pp. 161–162, 526; Vol. XXII, p. 497). There is only one way to develop the science "physics" in a truly systematic way and that is by founding it upon a plan which a priori comprises the whole and into which the individual facts and a posteriori laws can be integrated.

What Kant appears to be looking for here is a general over-all schema of physics, a systematic set of anticipations and conditions for the empirical investigation of nature. Mere empirical observation of the moving forces of nature cannot be called physical *science*. As a science, physics involves systematization and cannot be a mere aggregation of empirical data. But systematization takes place according to a priori principles which must furnish the guiding clues for empirical investigation, says Kant. For, from empirical intuition one can take nothing but what we ourselves have put there for physics

(*Ib.*, Vol. XXII, p. 323). Therefore, there must be a priori principles according to which moving forces are coordinated in relation to one another (that is, according to the formal element), while these forces in themselves (that is according to the material element or the object) are considered empirically (*Ib.*, Vol. XXI, p. 291). In the *Metaphysical First Principles of Natural Science* it becomes clear that one can deduce *some* definite truths merely a priori. But in addition to these there are also problematic anticipations of the empirical investigation of nature through which we know that this *or* that is the case, though experience alone can tell us which is the case.

In other words in Kant's view, physics, considered as concerned with the laws of the moving forces of matter as given in experience, presupposes something corresponding to a schematism of the a priori concepts brought to light in the *Metaphysical First Principles of Natural Science,* and which can constitute a bridge between the a priori concepts and the empirical representations (*Ib.*, Vol. XXI p. 525). The transition from the one science to the other must involve certain mediating concepts (*Zwisschenbegriffe*) which are given in the former and applied to the latter, and which belong both to the domain of the one and that of the other. Otherwise this transition would be, not a legitimate transition, but "a leap, in which one does not know where one will arrive, and after which, when one looks back, one does not really see the point of departure" (*Ib.*, pp. 525–526; Adickes, *op. cit.*, pp. 157–158).

Some scholars have envisioned mathematics as able to present physics with these necessary principles. Not so Kant, who was convinced that there cannot be mathematical principles of natural science because mathematics is unable to furnish physics with any insight into the essence and multiplicity of physical forces. Mathematics itself is not physics, nor is it a canon of physics, but rather a very powerful organon or instrument for use in investigations dealing with movements and their laws. Motions as such can be adequately explained mathematically because motion involves only spatial and temporal concepts which can be represented a priori in pure intuition. Moving forces, however, which act as causes of motions, and without which physics as a systematic science is inconceivable, need philosophical principles (*Op. Post.*, Vol. XXI, p. 209). It is in this connection that Kant inserts his polemic against the title of Newton's main work, *Philosophia naturalis principia mathematica.* In his view this title is absurd because it is equally impossible to speak about mathematical principles of philosophy as to talk about philosophical principles of mathematics (*Ib.*, pp. 204–205, 207–208). Mathematics is unable to show and explain the original forces characteristic of nature; presupposing these forces it merely describes and explains motions of things and the laws governing them. That is why the systematization of physics can be only the task of philosophy (*Ib.*, p. 163).

In this kind of metaphysics of natural science Kant presupposes that matter is known to us only via its affections upon our senses. These affections, however, are impossible without certain movements which, in turn, presuppose moving forces. Whereas in the *Metaphysical First Principles of Natural Science* the concept of matter plays the predominant role, in the *Opus Postumum* the concept of moving force, or more aptly matter insofar as it contains moving forces, plays the important part. This is why the "Science of the Transition" is so well suited to leading us from a pure metaphysics of natural science, in which the only predicate of matter permitted was that of "what moves in space," to physics; for the concept of "matter insofar as it contains moving forces" is a concept which is merely empirical on the one hand, but an a priori concept as well on the other. All these factors taken into account, it becomes clear why the treatise "On the Elementary System of Motor Forces" occupies such an important place in the *Opus Postumum*, and also why a study of ether appeared to be necessary to avoid action at a distance, in Kant's view absolutely impossible (Adickes, *op. cit.*, pp. 159–162, 363–365).

In view of the fact that, as far as philosophy of science as such is concerned, there is (*from our contemporary view*) no essential difference between the "Metaphysical Principles" and the "Science of the Transition," insofar as in both cases philosophy of science is conceived of as a philosophical discipline which is necessary for the *foundation* of physics as a *systematic* science, it was decided to present a selection from the first book only; however, the entire issue was deemed of sufficient importance to warrant at least a brief indication of the change in perspective from the first to the second book.

SELECTIVE BIBLIOGRAPHY

Kant, I., *Critique of Pure Reason*, trans. N. Kemp Smith. London, 1933.

———, *Prolegomena to Any Future Metaphysics That Will Be Able to Present Itself as a Science*, trans. P. G. Lucas. Manchester, 1953.

————, *Opus Postumum. Gesammelte Schriften* (herausgegeben von der Preussischen Akademie der Wissenschaften, 22 vols.). Berlin, 1902–1942. Vols. XXI and XXII.

————, *Kant's Prolegomena and Metaphysical Foundations of Natural Science,* trans. E. B. Bax. London, 1883.

Adickes E., *Kant's Opus Postumum.* Berlin, 1920.

————, *Kant als Naturforscher* (2 vols.). Berlin, 1924–1925.

Smith, N. Kemp, *A Commentary to Kant's "Critique of Pure Reason."* New York, 1962.

Lindsay, A. D., *Kant.* London, 1934.

Martin, G., *Kant's Metaphysics and Theory of Science,* trans. P. G. Lucas. Manchester, 1955.

Warnock, G. J., "Kant," in *A Critical History of Western Philosophy,* ed. D. J. O'Connell, New York, 1964. Pp. 296–318.

Ewing, A. C., *A Short Commentary on Kant's Critique of Pure Reason.* Chicago, 1950.

Weldon, T. D., *Introduction to Kant's Critique of Pure Reason.* Oxford, 1958.

PREFACE TO THE METAPHYSICAL
FOUNDATIONS OF NATURAL SCIENCE

*I*F the word Nature be merely taken in its *formal* signification, there may be as many natural sciences as there are specifically different things (for each must contain the inner principle special to the determinations pertaining to its existence), inasmuch as it [Nature] signifies the primal inner principle of all that belongs to the existence of a thing.[1] But Nature, regarded in its *material* significance, means not a quality, but the sum total of all things, in so far as they can be *objects of our senses,* and therefore of experience; in short, the totality of all phenomena—the sense-world, exclusive of all nonsensuous objects. Now Nature, in this sense of the word, has two main divisions, in accordance with the main distinction of our sensibility, one of which comprises the objects of the *outer,* the other the object of the *inner* sense; thus rendering possible a two-fold doctrine of Nature, the DOCTRINE OF BODY and the DOCTRINE OF SOUL, the first dealing with *extended,* and the second with *thinking,* Nature.

Every doctrine constituting a system, namely, a whole of cognition, is termed a science; and as its principles may be either axioms of the *empirical* or *rational* connection of cognitions in a whole, so natural science, whether it be doctrine of body or doctrine of soul, would have to be divided into *historical* and *rational* natural science, were it not that the word *nature* (as implying the deduction of the manifold pertaining to the existence of things, from its inner *principle*) necessitates a knowledge through reason of its system, if it is to deserve the name natural science. Hence, doctrine of nature may be

---

[1] Essence is the primal inner principle of all that belongs to the possibility of a thing. Hence one can only predicate an essence, but not a nature of geometrical figures (for nothing is contained in their conception expressive of an existence).

This article is from **Kant's Prolegomena and Metaphysical Foundations of Natural Science,** trans. E. Belfort Bax (London, 1883), pp. 137–149. Reprinted by permission of G. Bell & Sons, Ltd.

better divided into *historical doctrine of nature,* comprising nothing but systematically-ordered facts respecting natural things (which again would consist of *description of nature* as a system of classes according to resemblances, and *history of nature* as a systematic presentation of the same at different times and in different places), and *natural science.* Natural science, once more, would be either natural science *properly* or *improperly* so-called, of which the first would treat its subject wholly according to principles *à priori,* and the second according to laws derived from experience.

That only can be called science (*Wissenschaft*) *proper* whose certainty is apodictic: cognition that can merely contain empirical certainty is only improperly called science. A whole of cognition which is systematic is for this reason called *science,* and, when the connection of cognition in this system is a system of causes and effects, *rational* science. But when the grounds or principles it contains are in the last resort merely empirical, as, for instance, in chemistry, and the laws from which the reason explains the given facts are merely empirical laws, they then carry no consciousness of their *necessity* with them (they are not apodictically certain), and thus the whole does not in strictness deserve the name of science; chemistry indeed should be rather termed systematic art than science.

A rational doctrine of nature deserves the name of natural science only when the natural laws at its foundation are cognised *à priori,* and are not mere laws of experience. A natural cognition of the first kind is called *pure,* that of the second *applied,* rational cognition. As the word nature itself carries with it the conception of law, and this again the conception of the *necessity* of all the determinations of a thing appertaining to its existence, it is easily seen why natural science must deduce the legitimacy of its designation only from a *pure* part of it, [a part] namely, which contains the principles *à priori* of all remaining natural explanations, and why only by virtue of this portion it is properly science, in such wise, that, according to the demands of the reason, all natural knowledge must at last turn on natural science and there find its conclusion. This is because the above necessity of law inseparably attaches to the conception of nature, and hence must be thoroughly comprehended. For this reason the most complete explanation of particular phenomena upon chemical principles, invariably leaves an unsatisfactoriness behind it, because from these accidental laws, learnt by mere experience, no grounds *à priori* can be adduced.

Thus all natural science *proper* requires a *pure* portion, upon which the apodictic certainty required of it by the reason can be based; and inasmuch as this is in its principles wholly heterogeneous from those which are merely empirical, it is at once a matter of the utmost importance, indeed in the nature of the case, as regards method of indispensable duty, to expound this part separately and unmixed with the other, and as far as possible in its completeness; in order that we may be able to determine precisely what the reason can accomplish for itself, and where its capacity begins to require the assistance of empirical principles. Pure cognition of the reason from mere *conceptions* is called pure philosophy or metaphysics, while that which only bases its cognition on the *construction* of conceptions, by means of the presentation of the object in an *à priori* intuition, is termed mathematics.

What may be called natural science proper presupposes metaphysics of nature; for laws, i.e., principles of the necessity of that which belongs to the *existence* of a thing, are occupied with a conception which does not admit of construction, because its existence cannot be presented in any *à priori* intuition; natural science proper, therefore, presupposes metaphysics. Now this must indeed always contain exclusively principles of a non-empirical origin (for, for this reason it bears the name of metaphysics); but it may be *either* without reference to any definite object of experience, and therefore undetermined as regards the nature of this or that thing of the sense-world, and treat of the laws rendering possible the conception of nature in general, in which case it is the *transcendental* portion of the metaphysics of nature; *or* it may occupy itself with the particular nature of this or that kind of thing, of which an empirical conception is given, in such wise, that except what lies in this conception, no other empirical principle will be required for its cognition. For instance: it lays the empirical conception of a matter, or of a thinking entity, at its foundation, and searches the range of the cognition of which the reason is *à priori* capable respecting these objects; and thus, though such a science must always be termed a metaphysic of nature (namely, of corporeal or thinking nature), it is then not a univeral but a *particular* metaphysical natural science (physics and psychology), in which the above transcendental principles are applied to the two series of sense-objects. But I maintain that in every special natural doctrine only so much science *proper* is to be met with as mathematics; for, in accordance with the foregoing, science proper, especially [science] of nature, requires a pure portion, lying at the foundation of the empirical, and based upon an *à priori* knowledge of natural things. Now to cognise anything *à priori* is to cognise it from its mere possibility; but the possibility of determinate natural things cannot be known from mere conceptions; for from these the possibility of the thought (that it does not contradict itself) can indeed be known, but not of the object, as natural thing which can be given (as existent) outside the thought. Hence, to the possibility of a determinate natural thing, and therefore to cognise it *à priori*, is further requisite that the *intuition* corresponding *à priori* to the conception should be given; in other words, that the conception should be constructed. But cognition of the reason through construction of conceptions is mathematical. A pure philosophy of nature in general, namely, one that only investigates what constitutes a nature in general, may thus be possible without mathematics; but a pure doctrine of nature respecting *determinate* natural things (corporeal doctrine and mental doctrine), is only possible by means of mathematics; and as in every natural doctrine only so much science proper is to be met with therein as there is cognition *à priori*, a doctrine of nature can only contain so much science proper as there is in it of applied mathematics.

So long, therefore as no conception is discovered for the chemical effects of substances on one another, which admits of being constructed, that is, no law of the approach or retreat of the parts can be stated in accordance with which (as, for instance, in proportion to their densities) their motions, together with the consequences of these, can be intuited and presented *à priori* (a demand that will scarcely ever be fulfilled), chemistry will be nothing more than a systematic art or experimental doctrine, but never science proper, its principles being merely empirical and not admitting of any presentation *à priori;* as a consequence, the principles of chemical phenomena cannot make their possibility in the least degree conceivable, being incapable of the application of mathematics.

But still farther even than chemistry must empirical psychology be removed from the rank of what may be termed a natural science proper; firstly, because mathematics is inapplicable to the phenomena of the internal sense and its laws, unless indeed we consider merely the *law of permanence* in the flow of its internal changes; but this would be an extension of cognition, bearing much the same relation to that procured by the mathematics of corporeal knowledge, as the doctrine of the properties of the straight line does to the whole of geometry; for the pure internal intuition in which psychical phenomena are constructed is time, which has only one dimension. But not even as a systematic art of analysis, or experimental doctrine, can it ever approach chemistry, because in it the manifold of internal observation is only separated in thought, but cannot be kept separate and be connected again at pleasure; still less is another thinking subject amenable to investigations of this kind, and even the observation itself, alters and distorts the state of the object observed. It can never therefore be anything more than an historical, and as such, as far as possible systematic natural doctrine of the internal sense, i.e., a natural description of the soul, but not a science of the soul, nor even a psychological experimental doctrine. This is the reason why, in the title of this work, which, properly speaking, contains the axioms of corporeal doctrine, we have employed, in accordance with the usual custom, the general name of natural science, because this designation in the strict sense is applicable to it alone, and hence occasions no ambiguity.

But to render possible the application of mathematics to the doctrine of body, by which alone it can become natural science, principles of the *construction* of conceptions belonging to the possibility of matter in general must precede. Hence a complete analysis of the conception of a matter in general must be laid at its foundation; this is the business of pure philosophy, which for the purpose makes use of no special experiences, but only of those which it meets with in separate (although in themselves empirical) conceptions, with reference to pure intuitions in space and time (according to laws, essentially depending on the conception of nature in general), thus constituting it a real *metaphysic of corporeal nature*.

All natural philosophers, who wished to proceed mathematically in their work, have hence invariably (although unknown to themselves) made use of metaphysical principles, and must make use of such, it matters not how energetically they may otherwise repudiate any claim of metaphysics on their science. Without doubt by the latter they understood the illusion of manufacturing possibilities at pleasure, and playing with conceptions, perhaps quite incapable of being presented in intuition, and possessing no other guarantee of their objective reality than that they do not stand in contradiction with themselves. But all true metaphysics is taken from the essential nature of the thinking faculty itself, and therefore in nowise invented, since it is not borrowed from experience, but contains the pure operations of thought, that is, conceptions and principles *à priori*, which the manifold of *empirical presentations* first of all brings into legitimate connection, by which it can become empirical KNOWLEDGE, i.e., experience. These mathematical physicists were thus quite unable to dispense with such metaphysical principles, and amongst them, not even with that which makes the conception of their own special subject, namely, matter, available *à priori*, in its application to external experience (as the conception of motion, of the filling of space, of inertia, etc.). But to allow merely empirical principles to obtain in such a question, they rightly held as quite unsuited to the apodictic certainty they desired to give to their natural laws, and hence they preferred to postulate such, without investigating their sources *à priori*.

But it is of the utmost importance in the progress of the sciences, to sever heterogeneous principles from one another, to bring each into a special system, so that it may constitute a science of its own kind, and thereby to avoid the uncertainty springing from their confusion, owing to our not being able to distinguish to which of the two, on the one hand the limitations, and on the other the mistakes occurring in their use, are to be attributed. For this reason I have regarded it as necessary to present in one system the first principles of the pure portion of natural science (*physica generalis*) where mathematical constructions traverse one another, and at the same time the principles of the construction of these conceptions; in short, the possibility of a mathematical doctrine of nature itself. This separation, besides the uses already mentioned, has the special charm, which the unity of knowledge brings with it, if we take care that the boundaries of the sciences do not run into one another, but occupy properly their subdivided fields.

It may serve as a second ground for gauging this procedure, that in all that is called metaphysics the *absolute completeness* of the sciences may be hoped for, in such a manner as can be promised by no other species of knowledge, and therefore, just as in the metaphysics of nature generally, so here also, the completeness of corporeal nature may be confidently expected; the reason being, that in metaphysics the object is considered merely according to the universal laws of thought, but in other sciences as it must be presented according to *data* of intuition (empirical as well as pure). Hence the former, because the object must be invariably compared *with all* the necessary laws of thought, must furnish a *definite* number of cognitions, which can be fully exhausted; but the latter, because it offers an endless multiplicity of intuitions (pure or empirical), and therefore of objects of thought, can never attain to absolute completeness, but can be extended to infinity, as in pure mathematics and empirical natural knowledge. This metaphysical corporeal doctrine I believe myself to have, as far as it reaches, completely exhausted, but do not affect thereby to have achieved any great work.

The scheme for the completeness of a metaphysical system, whether of nature in general, or of corporeal nature in particular, is the table of the categories.[1] For there are not any more pure conceptions of the

[1] I find doubts expressed in the criticism of Professor Ulrich's *Institutiones Logicæ et Metaphysicæ*, in the 'Allgemeine Litteratur Zeitung' (1785), No. 295, not indeed respecting this table of the pure conceptions of the understanding, but the conclusions drawn therefrom as to the limitation of the whole faculty of the pure Reason, and therefore of all metaphysics, in which the learned critic expresses himself at one with his no less accurate author; doubts which, because they are supposed to touch the foundation-stone of my system, as put forward in the Critique, should be reasons for thinking that the latter did not by far carry that apodictic necessity with it, in respect of its main object, which is indispensable in compelling an unqualified acceptance. This foundation-stone is said to be a deduction expounded partly *there*, and partly in the *Prolegomena*, of the pure conceptions of the understanding, which in that part of the Critique, that should have been the clearest, is said to be the most obscure, or indeed, to move in a circle, etc. I direct my answer to these objections, only to their chief point, namely, *that without a completely clear and adequate deduction of the categories*, the system of the Critique of pure Reason would totter to its foundations. I maintain, on the contrary, that for those who subscribe to my propositions as to the sensibility of all our intuition, and the sufficiency of the table of the categories, as determinations of our consciousness borrowed from the logical functions of judgment in general (as the Reviewer does) the system of the Critique must carry with it apodictic certainty because it is built on the proposition, *that the whole speculative use of our Reason never reaches beyond objects of possible experience*. For if it can be proved *that* the categories of which the Reason must make use in all its cognition, can have no other employment whatever, except merely with reference to objects of experience (in such a way that only in them [viz. the categories] is the form of thought possible), the answer to the question, *how* they make such possible is indeed important enough, in order, as far as may be to *complete* this deduction, but in respect of the main object of the system, namely the determination of the boundary of the pure Reason is nowise *necessary*, but merely *desirable*. For in this respect, the deduction is already carried *far enough*, when it shows that the conceived categories are nothing but mere forms of the judgments, in so far as they are applied to intuitions (which are with us always sensuous), by which they first of all become objects and cognitions; because this already suffices to found the whole system of the Critique proper with complete certainty. Thus Newton's system of universal gravitation is established, although it carries with it the inexplicable difficulty of how attraction at a distance is possible; but *difficulties are not doubts*. That the foundation remains even without the complete deduction of the categories being established, I can prove, from what is conceded, thus:

Conceded: that the table of the categories contains all the pure conceptions of the understanding complete, as well as all the formal operations of the under-

Understanding, which concern the nature of things. Under the four classes of Quantity, Quality, Relation, and finally Modality, all the determinations of the universal conception of a matter in general, and, therefore, of all that can be thought *à priori* respecting it, that can be presented in mathematical construction, or given in experience

---

standing in judgments, from which they are deduced and differ in nothing, beyond that in the conception of the understanding an object is regarded as defined in respect of one or the other function of judgment (e.g., in the categorical judgment *the stone is hard;* the *stone* is employed as subject, and *hard* as predicate, so that it remains permissible to the understanding to turn the logical function of these conceptions round, and say, something hard is a stone: on the contrary, when I represent it to myself *in the object* as *determined*, that the *stone* (in every possible determination of an object, not of the mere conception) must be conceived only as subject, and the hardness only as predicate, the same logical functions become *pure conceptions of the understanding* of objects, namely, as *substance and accident*);

2, *Conceded:* that the understanding, by its nature, carries with it synthetic principles *à priori*, by which it subordinates to the foregoing categories all objects that may be given it; and therefore that there must be also intuitions *à priori*, containing the requisite conditions for the application of the above pure conceptions of the understanding, *because, without intuition there is no object* in respect of which the logical function can be determined as category, and hence no cognition of any object; and that without pure intuition, no axiom defining it *à priori* in this respect can obtain;

3, *Conceded:* that these pure intuitions can never be anything but mere forms of the phenomena of the external or internal sense (space and time), and consequently only of the *objects of possible experience:*

It follows, that no employment of the pure Reason can ever refer to anything but objects of experience, and, as in axioms *à priori*, nothing empirical can be the condition, they can be nothing more than principles of the *possibility of experience* generally. This alone is the true and adequate foundation of the determination of the boundary of the pure Reason, but not the solution of the problem; how experience is possible by means of these categories and only by means of them. The last problem, although even without it the structure would be firm, has meanwhile great importance, and, as I now see, equally great facility, since it can be solved well-nigh by a single conclusion from the precisely determined definition of a *judgment* in general (an act by which the given presentations first become cognitions of an object). The obscurity which, in this portion of the deduction attaches to my previous operations, and which I do not disclaim, is attributable to the usual fortune of the understanding in research, the shortest way being commonly not the first it is aware of, I shall, therefore, take the earliest opportunity of supplying this defect (which more concerns the style of exposition than the ground of explanation, which is given correctly enough, even there) without placing my acute critic in the, doubtless, to himself, unpleasant necessity of taking refuge in a pre-established harmony, by reason of the unaccountable agreement of the phenomena with the laws of the under-

as its definite object, must be capable of being brought. There is no more to do in the way of discovery or addition, although certainly, should there be anything lacking in clearness or thoroughness, it may be made better.

Hence the conception of matter had to be carried out through all the four functions of the conceptions of the understanding (in four divisions), in each of which a new determination of the same was added. The fundamental determination of a something that is to be an object of the external sense, must be motion, for thereby only can this sense be affected. The understanding leads all other predicates pertaining to the nature of matter back to this, and thus natural science is throughout either a pure or an applied *doctrine of motion.* The *metaphysical* foundations of natural science may thus be brought under *four* main divisions, of which the *first—motion* considered as pure *quantum,* according to its composition, without any *quality* of the movable, may be termed Phoronomy; the *second,* which regards it as belonging to the *quality* of the matter, under the name of an original moving force, may be called Dynamics; and the *third,* where matter with this quality is conceived as by its own reciprocal motion in *relation,* appears under the name of Mechanics; and the *fourth,* where its motion or rest [is conceived], merely in reference to the mode of presentation or *modality,* in other words as determined as phenomenon of the external sense, is called Phenomenology.

But besides the above internal necessity, whereby the metaphysical foundations of the doctrine of body are not only to be distinguished from physics, which employs empirical principles, but even from the

---

standing notwithstanding that the latter have sources quite distinct from the former—a remedy, by the way, far worse than the evil it is intended to cure, and against which it can really avail nothing at all. For the *objective necessity* in question, characterising the pure conceptions of the understanding (and the principles of their application to phenomena) cannot come out of this. For instance, in the conception of cause in connection with effect, everything remains *subjectively necessary,* but objectively simply chance combination, just as Hume has it, when he terms it mere illusion through custom. No system in the world can derive this necessity otherwise than from the pure *à priori* principles lying at the foundation of the possibility of *thought itself,* whereby alone the cognition of objects whose phenomenon is given us, that is, experience, is possible; and even supposing that the mode, *how* experience is thereby possible, were never adequately explained, it would remain indisputably certain *that* it is merely possible through these conceptions, and conversely that these conceptions are capable of no meaning of employment in any other reference than to objects of possible experience.

rational premises of the latter, in which the employment of mathematics is to be met with, there is an external, and, though only accidental, at the same time an important reason, for separating its thorough working-out from the general system of metaphysics, and for presenting it systematically as a special whole. For if it be permissible to indicate the boundaries of a science, not merely according to the construction of its object, and its specific kind of cognition, but also according to the aim that is kept in view as a further use of the science itself, and it is found that metaphysics has engaged so many heads, and will continue to engage them, not in order to extend natural knowledge (which could be done much more easily and certainly by observation, experiment, and the application of mathematics to external phenomena), but in order to attain to a knowledge of that which lies wholly beyond all the boundaries of experience, of God, Freedom, and Immortality; [in this case] one gains in the promotion of this object, if one liberates it from a shoot springing indeed from its own stem, but only detrimental to its regular growth, and plants this [shoot] apart, without thereby mistaking its origination, or ignoring its entire growth from the system of general metaphysics. This does not affect the completeness of the latter, but it facilitates the uniform progress of this science towards its goal, if in all cases where the universal doctrine of body is required, one can call to aid the separate system of such a science, without encumbering it with the larger system [viz. of metaphysics in general]. It is indeed very remarkable (though it cannot here be thoroughly entered into), that universal metaphysics, in all cases where it requires instances (intuitions) to procure significance for its pure conceptions of the understanding, must always take them from the universal doctrine of body; in other words, from the form and principle of external intuition; and if these are not found to hand in their entirety, it gropes uncertainly and tremblingly amid mere empty conceptions. Hence the well-known disputes, or at least the obscurity in questions, as to the possibility of an opposition of realities, of intensive quantity, &c., by which the understanding is only taught, through instances from corporeal nature, what the conditions are under which the above conceptions can alone have objective reality, that is, significance and truth. And thus a separate metaphysics of corporeal nature does excellent and indispensable service to the *universal* [metaphysics], in that it procures instances (cases *in concreto*) in which to realise the conceptions and doctrines of the latter (properly the transcendental philosophy), that is, to give to a mere form of thought sense and meaning.

I have in this treatise followed the mathematical method, if not with all strictness (for which more time would have been necessary than I had to devote to it), at least imitatively, not in order, by a display of profundity, to procure a better reception for it, but because I believe such a system to be quite capable of it, and that perfection may in time be obtained by a cleverer hand, if stimulated by this sketch, mathematical investigators of nature should find it not unimportant to treat the metaphysical portion, which anyway cannot be got rid of, as a special fundamental department of general physics, and to bring it into unison with the mathematical doctrine of motion.

Newton, in the preface to his mathematical principles of natural science (after having remarked that geometry only requires two of the mechanical actions which it postulates, namely, to describe a straight line and a circle) says: *geometry is proud of being able to achieve so much while taking so little from extraneous sources.*[1] One might say of metaphysics, on the other hand: *it stands astonished, that with so much offered it by pure mathematics it can effect so little.* In the meantime, this little is something which mathematics indispensably requires in its application to natural science, which, inasmuch as it must here necessarily borrow from metaphysics, need not be ashamed to allow itself to be seen in company with the latter.

[1] Gloria geometria, quod tam paucis principiis alicunde petitis tam multa praestet.
—*Newton,* Princ. Phil. Nat. Math. Praefat.

Chapter 2
SIR JOHN FREDERICK WILLIAM HERSCHEL
(1792–1871)

INTRODUCTION

SIR JOHN FREDERICK WILLIAM HERSCHEL was born in Slough, Bucks, England on March 7, 1792. His father, a well-known astronomer, had come to Slough from Hanover, Germany, in 1786. Herschel studied with a private teacher in addition to attending Eton and St. John's at Cambridge, where he concentrated on mathematics. He eventually produced several excellent essays on the subject. In 1814 he decided to become a lawyer and began to study in this field, but by 1816 he had been so influenced by W. H. Wollaston that he redirected his interest toward astronomy. Within a few years he had distinguished himself; he was awarded several gold medals and in 1831 had the honor of knighthood conferred upon him by William IV.

For a two-year period beginning in 1831 Herschel re-examined all the double stars and nebulae discovered by his father, then decided to go to the southern hemisphere to explore the skies from that location, using instruments which he, himself, had helped to perfect. In this venture to the Cape of Good Hope he was accompanied by his family; they remained for four years before returning to England to continue with the work there.

Herschel was an accomplished chemist as well as a famous astronomer. He made important contributions to the field of photography with his discovery (1819) of the solvent power of hyposulfite of soda on salts of silver, and his invention, independently of Fox Talbot, of the process of photography on sensitized paper. Herschel died in Collingwood in 1871.

In addition to the laborious *Cape Observations*, he wrote an important book on astronomy, *Outlines of Astronomy* (1849), and his famous *Discourse on the Study of Natural Philosophy* (1831). He was also the author of many scientific articles, and two collections: *Familiar Lectures on Scientific Subjects* (1866) and *Collected Addresses* (1857).

Herschel's *A Preliminary Discourse on the Study of Natural Phi-*

losophy, which John Stuart Mill was later to find an important source of inspiration for his new theory of induction, was the first attempt by a scientist to make the methods of science explicit and to formulate a view concerning the nature of the objects of the natural sciences, and the relation of these objects to the cognitive powers of the human mind. It is a work of three parts: (1) The general nature and advantages of the study of the physical sciences; (2) The principles on which physical science relies for its successful prosecution; (3) The subdivision of physics into distinct branches, and their mutual relations.

The second part of the book, undoubtedly the most important, consists of seven chapters which deal successively with problems connected with observation, experiment, classification, analysis of phenomena, causality, laws, the uniformity of nature, and Herschel's view of induction.

Generally speaking the book is written in the spirit of the empiricist tradition. The influence of philosophers such as Locke, Berkeley, and Hume is less explicit, whereas often open references are made to the *Novum Organum* of Francis Bacon. Although it was not Herschel's intention to develop an explicit epistemology, he was nonetheless brought to epistemological investigations inasmuch as the theory of scientific method is so intimately connected with epistemology. Actually his epistemological considerations are of very little significance in that they amount to somewhat amateurish improvisations of empiricist and associationist principles, not always free from contradictions, and certainly replete with mistakes and unsolved problems. His insights on observation, experiment, classification, hypothesis, and verification are of far greater importance. It is here that Herschel rethinks insights developed earlier by Bacon, clarifying them by means of concrete illustrations taken from the sciences of his era. Perhaps the most interesting part of the book is the section dealing with the "ten rules of philosophizing" in which Herschel clearly describes and concretely illustrates the four methods of agreement, difference, concomitant variations, and residues, as Mill was to call them later. For this part Herschel leans heavily upon Bacon, perhaps unaware of the historical origin of the methods and their further development by Hume and others.

On the subject of induction Herschel limited himself to a direct enunciation and clarification of the methodological precepts which must govern induction. He distinguished two stages: Science is first concerned with the discovery of "proximate causes" and of laws of the lowest degree of generality, as well as with the verification of these laws. At the second level science no longer deals with individual facts, but with those general facts, laws, and causes which were discovered in the first stage by the examination of individual facts. The results of investigations on this level consist in laws of higher generality which Herschel called "theories"; they must also be verified in experience.

With respect to the first level of induction Herschel is mainly interested in the discovery of causes and the proof of causal connections, which in his view consists in the verification of causal hypotheses by deduction of predictions and comparison of these predictions with observed facts. As for the second level of induction Herschel claims that "theories" are formed in a way analogous to the laws arrived at in the first stage. First one must try to form hypotheses; these are in regard to theories what presumed proximate causes are in regard to particular inductions of the first level. Then one has to examine whether or not the hypotheses cover all the facts and include all the laws to which observation and induction lead. Next they are to be verified in a way analogous to the verification of laws. In other words, theories are not invented, but are to be discovered, and thus in the final analysis obtained from experience.

The following selections are taken from the first two chapters of Part II of the *Discourse*, which deals with experience, laws, and induction.

SELECTIVE BIBLIOGRAPHY

Herschel, J. F. W., Sir, *A Preliminary Discourse on the Study of Natural Philosophy* (1831). London, 1842.

———, *Collected Addresses*. London, 1857.

———, *Familiar Lectures on Scientific Subjects*. London, 1866.

———, *Scientific Papers* (2 vols.). London, 1912.

Cannon, W. F., "John Herschel and the Idea of Science." *Journal of the History of Ideas*, 22 (1961) 215–239.

Clerke, A. M., *The Herschels and Modern Astronomy*. London, 1895.

Ducasse, C. J., "John F. W. Herschel's Methods of Experimental Inquiry," in *Studies in the History of Cultures: The Disciplines of the Humanities*. Menasha, Wis., 1942. (Also in *Theories of Scientific Method: The Renaissance Through the Nineteenth Century*, ed. E. H. Madden. Seattle, 1960. Pp. 153–182).

Metz, R., *Die philosophischen Strömungen der Gegenwart in Grossbritannien*. Leipzig, 1935.

OF THE PRINCIPLES ON WHICH PHYSICAL SCIENCE RELIES FOR ITS SUCCESSFUL PROSECUTION

And the Rules by Which a Systematic Examination of Nature Should Be Conducted, with Illustrations of Their Influence, As Exemplified in the History of Its Progress

OF EXPERIENCE AS THE SOURCE OF OUR KNOWLEDGE

—Of the Dismissal of Prejudices—Of the Evidence of our Senses

*I*NTO abstract science, as we have before observed, the notion of cause does not enter. The truths it is conversant with are *necessary* ones, and exist independent of cause. There may be no such real *thing* as a right-lined triangle marked out in space; but the moment we conceive one in our minds, we cannot refuse to admit the sum of its three angles to be equal to two right angles; and if, in addition, we conceive one of its angles to be a right angle, we cannot thenceforth refuse to admit that the sum of the squares on the two sides, including the right angle, is equal to the square on the side subtending it. To maintain the contrary, would be, in effect, to deny its being right-angled. No one *causes* or *makes* all the diameters of an ellipse to be bisected in its centre. To assert the contrary, would not be to rebel against a power, but to deny our own words. But in natural science, *cause* and *effect* are the ultimate relations we con-

This article is from **A Preliminary Discourse on the Study of Natural Philosophy** by J. F. W. Herschel (Philadelphia, 1831), pp. 57–79. Reprinted by permission of Lea & Febiger, Publishers.

template; and *laws*, whether imposed or maintained, which, for aught we can perceive, might have been other than they are. This distinction is very important. A clever man, shut up alone and allowed unlimited time, might reason out for himself all the truths of mathematics, by proceeding from those simple notions of space and number of which he cannot divest himself without ceasing to think. But he could never tell, by any effort of reasoning, what would become of a lump of sugar if immersed in water, or what impression would be produced on his eye by mixing the colors yellow and blue.

We have thus pointed out to us, as the great, and indeed only ultimate source of our knowledge of nature and its laws, EXPERIENCE; by which we mean, not the experience of one man only, or of one generation, but the accumulated experience of all mankind, in all ages, registered in books or recorded by tradition. But experience may be acquired in two ways: either, first, by noticing facts as they occur, without any attempt to influence the frequency of their occurrence, or to vary the circumstances under which they occur; this is OBSERVATION: or, secondly, by putting in action causes and agents over which we have control, and purposely varying their combinations, and noticing what effects take place; this is EXPERIMENT. To these two sources we must look as the fountains of all natural science. It is not intended, however, by thus distinguishing observation from experiment, to place them in any kind of contrast. Essentially they are much alike, and differ rather in degree than in kind; so that, perhaps, the terms *passive* and *active observation* might better express their distinction; but it is, nevertheless, highly important to mark the different states of mind in inquiries carried on by their respective aids, as well as their different effects in promoting the progress of science. In the former, we sit still and listen to a tale, told us, perhaps obscurely, piecemeal, and at long intervals of time, with our attention more or less awake. It is only by after-rumination that we gather its full import; and often, when the opportunity is gone by, we have to regret that our attention was not more particularly directed to some point which, at the time, appeared of little moment, but of which we at length appreciate the importance. In the latter, on the other hand, we cross-examine our witness, and, by comparing one part of his evidence with the other, while he is yet before us, and reasoning upon it in his presence, are enabled to put pointed and searching questions, the answer to which may at once enable us to make up our minds. Accordingly, it has been found invariably, that in those departments of physics where the phenomena are beyond our control, or into which experimental inquiry, from other causes, has not been carried, the progress of knowledge has been slow, uncertain, and irregular; while in such as admit of experiment, and in which mankind have agreed to its adoption, it has been rapid, sure, and steady. . . .

Experience once recognised as the fountain of all our knowledge of nature, it follows that, in the study of nature and its laws, we ought at once to make up our minds to dismiss as idle prejudices, or at least suspend as premature, any preconceived notion of what might or what ought to be the order of nature in any proposed case, and content ourselves with observing, as a plain matter of fact, what *is*. To experience we refer, as the only ground of all physical inquiry. But before experience itself can be used with advantage, there is one preliminary step to make, which depends wholly on ourselves; it is the absolute dismissal and clearing the mind of all prejudice, from whatever source arising, and the determination to stand and fall by the result of a direct appeal to facts in the first instance, and of strict logical deduction from them afterwards. Now, it is necessary to distinguish between two kinds of prejudices, which exercise very different dominion over the mind, and, moreover, differ extremely in the difficulty of dispossessing them, and the process to be gone through for that purpose. These are,—

1. Prejudices of opinion.
2. Prejudices of sense.

By prejudices of opinion, we mean opinions hastily taken up, either from the assertion of others, from our own superficial views, or from vulgar observation, and which, from being constantly admitted without dispute, have obtained the strong hold of habit on our minds. Such were the opinions once maintained that the earth is the greatest body in the universe, and placed immovable in its centre, and all the rest of the universe created for its sole use; that it is the nature of fire and of sounds to ascend; that the moonlight is cold; that dews *fall* from the air, &c.

To combat and destroy such prejudices we may proceed in two ways, either by demonstrating the falsehood of the facts alleged in their support, or by showing how the appearances, which seem to countenance them, are more satisfactorily accounted for without their admission. But it is unfortunately the nature of prejudices of opinion to adhere, in a certain degree, to every mind, and to some with pertinacious obstinacy, *pigris radicibus*, after all ground for their reasonable entertainment is destroyed. Against such a disposition the student of natural science must contend with all his power. Not that we are so unreasonable as to demand of him an instant and peremptory dismission of all his former opinions and judgments; all we require is, that he will hold them without bigotry, retain till he shall see reason to question them, and be ready to resign them when fairly proved untenable, and to doubt them when the weight of probability is shown to lie against them. If he refuse this, he is incapable of science.

Our resistance against the destruction of the other class of prejudices, those of sense, is commonly more violent at first, but less persistent, than in the case of those of opinion. Not to trust the evidence of our senses seems indeed a hard condition, and one which, if proposed, none would comply with. But it is not the direct evidence of our senses that we are in any case called upon to reject, but only the erroneous judgments we unconsciously form from them, and this only when they can be shown to be so *by counter evidence of the same sort;* when one sense is brought to testify against another, for instance; or the same sense against itself, and the obvious conclusions in the two cases disagree, so as to compel us to acknowledge that one or other must be wrong. For example, nothing at first can seem a more rational, obvious, and incontrovertible conclusion, than that the *color* of an object is an inherent quality, like its weight, hardness, &c., and that to *see* the object, and see it *of its own color,* when nothing intervenes between our eyes and it, are one and the same thing. Yet this is only a prejudice; and that it is so, is shown by bringing forward the same sense of vision which led to its adoption, as evidence on the other side; for, when the differently colored prismatic rays are thrown, in a dark room, in succession upon any object, whatever be the color we are in the habit of calling its own, it will appear of the particular hue of the light which falls upon it: a yellow paper, for instance, will appear scarlet when illuminated by red rays, yellow when by yellow, green by green, and blue by blue rays; its own (so called) proper color *not in the least degree mixing with that it so exhibits.* . . .

These, and innumerable instances we might cite, will convince us, that though we are never deceived in the *sensible impression* made by external objects on us, yet in forming our judgments of them we are greatly at the mercy of circumstances, which either modify the impressions actually received, or combine them, with adjuncts which have become habitually associated with different judgments; and, therefore, that, in estimating the degree of confidence we are to place in our conclusions, we must, of necessity, take into account these modifying or accompanying circumstances, whatever they may be. We do not, of course, here speak of deranged organization;

such as, for instance, a distortion of the eye, producing double vision, and still less of mental delusion, which absolutely perverts the meaning of sensible impressions.

As the mind exists not in the place of sensible objects, and is not brought into immediate relation with them, we can only regard sensible impressions as signals conveyed from them by a wonderful, and, to us, inexplicable mechanism, to our minds, which receives and reviews them, and, by habit and association, connects them with corresponding qualities or affections in the objects; just as a person writing down and comparing the signals of a telegraph might interpret their meaning. As, for instance, if he had constantly observed that the exhibition of a certain signal was sure to be followed next day by the announcement of the arrival of a ship at Portsmouth, he would connect the two facts by a link of the very same nature with that which connects the notion of a large wooden building, filled with sailors, with the impression of her outline on the retina of a spectator on the beach. . . .

OF THE ANALYSIS OF PHENOMENA

Phenomena, then, or appearances, as the word is literally rendered, are the sensible results of processes and operations carried on among external objects, or their constituent principles, of which they are only signals, conveyed to our minds as aforesaid. Now, these processes themselves may be, in many instances, rendered *sensible;* that is to say, analyzed, and shown to consist in the motions or other affections of sensible objects themselves. For instance, the phenomenon of the sound produced by a musical string, or a bell, when struck, may be shown to be the result of a process consisting in the rapid vibratory motion of its parts communicated to the air, and thence to our ears; though the immediate effect on our organs of hearing does not excite the least idea of such a motion. On the other hand, there are innumerable instances of sensible impressions which (at least at present) we are incapable of tracing beyond the mere sensation; for example, in the sensations of bitterness, sweetness, &c. These, accordingly, if we were inclined to form hasty decisions, might be regarded as ultimate qualities; but the instance of sounds, just adduced, alone would teach us caution in such decisions, and incline us to believe them mere results of some secret process going on in our organs of taste, which is too subtle for us to trace. A simple experiment will serve to set this in a clearer light. A solution of the salt called by chemists *nitrate of silver,* and another of the *hyposulphite of soda,* have each of them separately, when taken into the mouth, a disgustingly bitter taste; but if they be mixed, or if one be tasted before the mouth is thoroughly cleared of the other, the sensible impression is that of intense sweetness. Again, the salt called *tungstate of soda* when first tasted is sweet, but speedily changes to an intense and pure bitter, like quassia.[1]

How far we may even be enabled to attain a knowledge of the ultimate and inward processes of nature in the production of phenomena, we have no means of knowing; but to judge from the degree of obscurity which hangs about the only case in which we feel within ourselves a *direct* power to produce any one, there seems no great hope of penetrating so far. The case alluded to is the production of motion by the exertion of force. We are conscious of a power to move our limbs, and by their intervention other bodies; and that this effect is the result of a certain inexplicable process which we are aware of, but can no way describe in words, by which we exert *force.* And even when such exertion produces no visible effect (as when we press our two hands violently together, so as just to oppose each other's effort), we still perceive, by the fatigue and exhaustion, and by the impossibility of maintaining the effort long, that something is going on within us, of which the mind is the agent, and the will, the determining cause. This impression which we receive of the nature of force from our own effort and our sense of fatigue, is quite different from that which we obtain of it from seeing the effect of force exerted by others in producing *motion.* Were there no such thing as motion, had we been from infancy shut up in a dark dungeon, and every limb encrusted with plaster, this internal consciousness would give us a complete idea of *force;* but when set at liberty, habit alone would enable us to recognise its exertion by its *signal* motion, and *that* only by finding that the same action of the mind which in our confined state enables us to fatigue and exhaust ourselves by the tension of our muscles, puts it in our power, when at liberty, to move ourselves and other bodies. But how obscure is our knowledge of the process going on within us in the exercise of this important privilege, in virtue of which alone we act as direct *causes,* we may judge from this, that when we put any limb in motion, the seat of the exertion seems to us to be *in* the limb, whereas

[1] Thomson's First Principles of Chemistry, vol. II p. 68.

it is demonstrably no such thing, but either in the brain or in the spinal marrow; the proof of which is, that if a little fibre, called a nerve, which forms a communication between the limb and the brain, or spine, be divided in any part of its course, however we may make the effort, the limb will not move.

This one instance of the obscurity which hangs about the only act of direct *causation* of which we have an immediate consciousness, will suffice to show how little prospect there is that, in our investigation of nature, we shall ever be able to arrive at a knowledge of ultimate causes, and will teach us to limit our views to that of *laws*, and to the analysis of complex phenomena by which they are resolved into simpler ones, which appearing to us incapable of further analysis, we must consent to regard as causes. Nor let any one complain of this as a limitation of his faculties. We have here "ample room and verge enough" for the full exercise of all the powers we possess; and, besides, it does so happen, that we are actually able to trace up a very large portion of the phenomena of the universe to this one *cause*, viz. the exertion of mechanical *force;* indeed, so large a portion, that it has been made a matter of speculation whether this is not the only one that is capable of acting on material beings.

What we mean by the analysis of complex phenomena into simpler ones, will best be understood by an instance. Let us, therefore, take the phenomenon of sound, and, by considering the various cases in which sounds of all kinds are produced, we shall find that they all agree in these points:—1st, The excitement of a motion in the sounding body. 2dly, The communication of this motion to the air or other intermedium which is interposed between the sounding body and our ears. 3dly, The propagation of such motion from particle to particle of such intermedium in due succession. 4thly, Its communication, from the particles of the intermedium adjacent to the ear, to the ear itself. 5thly, Its conveyance in the ear, by a certain mechanism, to the auditory nerves. 6thly, The excitement of sensation. Now, in this analysis, we perceive that two principal matters must be understood, before we can have a true and complete knowledge of sound:—1st, The excitement and propagation of motion. 2dly, The production of sensation. These, then, are two other phenomena, of a simpler, or, it would be more correct to say, of a more general or elementary order, into which the complex phenomenon of sound resolves itself. But again, if we consider the communication of motion from body to body, or from one part to another of the same, we shall perceive that it is again resolvable into several other phenomena:—1st, The original setting in motion of a material body, or any part of one. 2dly, The behavior of a particle set in motion, when it meets another lying in its way, or is otherwise impeded or influenced by its connection with surrounding particles. 3dly, The behavior of the particles so impeding or influencing it under such circumstances; besides which, the last two point out another phenomenon, which it is necessary also to consider, viz. the phenomenon of the connection of the parts of material bodies in masses, by which they form aggregates, and are enabled to influence each other's motions.

Thus, then, we see that an analysis of the phenomenon of sound leads to the inquiry, 1st, of two *causes*, viz. the cause of motion, and the cause of sensation, these being phenomena which (at least as human knowledge stands at present) we are unable to analyze further; and, therefore, we set them down as simple, elementary, and referable, for any thing we can see to the contrary, to the immediate action of their causes. 2dly, Of several questions relating to the connection between the motion of material bodies and its cause, such as, *What will happen* when a moving body is surrounded on all sides by others not in motion? *What will happen* when a body not in motion is advanced upon by a moving one? It is evident that the answers to such questions as these can be no other than *laws of motion*, in the sense we have above attributed to laws of nature, viz. a statement in words of what will happen in such and such proposed general contingencies. Lastly, we are led, by pursuing the analysis, and considering the phenomenon of the aggregation of the parts of material bodies, and the way in which they influence each other, to two other general phenomena, viz. the cohesion and elasticity of matter; and these we have no means of analyzing further, and must, therefore, regard them (till we see reasons to the contrary) as *ultimate phenomena*, and referable to the direct action of causes, viz. an attractive and a repulsive *force*.

Of force, as counterbalanced by opposing force, we have, as already said, an internal consciousness; and though it may seem strange to us that matter should be capable of exerting on matter the same kind of effort, which, judging alone from this consciousness, we might be led to regard as a mental one; yet we cannot refuse the direct evidence of our senses, which shows us that when we keep a spring stretched with one hand, we feel an effort opposed exactly in the same way as if we had ourselves opposed it with the other hand, or as it would be by that of another person. The inquiry, therefore, into the aggregation of matter resolves itself into the general question,

What will be the behavior of material particles under the mutual action of opposing forces capable of counterbalancing each other? and the answer to this question can be no other than the announcement of the *law* of equilibrium, whatever law that may be.

With regard to the cause of sensation, it must be regarded as much more obscure than that of motion, inasmuch as we have no conscious knowledge of it, *i.e.*, we have no power, by any act of our minds and will, to call up a sensation. It is true, we are not destitute of an approach to it, since by an effort of memory and imagination, we can produce in our minds an impression, or idea, of a sensation, which, in peculiar cases may even approach in vividness to actual reality. In dreams, too, and in some cases of disordered nerves, we have sensations without objects. But if force, as a cause of motion, is obscure to us, even while we are in the act of exercising it, how much more so is this other cause, whose exercise we can only imitate imperfectly by any voluntary act, and of whose purely internal action we are only fully conscious when in a state that incapacitates us from reasoning, and almost from observation!

Dismissing, then, as beyond our reach, the inquiry into causes, we must be content at present to concentrate our attention on the laws which prevail among phenomena, and which seem to be their immediate results. From the instance we have just given, we may perceive that every inquiry into the intimate nature of a complex phenomenon branches out into as many different and distinct inquiries as there are simple or elementary phenomena into which it may be analyzed; and that, therefore, it would greatly assist us in our study of nature, if we could, by any means, ascertain what *are* the ultimate phenomena into which all the composite ones presented by it may be resolved. There is, however, clearly no way by which this can be ascertained *à priori*. We must go to nature itself, and be guided by the same kind of rule as the chemist in his analysis, who accounts every ingredient an *element* till it can be decompounded and resolved into others. So, in natural philosophy, we must account every phenomenon an elementary or simple one till we can analyze it, and show that it is the result of others, which in their turn become elementary. Thus, in a modified and relative sense, we may still continue to speak of causes, not intending thereby those ultimate principles of action on whose exertion the whole frame of nature depends, but those proximate links which connect phenomena with others of a simpler, higher, and more general or elementary kind. For example: we may regard the vibration of a musical string as the proximate cause of the sound it yields, receiving it, so far, as an ultimate fact, and waiving or deferring inquiry into the cause of vibrations, which is of a higher and more general nature.

Moreover, as in chemistry we are sometimes compelled to acknowledge the existence of elements different from those already identified and known, though we cannot insulate them, and to perceive that substances have the characters of compounds, and must therefore be susceptible of analysis, though we do not see how it is to be set about; so in physics, we may perceive the complexity of a phenomenon, without being able to perform its analysis. For example, in magnetism, the agency of electricity is clearly made out, and they are shown to stand to one another in the relation of effect and cause, at least in so far as that all the phenomena of magnetism are producible by electricity, but no electric phenomena have hitherto ever been produced by magnetism. But the analysis of magnetism, in its relation to particular metals, is not yet quite satisfactorily performed; and we are compelled to admit the existence of some cause, whether proximate or ultimate, whose presence in the one and not in the other phenomenon determines their difference. Cases like these, of all which science presents, offer the highest interest. They excite inquiry, like the near approach to the solution of an enigma; they show us that there is light, could only a certain veil be drawn aside.

In pursuing the analysis of any phenomenon, the moment we find ourselves stopped by one of which we perceive no analysis, and which, therefore, we are forced to refer (at least provisionally) to the class of ultimate facts, and to regard as elementary, the study of that phenomenon and of its laws becomes a separate branch of science. If we encounter the same elementary phenomenon in the analysis of several composite ones, it becomes still more interesting, and assumes additional importance; while at the same time we acquire information respecting the phenomenon itself, by observing those with which it is habitually associated, that may help us at length to its analysis. It is thus that sciences increase, and acquire a mutual relation and dependency. It is thus, too, that we are at length enabled to trace parallels and analogies between great branches of science themselves, which at length terminate in a perception of their dependence on some common phenomenon of a more general and elementary nature than that which forms the subject of either separately. It was thus, for example, that, previous to Oërsted's great discovery of electro-magnetism, a general resemblance between the two sciences of electricity and magnetism was recognised, and many of the chief phenomena in each were ascertained to have their parallels, *mutatis mutandis*, in the other. It was thus, too, that an analogy subsisting between sound and light has been gradually traced into a closeness of agreement, which can hardly leave any reasonable doubt of their ultimate coincidence in one common phenomenon, the vibratory motion of an elastic medium. If it be allowed to pursue our illustration from chemistry, and to ground its application, not on what has been, but on what may one day be done, it is thus that the general family resemblance between certain groups of bodies, now regarded as elementary (as nickel and cobalt, for instance, chlorine, iode and brome), will, perhaps, lead us hereafter to perceive relations between them of a more intimate kind than we can at present trace.

On those phenomena which are most frequently encountered in an analysis of nature, and which most decidedly resist further decomposition, it is evident that the greatest pains and attention ought to be bestowed, not only because they furnish a key to the greatest number of inquiries, and serve to group and classify together the greatest range of phenomena, but by reason of their higher nature, and because it is in these that we must look for the direct action of causes, and the most extensive and general enunciation of the laws of nature. These, once discovered, place in our power the explanation of all particular facts, and become grounds of reasoning, independent of particular trial: thus playing the same part in natural philosophy that axioms do in geometry; containing, in a refined and condensed state, and as it were in a quintessence, all that our reason has occasion to draw from experience, to enable it to follow out the truths of physics by the mere application of logical argument. Indeed, the axioms of geometry themselves may be regarded as in some sort an appeal to experience, not corporeal, but mental. When we say, the whole is greater than its part, we announce a general fact, which rests, it is true, on our ideas of whole and part, but, in abstracting these notions, we begin by considering them as subsisting in space, and time, and body, and again, in linear, and superficial, and solid space. Again, when we say, the equals of equals are equal, we mentally make comparisons, in equal spaces, equal times, &c.; so that these axioms, however self-evident, are still general propositions so far of the inductive kind, that independently of experience they would not present themselves to the mind. The only difference between these and axioms obtained from extensive induction is this, that, in raising the axioms of geometry, the instances offer themselves spontaneously, and without the trouble of search, and are few and simple; in raising those of nature, they are infinitely numerous, complicated and remote: so that the most diligent research and the utmost acuteness are required to unravel their web, and place their meaning in evidence.

By far the most general phenomenon with which we are acquainted, and that which occurs most constantly, in every inquiry into which we enter, is motion, and its communication. Dynamics, then, or the science of force and motion, is thus placed at the head of all the sciences; and, happily for human knowledge, it is one in which the highest certainty is attainable, a certainty no way inferior to mathematical demonstration. As its axioms are few, simple, and in the highest degree distinct and definite, so they have at the same time an immediate relation to geometrical quantity, space, time, and direction, and thus accommodate themselves with remarkable facility to geometrical reasoning. Accordingly, their consequences may be pursued, by arguments purely mathematical, to any extent, insomuch that the limit of our knowledge of dynamics is determined only by that of pure mathematics, which is the case in no other branch of physical science.

But, it will now be asked, how we are to proceed to analyze a composite phenomenon into simpler ones, and whether any general rules can be given for this important process? We answer, None; any more than (to pursue the illustration we have already had recourse to) general rules can be laid down by the chemist for the analysis of substances of which all the ingredients are unknown. Such rules, could they be discovered, would include the whole of natural science; but we are very far, indeed, from being able to propound them. However, we are to recollect that the analysis of phenomena, philosophically speaking, is principally useful, as it enables us to recognise, and mark for special investigation, those which appear to us simple; to set methodically about determining their laws, and thus to facilitate the work of raising up general axioms, or forms of words, which shall include the whole of them; which shall, as it were, transplant them out of the external into the intellectual world, render them creatures of pure thought, and enable us to reason them out *à priori*. And what renders the power of doing this so eminently desirable is, that, in thus reasoning back from generals to particulars, the propositions at which we arrive apply to an immense multitude of combinations and cases, which were never individually contemplated in the mental process by which our axioms were first discovered; and that, consequently, when our reasonings are pushed to the utmost limit of particularity, their results appear in the form of *individual facts,* of which we might have had no knowledge from immediate experience; and thus we are not only furnished with the explanation of all known facts, but with the actual discovery of such as were before unknown. A remarkable example of this has already been mentioned in Fresnel's *à priori* discovery of the extraordinary refraction of both rays in a doubly refracting medium. To give another example:—The law of gravitation is a physical axiom of a very high and universal kind, and has been raised by a succession of inductions and abstractions drawn from the observation of numerous facts and subordinate laws in the planetary system. When this law is taken for granted, and laid down as a basis of reasoning, and applied to the actual condition of our own planet, one of the consequences to which it leads is, that the earth, instead of being an exact sphere, must be compressed or flattened in the direction of its polar diameter, the one diameter being about thirty miles shorter than the other; and this conclusion, deduced at first by mere reasoning, has been since found to be true in fact. All astronomical predictions are examples of the same thing.

In the important business of raising these axioms of nature, we are not, as in the analysis of phenomena, left wholly without a guide. The nature of abstract or general reasoning points out in a great measure the course we must pursue. A law of nature, being the statement of what will happen in certain general contingencies, may be regarded as the announcement, in the same words, of a whole group or class of phenomena. Whenever, therefore, we perceive that two or more phenomena agree in so many or so remarkable points, as to lead us to regard them as forming a class or group, if we lay out of consideration, or *abstract,* all the circumstances in which they disagree, and retain in our minds those only in which they agree, and then, under this kind of mental convention, frame a definition or statement of one of them, in such words that it shall apply equally to them all, such statement will appear in the form of a general proposition, having so far at least the character of a law of nature.

For example: a great number of transparent substances, when exposed, in a certain particular manner, to a beam of light which has been prepared by undergoing certain reflexions or refractions (and has thereby acquired peculiar properties, and is said to be "*polarized*"), exhibit very vivid and beautiful colors, disposed in streaks, bands, &c. of great regularity, which seem to arise within the substance, and which, from a certain regular succession observed in their appearance, are called "periodical colors." Among the substances which exhibit periodical colors occur a great variety of transparent solids, but no fluids and no opaque solids. Here, then, there seems to be sufficient community of nature to enable us to use a general term, and to state the proposition as a law, viz. *transparent solids* exhibit periodical colors by exposure to polarized light. However, this, though true of many, does not apply to *all* transparent solids, and therefore we cannot state it as a general truth or law of nature in this form; although the reverse proposition, that all solids which exhibit such colors in such circumstances are *transparent,* would be correct and general. It becomes necessary, then, to make a list of those to which it does apply; and thus a great number of substances of all kinds become grouped together, in a class linked by this common property. If we examine the individuals of this group, we find among them the utmost variety of color, texture, weight, hardness, form and composition; so that, in these respects, we seem to have fallen upon an assemblage of contraries. But when we come to examine them closely, in all properties, we find they have all one point of agreement, in the property of double refraction, . . . and therefore we may describe them all truly as *doubly refracting substances.* We may, therefore, state the fact in the form, "Doubly refracting substances exhibit periodical colors by exposure to polarized light;" and in this form it is found, on further examination, to be true, not only for those particular instances which we had in view when we first propounded it, but in all cases which have since occurred on further inquiry, without a single exception; so that the proposition is general, and entitled to be regarded as a law of nature.

We may therefore regard a law of nature either, 1st, as a general proposition, announcing, in abstract terms, a whole group of particular facts relating to the behavior of natural agents in proposed circumstances; or, 2dly, as a proposition announcing that a whole class of individuals agreeing in one character agree also in another. For example: in the case before us, the law arrived at includes, in its general announcement, among others, the particular facts, that rock crystal and saltpetre exhibit periodical colors; for these are both of them doubly refracting substances. Or, it may be regarded as announcing a relation between the two phenomena of double refraction, and the exhibition of periodical colors; which in the actual case is one of the most important, viz. the relation of *constant association,* inasmuch as it asserts that in whatever individual the one character is found, the other will invariably be found also.

These two lights, in which the announcement of a general law may be regarded, though at bottom they come to the same thing, yet differ widely in their influence on our minds. The former exhibits a law as little more than a kind of artificial memory; but in the latter it becomes a step in philosophical investigation, leading directly to the consideration of a proximate, if not an ultimate cause; inasmuch as, whenever two phenomena are observed to be invariably connected together, we conclude them to be related to each other, either as cause and effect, or as common effects of a single cause.

There is still another light in which we may regard a law of the kind in question, viz. as a proposition asserting the mutual connection, or in some cases the entire identity, of two classes of individuals (whether individual objects or individual facts); and this is, perhaps, the simplest and most instructive way in which it can be conceived, and that which furnishes the readiest handle to further generalization in the raising of yet higher axioms. For example: in the case above mentioned, if observation had enabled us to establish the existence of a class of bodies possessing the property of double refraction, and observations of another kind had, independently of the former, led us to recognise a class possessing that of the exhibition of periodical colors in polarized light, a mere comparison of lists would at once demonstrate the identity of the two classes, or enable us to ascertain whether one was or was not included in the other.

It is thus we perceive the high importance in physical science of just and accurate classifications of particular facts, or individual objects, under general well considered heads or points of agreement (for which there are none better adapted than the simple phenomena themselves into which they can be analyzed in the first instance); for by so doing, each of such phenomena, or heads of classification, becomes, not a particular, but a general fact; and when we have amassed a great store of such *general facts*, they become the objects of another and higher species of classification, and are themselves included in laws which, as they dispose of groups, not individuals, have a far superior degree of generality, till at length, by continuing the process, we arrive at *axioms* of the highest degree of generality of which science is capable.

This process is what we mean by induction; and, from what has been said, it appears that induction may be carried on in two different ways,—either by the simple juxtaposition and comparison of ascertained classes, and marking their agreements and disagreements; or by considering the individuals of a class, and casting about, as it were, to find in what particular they all agree, besides that which serves as their principle of classification.

Either of these methods may be put in practice, as one or the other may afford facilities in any case; but it will naturally happen that, where facts are numerous, well observed, and methodically arranged, the former will be more applicable than in the contrary case: the one is better adapted to the maturity, the other to the infancy of science: the one employs, as an engine, the division of labor; the other mainly relies on individual penetration, and requires a union of many branches of knowledge in one person.

It is to our immortal countryman Bacon that we owe the broad announcement of this grand and fertile principle; and the development of the idea, that the whole of natural philosophy consists entirely of a series of inductive generalizations, commencing with the most circumstantially stated particulars, and carried up to universal laws, or axioms, which comprehend in their statements every subordinate degree of generality, and of a corresponding series of inverted reasoning from generals to particulars by which these axioms are traced back into their remotest consequences, and all particular propositions deduced from them; as well those by whose immediate consideration we rose to their discovery, as those of which we had no previous knowledge. In the course of this descent to particulars, we must of necessity encounter all those facts on which the arts and works that tend to the accommodation of human life depend, and acquire thereby the command of an unlimited practice, and a disposal of the powers of nature co-extensive with those powers themselves. A noble promise, indeed, and one which ought, surely, to animate us to the highest exertion of our faculties; especially since we have already such convincing proof that it is neither vain nor rash, but, on the contrary, has been, and continues to be fulfilled, with a promptness and liberality which even its illustrious author in his most sanguine mood would have hardly ventured to anticipate.

Chapter 3
WILLIAM WHEWELL
(1794–1866)

INTRODUCTION

W<small>ILLIAM</small> W<small>HEWELL</small> was born in Lancaster, England, on May 24, 1794. He studied at Trinity College in Cambridge where he later became a fellow and tutor; at twenty-six he was already a fellow of the Royal Society. Whewell's work in the realm of mechanics and dynamics was so important that it led to an appointment as professor of mineralogy at Cambridge in 1828. In 1838 he received an appointment as professor of moral philosophy. By 1841 he had become master of Trinity College; one year later he assumed the vice-chancellorship of Cambridge University.

At the beginning of the thirties Whewell began to study the history of empirical sciences; the results of these investigations were published in his three-volume *History of the Inductive Sciences* in 1837. Three years later he published *Philosophy of the Inductive Sciences* which he later expanded first into two, then three, volumes under the titles *The History of Scientific Ideas* (2 vols.), and *Novum Organon Renovatum. On the Philosophy of Discovery* appeared in 1860. Whewell's other publications were in the areas of natural science and moral philosophy: *Astronomy and General Physics Considered in Reference to Natural Theology* (1834), *Mechanical Euclid* (1837), *Elements of Morality* (2 vols., 1845), *Lectures on Systematic Morality* (1846), and *Lectures on Moral Philosophy in England* (1852). He died at Cambridge on March 6, 1866.

Whewell made his most important contribution in the area of history and philosophy of natural science: in fact his history of inductive logic furnished John Stuart Mill with most of the material for his third book of logic, a work Mill claimed could never have been written without Whewell's preparatory work. Notwithstanding, Whewell's theory of induction stems from a criterion more or less antithetical to the stand Mill adopted, and in fact closer to Kant's doctrine of the a priori. Most historians agree that Whewell's familiarity with Kant's philosophy gave him many advantages over the other philosophers of his time, but they also recognize it as a reason why Whewell's work did not immediately receive its well-deserved recognition.

In Whewell's view induction does not consist in a collection of facts. Facts for him are confused and without order, without necessary connection; their simple observation by man does not make them an object of laws if man is not somehow already guided in his observations by a universal form of his understanding. This universal form a priori is that upon which scientific knowledge is to be founded. In short, in all scientific statements one must make a distinction between a "formal" element, which as a priori is able to account for the necessity of that statement, and an "objective" element, which has its origin in experience. The "sources of necessary truths" (which are to be found in the human mind) are called "fundamental ideas." The latter are, for example, space, time, and causality, as well as many others. Regarding space, time, and causality Whewell followed quite closely Kant's conception as explained in the *Critique of Pure Reason*, but for all of his other fundamental ideas he developed new insights, many the fruits of his historical investigations in the realm of inductive science. Whewell believed that the fundamental ideas cannot be discovered a priori as Kant thought, but that they must be discovered in, their nature and character determined by, an analysis of the history of the different sciences to which those ideas are essential.

Whewell gives a brief outline of his views on the general nature of our knowledge in Book I of the *Philosophy of the Inductive Sciences*. In his opinion all forms of human knowledge essentially involve two opposite elements, one of which is given by experience and the other superimposed by our understanding upon what we experience. This fundamental opposition can be expressed by different pairs of terms all of which refer to one or another aspect of this opposition: thought and thing, deduction and induction, theory and fact, ideas and sensations, the subjective and the objective, form and matter. The latter pair of terms refers to our ideas as the form, and our sensations as the matter of our knowledge. This pair of terms shows more than any of the others "that the two elements of an antithesis which cannot be separated in fact, may be advantageously separated in our reasonings."

The balance of Book I deals with a detailed comparison of necessary and experiential truths, and with an account of the foundation of the former. This issue constitutes the subject matter of the first Whewell selection to follow. At its conclusion Whewell posits that philosophy of science for the greater part consists in an accurate analysis of the basic ideas which govern the different sciences. In one of the other books (on mechanics) he argues that the concept of force is fundamental in mechanics. In view of the fact that force is a form of cause he feels that he must first explain what is to be understood by cause before he can establish a founded view on forces. This theme is taken up in the second selection.

In the second volume of the *Philosophy of the Inductive Sciences* the process of scientific discovery is analyzed. In Whewell's view this process implies three major steps: the clarification of the elements of knowledge by means of analysis, the colligation of facts with the help of a determinate conception, and finally the verification of the colligation. The last selection presented here deals with the second of these phases, as Whewell turns briefly to the problem of induction. The material is from his *Novum Organon Renovatum,* which contains short aphorisms in addition to the original text found in *Philosophy of the Inductive Sciences.* Although these aphorisms may be somewhat obscure without the commentary, the reader will soon see the essential point Whewell is making.

For purposes here passages in which Whewell refers to earlier or later parts of his book have been omitted; their absence in no way affects the strength of Whewell's arguments, nor the reader's possibility of understanding them.

SELECTIVE BIBLIOGRAPHY

Whewell, W., *History of the Inductive Sciences* (3 vols.). London, 1837.

———, *The Philosophy of the Inductive Sciences Founded Upon Their History.* London, 1840. This work was expanded later into two volumes (1847) and into three (1858) under the titles: *The History of Scientific Ideas* (2 vols.). London, 1847; *Novum Organon Renovatum.* London, 1858.

———, *On the Philosophy of Discovery.* London, 1860.

Tothunter, J., *William Whewell* (2 vols.). London, 1876.

Douglas, S., *Life and Selected Correspondence of William Whewell.* London, 1881.

Stoll, M. R., *Whewell's Philosophy of Induction*. Lancaster, 1929.

Blanchée, R., *Le rationalisme de Whewell*. Paris, 1935.

Steward, C. G., *Die theoretische Philosophie William Whewells und der Kantische Einfluss*. Tübingen, 1938.

Ducasse, C. J., "Whewell's Philosophy of Scientific Discovery," in *Phil. Rev.*, **60** (1951) 56–69, 213–234.

Heathcote, A. W., "William Whewell's Philosophy of Science," *Brit. J. Phil. Sci.* (1953–1954) 302–312.

ON THE NATURE AND CONDITIONS
OF INDUCTIVE SCIENCE

A. INTRODUCTION

*T*HE PHILOSOPHY OF SCIENCE, if the phrase were to be understood in the comprehensive sense which most naturally offers itself to our thoughts, would imply nothing less than a complete insight into the essence and conditions of all real knowledge, and an exposition of the best methods for the discovery of new truths. We must narrow and lower this conception, in order to mould it into a form in which we may make it the immediate object of our labours with a good hope of success; yet still it may be a rational and useful undertaking, to endeavour to make some advance towards such a Philosophy, even according to the most ample conception of it which we can form. The present work has been written with a view of contributing, in some measure, however small it may be, towards such an undertaking.

But in this, as in every attempt to advance beyond the position which we at present occupy, our hope of success must depend mainly upon our being able to profit, to the fullest extent, by the progress already made. We may best hope to understand the nature and conditions of real knowledge, by studying the nature and conditions of the most certain and stable portions of knowledge which we already possess: and we are most likely to learn the best methods of discovering truth, by examining how truths, now universally recognized, have really been discovered. Now there do exist among us doctrines of solid and acknowledged certainty, and truths of which the discovery has been received with universal applause. These constitute what we commonly term *Sciences;* and of these bodies of exact and enduring knowledge, we have within our reach so large and varied a collection,

This article is from **The Philosophy of the Inductive Sciences Founded Upon Their History** by William Whewell (London, 1840), pp. 1–2. Reprinted by permission of Longmans, Green & Company, Ltd.

that we may examine them, and the history of their formation, with a good prospect of deriving from the study such instruction as we seek. We may best hope to make some progress towards the Philosophy of Science, by employing ourselves upon THE PHILOSOPHY OF THE SCIENCES.

The *Sciences* to which the name is most commonly and unhesitatingly given, are those which are concerned about the material world; whether they deal with the celestial bodies, as the sun and stars, or the earth and its products, or the elements; whether they consider the differences which prevail among such objects, or their origin, or their mutual operation. And in all these Sciences it is familiarly understood and assumed, that their doctrines are obtained by a common process of collecting general truths from particular observed facts, which process is termed *Induction*. It is further assumed that both in these and in other provinces of knowledge, so long as this process is duly and legitimately performed, the results will be real substantial truth. And although this process, with the conditions under which it is legitimate, and the general laws of the formation of Sciences, will hereafter be subjects of discussion in this work, I shall at present so far adopt the assumption of which I speak, as to give to the Sciences from which our lessons are to be collected the name of *Inductive* Sciences. And thus it is that I am led to designate my work as THE PHILOSOPHY OF THE INDUCTIVE SCIENCES.

B. OF FUNDAMENTAL IDEAS

Of Necessary Truths

Every advance in human knowledge consists . . . in adapting new ideal conceptions to ascertained facts, and thus in superinducing the Form upon the Matter, the active upon the passive processes of our minds. Every such step introduces into our knowledge an additional portion of the ideal element, and of those relations which flow from the nature of Ideas. It is, therefore, important for our purpose to examine more closely this element, and to learn what the relations are which may thus come to form part of our knowledge. An inquiry

This article is from **The Philosophy of the Inductive Sciences Founded Upon Their History** by William Whewell (London, 1840), pp. 54–79. Reprinted by permission of Longmans, Green & Company, Ltd.

into those Ideas which form the foundations of our sciences;—into the reality, independence, extent, and principal heads of the knowledge which we thus acquire;—is a task on which we must now enter. . . .

In this inquiry our object will be to pass in review all the most important Fundamental Ideas which our sciences involve; and to prove more distinctly in reference to each, what we have already asserted with regard to all, that there are everywhere involved in our knowledge acts of the mind as well as impressions of sense; and that our knowledge derives, from these acts, a generality, certainty, and evidence which the senses could in no degree have supplied. But before I proceed to do this in particular cases, I will give some account of the argument in its general form.

We have already considered the separation of our knowledge into its two elements,—Impressions of Sense and Ideas,—as evidently indicated by this; that all knowledge possesses characters which neither of these elements alone could bestow. Without our ideas, our sensations could have no connexion; without external impressions, our ideas would have no reality; and thus both ingredients of our knowledge must exist.

There is another mode in which the distinction of the two elements of knowledge appears, . . . namely in the distinction of *necessary* and *contingent* or *experiential* truths. For of these two classes of truths, the difference arises from this;—that the one class derives its nature from the one, and the other from the other, of the two elements of knowledge. . . . The former are true necessarily and universally: the latter are learnt from experience and limited by experience. Now with regard to the former kind of truths, I wish to show that the universality and necessity which distinguish them can by no means be derived from experience; that these characters do in reality flow from the ideas which these truths involve; and that when the necessity of the truth is exhibited in the way of logical demonstration, it is found to depend upon certain fundamental principles, (Definitions and Axioms,) which may thus be considered as expressing, in some measure, the essential characters of our ideas. These fundamental principles I shall afterwards proceed to discuss and to exhibit in each of the principal departments of science.

I shall begin by considering Necessary Truths more fully than I have yet done. As I have already said, necessary truths are those in which we not only learn that the proposition *is* true, but see that it *must be* true; in which the negation of the truth is not only false, but impossible; in which we cannot, even by an effort of imagination, or in a supposition, conceive the reverse of that which is asserted.

That there are such truths cannot be doubted. We may take, for example, all relations of number. Three and Two added together make Five. We cannot conceive it to be otherwise. We cannot, by any freak of thought, imagine Three and Two to make Seven.

It may be said that this assertion merely expresses what we mean by our words; that it is a matter of definition; that the proposition is an identical one.

But this is by no means so. The definition of Five is not Three and Two, but Four and One. How does it appear that Three and Two is the same number as Four and One? It is evident that it is so; but *why* is it evident?—not because the proposition is identical; for if that were the reason, all numerical propositions must be evident for the same reason. . . . If we divide 123375 by 987 according to the process taught us at school, how are we assured that the result is correct, and that the number 125 thus obtained is really the number of times one number is contained in the other?

The correctness of the rule, it may be replied, can be rigorously demonstrated. It can be shewn that the process must inevitably give the true quotient.

Certainly this can be shown to be the case. And precisely because it *can* be shown that the result must be true, we have here an example of a necessary truth; and this truth, it appears, is not *therefore* necessary because it is itself evidently identical, however it may be possible to prove it by reducing it to evidently identical propositions. And the same is the case with all other numerical propositions; for, as we have said, the nature of all of them is the same.

Here, then, we have instances of truths which are not only true, but demonstrably and necessarily true. Now such truths are, in this respect at least, altogether different from truths, which, however certain they may be, are learnt to be so only by the evidence of observation, interpreted, as observation must be interpreted, by our own mental faculties. There is no difficulty in finding examples of these merely observed truths. We find that sugar dissolves in water, and forms a transparent fluid, but no one will say that we can see any reason beforehand why the result *must* be so. We find that all animals which chew the cud have also the divided hoof; but could any one have predicted that this would be universally the case? or supposing the truth of the rule to be known, can any one say that he cannot conceive the facts as occurring otherwise? Water expands when it crystallizes, some other substances contract in the same circumstances; but can any one know that this will be so otherwise than by observation? We have here propositions *rigorously* true, (we will assume,) but can any one say they are *necessarily* true? These, and the great mass of the doctrines established by induction, are actual, but so far as we can see, accidental laws; results determined by some unknown selection, not demonstrable consequences of the essence of things, inevitable and perceived to be inevitable. Accord' ing to the phraseology which has been frequently used by philosophical writers, they are *contingent*, not necessary truths.

It is requisite to insist upon this opposition, because no insight can be obtained into the true nature of knowledge, and the mode of arriving at it, by any one who does not clearly appreciate the distinction. The separation of truths which are learnt by observation, and truths which can be seen to be true by a pure act of thought, is one of the first and most essential steps in our examination of the nature of truth, and the mode of its discovery. . . .

As I have already said, one mode in which we may express the difference of necessary truths and truths of experience, is, that necessary truths are those of which we cannot distinctly conceive the contrary. We can very readily conceive the contrary of experiential truths. We can conceive the stars moving about the pole or across the sky in any kind of curves with any velocities; we can conceive the moon always appearing during the whole month as a luminous disk, as she might do if her light were inherent and not borrowed. But we cannot conceive one of the parallelograms on the same base and between the same parallels larger than the other; for we find that, if we attempt to do this, when we separate the parallelograms into parts, we have to conceive one triangle larger than another, both having all their parts equal; which we cannot conceive at all, if we conceive the triangles distinctly. We make this impossibility more clear by conceiving the triangles to be placed so that two sides of the one coincide with two sides of the other; and it is then seen, that in order to conceive the triangles unequal, we must conceive the two bases which have the same extremities both ways, to be different lines, though both straight lines. This it is impossible to conceive: we assent to the impossibility as an axiom, when it is expressed by saying, that two straight lines cannot inclose a space; and thus we cannot distinctly conceive the contrary of the proposition just mentioned respecting parallelograms.

But it is necessary, in applying this distinction, to bear in mind the terms of it;—that we cannot *distinctly* conceive the contrary of a necessary truth. For in a certain loose, indistinct way, persons conceive the contrary of necessary geometrical truths, when they erroneously conceive false propositions to be true. . . .

I have taken examples of necessary truths from the properties of number and space; but such truths exist no less in other subjects, although the discipline of thought which is requisite to perceive them distinctly, may not be so usual among men with regard to the sciences of mechanics and hydrostatics, as it is with regard to the sciences of geometry and arithmetic. Yet every one may perceive that there are such truths in mechanics. If I press the table with my hand, the table presses my hand with an equal force: here is a self-evident and necessary truth. In any machine, constructed in whatever manner to increase the force which I can exert, it is certain that what I gain in force I must lose in the velocity which I communicate. This is not a contingent truth, borrowed from and limited by observation; for a man of sound mechanical views applies it with like confidence, however novel be the construction of the machine. When I come to speak of the ideas which are involved in our mechanical knowledge, I may, perhaps, be able to bring more clearly into view the necessary truth of general propositions on such subjects. That reaction is equal and opposite to action, is as necessarily true as that two straight lines cannot inclose a space; it is impossible theoretically to make a perpetual motion by mere mechanism as to make the diagonal of a square commensurable with the side.

Necessary truths must be *universal* truths. If any property belong to a right-angled triangle *necessarily*, it must belong to *all* right-angled triangles. And it shall be proved in the following Chapter, that truths possessing these two characters, of Necessity and Universality, cannot possibly be the mere results of experience.

Of Experience

I here employ the term Experience in a more definite and limited sense than that which it possesses in common usage; for I restrict it to matters belonging to the domain of science. In such cases, the knowledge which we acquire, by means of experience, is of a clear and precise nature; and the passions and feelings and interests, which make the lessons of experience in practical matters so difficult to read aright, no longer disturb and confuse us. We may, therefore, hope, by attending to such cases, to learn what efficacy experience really has, in the discovery of truth.

That from *experience* (including intentional experience, or *observation,*) we obtain much knowledge which is highly important, and which could not be procured from any other source, is abundantly clear. We have already taken several examples of such knowledge. We know by experience that animals which ruminate are cloven-hoofed; and we know this in no other manner. We know, in like manner, that all the planets and their satellites revolve round the sun from west to east. It has been found by experience that all meteoric stones contain chrome. Many similar portions of our knowledge might be mentioned.

Now what we have here to remark is this;—that in no case can experience prove a proposition to be *necessarily* or *universally* true. However many instances we may have observed of the truth of a proposition, yet if it be known merely by observation, there is nothing to assure us that the next case shall not be an exception to the rule. If it be strictly true that every ruminant animal yet known has cloven hoofs, we still cannot be sure that some creature will not hereafter be discovered which has the first of these attributes without having the other. When the planets and their satellites, as far as Saturn, had been all found to move round the sun in one direction, it was still possible that there might be other such bodies not obeying this rule; and, accordingly, when the satellites of Uranus were detected, they appeared to offer an exception of this kind. Even in the mathematical sciences, we have examples of such rules suggested by experience, and also of their precariousness. However far they may have been tested, we cannot depend upon their correctness, except we see some reason for the rule. . . .

Experience must always consist of a limited number of observations. And, however numerous these may be, they can show nothing with regard to the infinite number of cases in which the experiment has not been made. Experience being thus unable to prove a fact to be universal, is, as will readily be seen, still more incapable of proving a truth to be necessary. Experience cannot, indeed, offer the smallest ground for the necessity of a proposition. She can observe and record what has happened; but she cannot find, in any case, or in any accumulation of cases, any reason for what *must* happen. She may see objects side by side; but she cannot see a reason why they must ever be side by side. She finds certain events to occur in succession; but the succession supplies, in its occurrence, no reason for its recurrence. She contemplates external objects; but she cannot detect any internal bond, which indissolubly connects the future with the past, the possible with the real. To learn a proposition by experience, and to see it to be necessarily true, are two altogether different processes of thought.

But it may be said, that we do learn by means of observation and experience many universal truths; indeed, all the general truths of which science consists. Is not the doctrine of universal gravitation learnt by experience? Are not the laws of motion, the properties of light, the general principles of chemistry, so learnt? How, with these examples before us, can we say that experience teaches no universal truths?

To this we reply, that these truths can only be known to be general, not universal, if they depend upon experience alone. Experience cannot bestow that universality which she herself cannot have, and that necessity of which she has no comprehension. If these doctrines *are* universally true, this universality flows from the *ideas* which we apply to our experience, and which are, as we have seen, the real sources of necessary truth. How far these ideas can communicate their universality and necessity to the results of experience, it will hereafter be our business to consider. It will then appear, that when the mind collects from observation truths of a wide and comprehensive kind, which approach to the simplicity and universality of the truths of pure science; she gives them this character by throwing upon them the light of her own Fundamental Ideas.

But the truths which we discover by observation of the external world, even when most strikingly simple and universal, are not necessary truths. Is the doctrine of universal gravitation necessarily true? It was doubted by Clairaut (so far as it refers to the moon), when the progression of the apogee in fact appeared to be twice as great as the theory admitted. It has been doubted, even more recently, with respect to the planets, their mutual perturbations appearing to indicate a deviation from the law. It is doubted still, by some persons, with respect to the double stars. But suppose all these doubts to be banished, and the law to be universal; is it then proved to be necessary? Manifestly not: the very existence of these doubts proves that it is not so. For the doubts were dissipated by reference to observation and calculation, not by reasoning on the nature of the law. Clairaut's difficulty was removed by a more exact calculation of the effect of the sun's force on the motion of the apogee. The suggestion of Bessel, that the intensity of gravitation might be different for different planets, was found to be unnecessary, when Professor Airy gave a more accurate determination of the mass of Jupiter. . . .

Thus no knowledge of the necessity of any truths can result from the observation of what really happens. This being clearly understood, we are led to an important inquiry.

The characters of universality and necessity in the truths which form part of our knowledge, can never be derived from experience, by which so large a part of our knowledge is obtained. But since, as we have seen, we really do possess a large body of truths which are necessary, and because necessary, therefore universal, the question still recurs, from what source these characters of universality and necessity are derived. . . .

Of the Grounds of Necessary Truths

To the question just stated, I reply, that the necessity and universality of the truths which form a part of our knowledge, are derived from the *Fundamental Ideas* which those truths involve. These ideas entirely shape and circumscribe our knowledge; they regulate the active operations of our minds, without which our passive sensations do not become knowledge. They govern these operations, according to rules which are not only fixed and permanent, but which may be expressed in plain and definite terms; and these rules, when thus expressed, may be made the basis of demonstrations by which the necessary relations imparted to our knowledge by our Ideas may be traced to their consequences in the most remote ramifications of scientific truth.

These enunciations of the necessary and evident conditions imposed upon our knowledge by the Fundamental Ideas which it involves, are termed *Axioms*. Thus the Axioms of Geometry express the necessary conditions which result from the Idea of Space; the Axioms of Mechanics express the necessary conditions which flow from the Ideas of Force and Motion; and so on.

It will be the office of several of the succeeding Books of this work to establish and illustrate in detail what I have thus stated in general terms. I shall there pass in review many of the most important fundamental ideas on which the existing body of our science depends; and I shall endeavour to show, for each such idea in succession, that knowledge involves an active as well as a passive element; that it is not possible without an act of the mind, regulated by certain laws. I shall further attempt to enumerate some of the principal fundamental relations which each idea thus introduces into our thoughts, and to express them by means of definitions and axioms, and other suitable forms.

I will only add a remark or two to illustrate further this view of the ideal grounds of our knowledge.

To persons familiar with any of the demonstrative sciences, it will be apparent that if we state all the Definitions and Axioms which are employed in the demonstrations, we state the whole basis on which those reasonings rest. For the whole process of demonstrative or deductive reasoning in any science, (as in geometry, for instance,) consists entirely in combining some of these first principles so as to obtain the simplest propositions of the science; then combining these so as to obtain other propositions of greater complexity; and so on, till we advance to the most recondite demonstrable truths; these last, however, intricate and unexpected, still involving no principles except the original definitions and axioms. Thus, by combining the Definition of a triangle, and the Definitions of equal lines and equal angles, namely, that they are such as when applied to each other, coincide, with the Axiom respecting straight lines (that two such lines cannot inclose a space,) we demonstrate the equality of triangles, under certain assumed conditions. Again, by combining this result with the Definition of parallelograms, and with the Axiom that if equals be taken from equals the wholes are equal, we prove the equality of parallelograms between the same parallels and upon the same base. From this proposition, again, we prove the equality of the square on the hypotenuse of a triangle to the squares on the two sides containing the right angle. But in all this there is nothing contained which is not rigorously the result of our geometrical Definitions and Axioms. All the rest of our treatises of geometry consists only of terms and phrases of reasoning, the object of which is to connect those first principles, and to exhibit the effects of their combination in the shape of demonstration.

This combination of first principles takes place according to the forms and rules of *Logic*. All the steps of the demonstration may be stated in the shape of which logicians are accustomed to exhibit processes of reasoning in order to show their conclusiveness, that is, in *Syllogisms*. Thus our geometrical reasonings might be resolved into such steps as the following:—

All straight lines drawn from the centre of a circle to its circumference are equal:

But the straight lines AB, AC, are drawn from the centre of a circle to its circumference:

Therefore the straight lines AB, AC, are equal.

Each step of geometrical, and all other demonstrative reasoning, may be resolved into three such clauses as these; and these three clauses are termed respectively, the *major premiss*, the *minor premiss*, and the *conclusion;* or, more briefly, the *major*, the *minor*, and the *conclusion*.

The principle which justifies the reasoning when exhibited in this syllogistic form, is this:—that a truth which can be asserted as generally, or rather as universally true, can be asserted as true also in each particular case. The *minor* only asserts a certain particular case to be an example of such conditions as are spoken of in the *major;* and hence the conclusion, which is true of the major by supposition, is true of the minor by consequence; and thus we proceed from syllogism to syllogism, in each one employing some general truth in some particular instance. Any proof which occurs in geometry, or any other science of demonstration, may thus be reduced to a series of processes, in each of which we pass from some general proposition to the narrower and more special propositions which it includes. And this process of deriving truths by the mere combination of general principles, applied in particular hypothetical cases, is called *deduction;* being opposed to *induction*, in which . . . a new general principle is introduced at every step.

Now we have to remark that, this being so, however far we follow such deductive reasoning, we can never have, in our conclusion any truth which is not virtually included in the original principles from which the reasoning started. For since at any step we merely take out of a general proposition something included in it, while at the preceding step we have taken this general proposition out of one more general, and so on perpetually, it is manifest that our last result was really included in the principle or principles with which we began. I say *principles*, because, although our logical conclusion can only exhibit the legitimate issue of our first principles, it may, nevertheless, contain the result of the combination of several such principles, and may thus assume a great degree of complexity, and may appear so far removed from the parent truths, as to betray at first sight hardly any relationship with them. Thus the proposition which has already been quoted respecting the squares on the sides of a right-angled triangle, contains the results of many elementary principles; as, the definitions of parallels, triangle, and square; the axioms respecting straight lines, and respecting parallels; and, perhaps, others. The conclusion is complicated by containing the effects of the combination of all these elements; but it contains nothing, and can contain nothing, but such elements and their combinations. . . .

In this manner the whole substance of our geometry is reduced to the Definitions and Axioms which we employ in our elementary reasonings; and in like manner we reduce the demonstrative truths of any other science to the definitions and axioms which we there employ.

But in reference to this subject, it has sometimes been said that demonstrative sciences do in reality depend upon Definitions only; and that no additional kind of principle, such as we have supposed Axioms to be, is absolutely required. It has been asserted that in geometry, for example, the source of the necessary truth of our propositions is this, that they depend upon definitions alone, and consequently merely state the identity of the same thing under different aspects.

That in the sciences which admit of demonstration, as geometry, mechanics, and the like, Axioms as well as Definitions are needed, in order to express the grounds of our necessary convictions, must be shown hereafter by an examination of each of these sciences in particular. But that the propositions of these sciences, those of geometry for example, do not merely assert the identity of the same thing, will, I think, be generally allowed. . . .

Thus the real logical basis of every body of demonstrated truths are the Definitions and Axioms which are the first principles of the reasonings. But when we are arrived at this point, the question further occurs, what is the ground of the truth of these Axioms? It is not the logical, but the philosophical, not the formal, but the real foundation of necessary truth, which we are seeking. Hence this inquiry necessarily comes before us, What is the ground of the Axioms of Geometry, of Mechanics, and of any other demonstrable science?

The answer which we are led to give, by the view which we have taken of the nature of knowledge, has already been stated. The ground of the axioms belonging to each science is the *Idea* which the axiom involves. The ground of the Axioms of Geometry is the *Idea of Space:* the ground of the Axioms of Mechanics is the *Idea of Force,* of *Action* and *Reaction,* and the like. And hence these Ideas are Fundamental Ideas; and since they are thus the foundations, not only of demonstration but of truth, an examination into their real import and nature is of the greatest consequence to our purpose.

Not only the Axioms, but the Definitions which form the basis of our reasonings, depend upon our Fundamental Ideas. And the

Definitions are not arbitrary definitions, but are determined by a necessity no less rigorous than the Axioms themselves. . . .

These principles,—Definitions, and Axioms,—thus exhibiting the primary developments of a fundamental idea, do in fact express the idea, so far as its expression in words forms part of our science. They are different views of the same body of truth; and though each principle, by itself, exhibits only one aspect of this body, taken together they convey a sufficient conception of it for our purposes. The Idea itself cannot be fixed in words; but these various lines of truth proceeding from it, suggest sufficiently to a fitly-prepared mind, the place where the idea resides, its nature, and its efficacy.

It is true that these principles,—our elementary Definitions and Axioms,—even taken altogether, express the Idea incompletely. Thus the Definitions and Axioms of Geometry, as they are stated in our elementary works, do not fully express the Idea of Space as it exists in our minds. For, in addition to these, other Axioms, independent of these, and no less evident, can be stated . . . And thus the Idea is disclosed but not fully revealed, imparted but not transfused, by the use we make of it in science. . . .

The Fundamental Ideas Are Not Derived From Experience

By the course of speculation contained in the last three Chapters, we are again led to the conclusion which we have already stated, that our knowledge contains an ideal element, and that this element is not derived from experience. For we have seen that there are propositions which are known to be necessarily true; and that such knowledge is not, and cannot be, obtained by mere observation of actual facts. It has been shown, also, that these necessary truths are the results of certain fundamental ideas, such as those of space, number, and the like. Hence it follows inevitably that these ideas and others of the same kind are not derived from experience. For these ideas possess a power of infusing into their developments that very necessity which experience can in no way bestow. This power they do not borrow from the external world, but possess by their own nature. Thus we unfold out of the Idea of Space the propositions of geometry, which are plainly truths of the most rigorous necessity and universality. But if the idea of space were merely collected from observation of the external world, it could never enable or entitle us to assert such propositions: it could never authorize us to say that not merely some lines, but *all lines*, not only have, but *must* have, those properties which geometry teaches. Geometry in every proposition speaks a language which experience never dares to utter; and indeed of which she but half comprehends the meaning. Experience sees that the assertions are true, but she sees not how profound and absolute is their truth. She unhesitatingly assents to the laws which geometry delivers, but she does not pretend to see the origin of their obligation. She is always ready to acknowledge the sway of pure scientific principles as a matter of fact, but she does not dream of offering her opinion on their authority as a matter of right; still less can she justly claim to be herself the source of that authority.

David Hume asserted,[1] that we are incapable of seeing in any of the appearances which the world presents anything of necessary connexion; and hence he inferred that our knowledge cannot extend to any such connexion. It will be seen from what we have said that we assent to his remark as to the fact, but we differ from him altogether in the consequence to be drawn from it. Our inference from Hume's observation is, not the truth of his conclusion, but the falsehood of his premisses;—not that, therefore, we can know nothing of natural connexion, but that, therefore, we have some other source of knowledge than experience:—not, that we can have no idea of connexion or causation, because, in his language, it cannot be the copy of an impression; but that since we have such an idea, our ideas are not the copies of our impressions.

Since it thus appears that our fundamental ideas are not acquired from the external world by our senses, but have some separate and independent origin, it is important for us to examine their nature and properties, as they exist in themselves; and this it will be our business to do through a portion of the following pages. But it may be proper first to notice one or two objections which may possibly occur to some readers.

It may be said that without the use of our senses, of sight and touch, for instance, we should never have any idea of space; that this idea, therefore, may properly be said to be derived from those senses. And to this I reply, by referring to a parallel instance. Without light we should have no perception of visible figure; yet the power of perceiving visible figure cannot be said to be derived from the light, but resides in the structure of the eye. If we had never seen objects in the light, we should be quite unaware that we possessed a power of vision; yet we should not possess it the less on that ac-

[1] *Essays*, Vol. II. p. 70.

count. If we had never exercised the senses of sight and touch (if we can conceive such a state of human existence) we know not that we should be conscious of an idea of space. But the light reveals to us at the same time the existence of external objects and our power of seeing. And in a very similar manner, the exercise of our senses discloses to us, at the same time, the external world, and our own ideas of space, time, and other conditions, without which the external world can neither be observed nor conceived. That light is necessary to vision, does not, in any degree, supersede the importance of a separate examination of the laws of our visual powers, if we would understand the nature of our own bodily faculties and the extent of the information they can give us. In like manner, the fact that intercourse with the external world is necessary for the conscious employment of our ideas, does not make it the less essential for us to examine those ideas in their most intimate structure, in order that we may understand the grounds and limits of our knowledge. Even before we see a single object, we have a faculty of vision; and in like manner, if we can suppose a man who has never contemplated an object in space or time, we must still assume him to have the faculties of entertaining the ideas of space and time, which faculties are called into play on the very first occasion of the use of the senses.

In answer to such remarks as the above, it has sometimes been said that to assume separate faculties in the mind for so many different processes of thought, is to give a mere verbal explanation, since we learn nothing concerning our idea of space by being told that we have a faculty of forming such an idea. It has been said that this course of explanation leads to an endless multiplication of elements in man's nature, without any advantage to our knowledge of his true constitution. We may, it is said, assert man to have a faculty of walking, of standing, of breathing, of speaking; but what, it is asked, is gained by such assertions? To this I reply, that we undoubtedly have such faculties as those just named; that it is by no means unimportant to consider them; and that the main question in such cases is, whether they are separate and independent faculties, or complex and derivative ones; and, if the latter be the case, what are the simple and original faculties by the combination of which the others are produced. . . .

Of the Philosophy of the Sciences

We proceed, in the ensuing Books, to the closer examination of a considerable number of those Fundamental Ideas on which the sciences, hitherto most successfully cultivated, are founded. In this task, our objects will be to explain and analyze such Ideas so as to bring into view the Definitions and Axioms, or other forms, in which we may clothe the conditions to which our speculative knowledge is subjected. I shall also try to prove, for some of these Ideas in particular, what has been already urged respecting them in general, that they are not derived from observation, but necessarily impose their conditions upon that knowledge of which observation supplies the materials. I shall further, in some cases, endeavour to trace the history of these Ideas as they have successively come into notice in the progress of science; the gradual development by which they have arrived at their due purity and clearness; and, as a necessary part of such a history, I shall give a view of some of the principal controversies which have taken place with regard to each portion of knowledge.

An exposition and discussion of the Fundamental Ideas of each Science may, with great propriety, be termed the PHILOSOPHY OF SUCH SCIENCE. These ideas contain in themselves the elements of those truths which the science discovers and enunciates; and in the progress of the sciences, both in the world at large and in the mind of each individual student, the most important steps consist in apprehending these ideas clearly, and in bringing them into accordance with the observed facts. . . .

C. PHILOSOPHY OF THE MECHANICAL SCIENCES

Of the Idea of Cause

We see in the world around us a constant succession of causes and effects connected with each other. The laws of this connexion we learn in a great measure from experience, by observation of the occurrences which present themselves to our notice, succeeding one another. But in doing this, and in attending to this succession of appearances, of which we are aware by means of our senses, we supply from our own minds the Idea of Cause. This Idea . . . is not derived

This article is from **The Philosophy of the Inductive Sciences Founded Upon Their History** by William Whewell (London, 1840), pp. 165–184. Reprinted by permission of Longmans, Green & Company, Ltd.

from experience, but has its origin in the mind itself;—is introduced into our experience by the active, and not by the passive part of our nature.

By Cause we mean some quality, power, or efficacy, by which a state of things produces a succeeding state. Thus the motion of bodies from rest is produced by a cause which we call *Force:* and in the particular case in which bodies fall to the earth, this force is termed *Gravity.* In these cases, the Conceptions of Force and Gravity receive their meaning from the Idea of Cause which they involve: for Force is conceived as the Cause of Motion. That this Idea of Cause is not derived from experience, we prove . . . by this consideration; that we can make assertions, involving this idea, which are rigorously necessary and universal; whereas knowledge derived from experience can only be true as far as experience goes, and can never contain in itself any evidence whatever of its necessity. We assert that "Every event must have a cause:" and this proposition we know to be true, not only probably, and generally, and as far as we can see; but we cannot suppose it to be false in any single instance. We are as certain of it as of the truths of arithmetic or geometry. We cannot doubt that it must apply to all events past and future, in every part of the universe, just as truly as to those occurrences which we have ourselves observed. *What* causes produce what effects;—what is the cause of any particular event;—what will be the effect of any peculiar process; —these are points on which experience may enlighten us. Observation and experience may be requisite, to enable us to judge respecting such matters. But that every event has *some* cause, Experience cannot prove any more than she can disprove. She can add nothing to the evidence of the truth, however often she may exemplify it. This doctrine, then, cannot have been acquired by her teaching; and the Idea of Cause, which the doctrine involves, and on which it depends, cannot have come into our minds from the region of observation.

That we do, in fact, apply the Idea of Cause in a more extensive manner than could be justified, if it were derived from experience only, is easily shown. For from the principle that everything must have a cause, we not only reason concerning the succession of the events which occur in the progress of the world, and which form the course of experience; but we infer that the world itself must have a cause;—that the chain of events connected by common causation, must have a First Cause of a nature different from the events themselves. This we are entitled to do, if our Idea of Cause be independent of, and superior to, experience: but if we have no Idea of Cause except such as we gather from experience, this reasoning is altogether baseless and unmeaning.

Again; by the use of our powers of observation, we are aware of a succession of appearances and events. But none of our senses or powers of external observation, can detect in these appearances the power or quality which we call Cause. Cause is that which connects one event with another; but no sense or perception discloses to us, or can disclose, any connexion among the events which we observe. We see that one occurrence follows another, but we can never see anything which shows that one occurrence *must* follow another. . . . One ball strikes another and causes it to move forward. But by what compulsion? Where is the necessity? If the mind can see any circumstance in this case which makes the result inevitable, let this circumstance be pointed out. But, in fact, there is no such discoverable necessity; for we can conceive this event not to take place at all. The struck ball may stand still, for aught we can see. "But the laws of motion will not allow it to do so." Doubtless they will not. But the laws of motion are learnt from experience, and therefore can prove no necessity. Why should not the laws of motion be other than they are? Are they necessarily true? That they are necessarily such as do actually regulate the impact of bodies, is at least no obvious truth; and therefore this necessity cannot be, in common minds, the ground of connecting the impact of one ball with the motion of another. And assuredly, if this fails, no other ground of such necessary connexion can be shown. In this case, then, the events are not seen to be necessarily connected. But if this case, where one ball moves another by impulse, be not an instance of events exhibiting a necessary connexion, we shall look in vain for any example of such a connexion. There is, then, no case in which events can be observed to be necessarily connected: our idea of causation, which implies that the event is necessarily connected with the cause, cannot be derived from observation.

But it may be said, we have not any such Idea of Cause, implying necessary connexion with effect, and a quality by which this connexion is produced. We see nothing but the succession of events; and by *cause* we mean nothing but a certain succession of events;— namely, a constant, unvarying succession. Cause and effect are only two events of which the second invariably follows the first. We delude ourselves when we imagine that our idea of causation involves anything more than this.

To this I reply by asking, what then is the meaning of the maxim above quoted, and allowed by all to be universally and necessarily true, that every event must have a cause? Let us put this maxim into the language of the explanation just noticed; and it becomes this:— "Every event must have a certain other event invariably preceding it." But why must it? Where is the necessity? Why must like events always be preceded by like, except so far as other events interfere? That there is such a necessity, no one can doubt. All will allow that if a stone ascend because it is thrown upwards in one case, a stone which ascends in another case has also been thrown upwards, or has undergone some equivalent operation. All will allow that in this sense, every kind of event must have some other specific kind of event preceding it. But this turn of men's thoughts shows that they see in events a connexion which is not mere succession. They see in cause and effect, not merely what does, often or always, precede and follow, but what *must* precede and follow. The events are not only conjoined, they are connected. The cause is more than the prelude, the effect is more than the sequel, of the fact. The cause is conceived not as a mere occasion; it is a power, an efficacy, which has a real operation.

Thus we have drawn from the maxim, that Every Effect must have a Cause, arguments to show that we have an Idea of Cause which is not borrowed from experience, and which involves more than mere succession. Similar arguments might be derived from any other maxims of universal and necessary validity, which we can obtain concerning Cause: as for example, the maxims that Causes are measured by their Effects, and that Reaction is equal and opposite to Action. These maxims we shall soon have to examine; but we may observe here, that the necessary truth which belongs to them, shows that they, and the Ideas which they involve, are not the mere fruits of observation; while their meaning, including, as it does, something quite different from the mere conception of succession of events, proves that such a conception is far from containing the whole import and signification of our Idea of Cause. . . .

Of the Axioms Which Relate to the Idea of Cause

Causes are abstract Conceptions—We have now to express, as well as we can, the fundamental character of that Idea of Cause, of which we have just proved the existence. This may be done, at least for purposes of reasoning, in this as in former instances, by means of axioms. I shall state the principal axioms which belong to this subject, referring the reader to his own thoughts for the axiomatic evidence which belongs to them.

But I must first observe, that in order to express general and abstract truths concerning cause and effect, these terms, *cause* and *effect*, must be understood in a general and abstract manner. When one event gives rise to another, the first *event* is, in common language, often called the cause, and the second the effect. Thus the meeting of two billiard balls may be said to be the cause of one of them turning aside out of the path in which it was moving. For our present purposes, however, we must not apply the term cause to such occurrences as this meeting and turning, but to a certain conception, *force*, abstracted from all such special events, and considered as a quality or property by which one body affects the motion of the other. And in like manner in other cases, cause is to be conceived as some abstract quality, power, or efficacy, by which change is produced; a quality not identical with the events, but disclosed by means of them. Not only is this abstract mode of conceiving force and cause useful in expressing the fundamental principles of science; but it supplies us with the only mode by which such principles can be stated in a general manner, and made to lead to substantial truth and real knowledge.

Understanding *cause*, therefore, in this sense, we proceed to our Axioms.

First Axiom. *Nothing can take place without a Cause.*

Every event, of whatever kind, must have a Cause in the sense of the term which we have just indicated; and that it must, is a universal and necessary proposition to which we irresistibly assent as soon as it is understood. We believe each appearance to come into existence, —we conceive every change to take place,—not only with something preceding it, but something by which it is made to be what it is. An effect without a cause;—an event without a preceding condition involving the efficacy by which the event is produced;—are suppositions which we cannot for a moment admit. That the connexion of effect with cause is universal and necessary, is a universal and constant conviction of mankind. It persists in the minds of all men, undisturbed by all the assaults of sophistry and skepticism and . . . remains unshaken, even when its foundations seem to be ruined. This axiom expresses, to a certain extent, our Idea of Cause; and when that idea is clearly apprehended, the axiom requires no proof, and indeed admits of none which makes it more evident. That notwithstanding its simplicity, it is of use in our speculations, we shall hereafter see; but in the first place, we must consider the other axioms belonging to this subject.

Second Axiom. *Effects are proportional to their Causes, and Causes are measured by their Effects.*

We have already said that *cause* is that quality or power, in the circumstances of each case, by which the effect is produced; and this power, an abstract property of the condition of things to which it belongs, can in no way fall directly under the cognizance of the senses. Cause, of whatever kind, is not apprehended as including objects and events which share its nature by being co-extensive with certain portions of it, as space and time are. It cannot therefore, like them, be measured by repetition of its own parts, as space is measured by repetition of inches, and time by repetition of minutes. Causes may be greater or less; as, for instance, the force of a man is greater than the force of a child. But how much is the one greater than the other? How are we to compare the abstract conception, force, in such cases as these?

To this, the obvious and only answer is, that we must compare causes by means of their effects;—that we must compare force by something which force can do. The child can lift one fagot; the man can lift ten such fagots: we have here a means of comparison. And whether or not the rule is to be applied in this manner, that is, by the number of the things operated on, (a question which we shall have to consider hereafter,) it is clear that this form of rule, namely, a reference to some effect or other as our measure, is the right, because the only possible form. The cause determines the effect. The cause being the same, the effect must be the same. The connexion of the two is governed by a fixed and inviolable rule. It admits of no ambiguity. Every degree of intensity in the cause has some peculiar modification of the effect corresponding to it. Hence the effect is an unfailing index of the amount of the cause; and if it be a measurable effect, gives a measure of the cause. We can have no other measure; but we need no other, for this is exact, sufficient, and complete.

It may be said, that various effects are produced by the same cause. The sun's heat melts wax and expands quicksilver. The force of gravity causes bodies to move downwards if they are free, and to press down upon their supports if they are supported. Which of the effects is to be taken as the measure of heat, or of gravity, in these cases? To this we reply, that if we had merely different states of the same cause to compare, any of the effects might be taken. The sun's heat on different days might be measured by the expansion of quicksilver, or by the quantity of wax melted. The force of gravity, if it were different at different places, might be measured by the spaces through which a given weight would bend an elastic support, or by the spaces through which a body would fall in a given time. All these measures are consistent with the general character of our idea of cause.

Limitation of the Second Axiom—But there may be circumstances in the nature of the case which may further determine the kind of effect which we must take for the measure of the cause. For example, if causes are conceived to be of such a nature as to be capable of addition, the effects taken as their measure must conform to this condition. This is the case with mechanical causes. The weights of two bodies are the causes of the pressure which they exert downwards; and these weights are capable of addition. The weight of the two is the sum of the weight of each. We are therefore not at liberty to say that weights shall be measured by the spaces through which they bend a certain elastic support, except we have first ascertained that the whole weight bends it through a space equal to the sum of the inflections produced by the separate weights. Without this precaution, we might obtain inconsistent results. . . .

Causes which are thus capable of addition are to be measured by the repeated addition of equal quantities. Two such causes are *equal* to each other when they produce exactly the same effect. So far our axiom is applied directly. But these two causes can be *added* together; and being thus added, they are *double* of one of them; and the cause composed by addition of *three* such, is *three* times as great as the first; and so on for any measure whatever. By this means, and by this means only, we have a complete and consistent measure of those causes which are so conceived as to be subject to this condition of being added and multiplied.

Causes are, in the present chapter, to be understood in the widest sense of the term; and the axiom now under our consideration applies to them, whenever they are of such a nature as to admit of any measure at all. But the cases which we have more particularly in view are *mechanical* causes, the causes of the motion and of the equilibrium of bodies. In these cases, forces are conceived as capable of addition; and what has been said of the measure of causes in such cases, applies peculiarly to mechanical forces. Two weights; placed together, may be considered as a single weight, equal to the *sum* of the two. Two pressures, pushing a body in the same direction at the same point, are identical in all respects with some single pressure, their *sum*, pushing in like manner; and this is true whether or not they put the body in motion. In the cases of mechanical forces, therefore, we take some certain effect, velocity generated or weight supported, which may fix the *unit* of force; and we then measure all other forces by the successive repetition of this unit, as we measure all spaces by the successive repetition of our unit of lineal measure. . . .

Third Axiom. *Reaction is equal and opposite to Action.*

In the case of mechanical forces, the action of a cause often takes place by an operation of one body upon another; and in this case, the action is always and inevitably accompanied by an *opposite* action. If I press a stone with my hand, the stone presses my hand in return. If one ball strike another and put it in motion, the second ball diminishes the motion of the first. In these cases the operation is mutual; the Action is accompanied by a Reaction. And in all such cases the Reaction is a force of exactly the same nature as the Action, exerted in an opposite direction. A pressure exerted upon a body at rest is resisted and balanced by another pressure; when the pressure of one body puts another in motion, the body, though it yields to the force, nevertheless exerts upon the pressing body a force like that which it suffers.

Now the axiom asserts further, that this Reaction is *equal*, as well as opposite, to the Action. For the Reaction is an effect of the Action, and is determined by it. And since the two, Action and Reaction, are forces of the same nature, each may be considered as cause and as effect; and they must, therefore, determine each other by a common rule. But this consideration leads necessarily to their equality: for since the rule is mutual, if we could for an instant suppose the Reaction to be less than the Action, we must by the same rule, suppose the Action to be less than the Reaction. And thus Action and Reaction, in every such case, are rigorously equal to each other.

It is easily seen that this axiom is not a proposition which is, or can be, proved by experience; but that its truth is anterior to special observation, and depends on our conception of Action and Reaction. Like our other axioms, this has its source in an Idea; namely, the Idea of Cause, under that particular condition in which cause and effect are mutual. The necessary and universal truth which we cannot help ascribing to the axiom, shows that it is not derived from the stores of experience, which can never contain truths of this character. Accordingly, it was asserted with equal confidence and generality by those who did not refer to experience for their principles, and by those who did. . . .

Extent of the Third Axiom—It may naturally be asked whether this third Axiom respecting causation extends to any other cases than those of mechanical action, since the notion of Cause in general has certainly a much wider extent. For instance, when a hot body heats a cold one, is there necessarily an equal reaction of the second body upon the first? Does the snowball cool the boy's hand exactly as much as the hand heats the snow? To this we reply, that, in every case in which one body acts upon another by its physical qualities, there must be some reaction. No body can effect another without being itself also affected. But in any physical change the *action* exerted is an abstract term which may be variously understood. The hot hand may *melt* a cold body, or may *warm* it: which kind of effect is to be taken as action? This remains to be determined by other considerations.

In all cases of physical change produced by one body in another, it is generally possible to assume such a meaning of action, that the reaction shall be of the same nature as the action; and when this is done, the third axiom of causation, that reaction is equal to action, is universally true. Thus if a hot body heats a cold one, the change may be conceived as the transfer of a certain substance, *heat* or *caloric*, from the first body to the second. On this supposition, the first body *loses* just as much heat as the other *gains;* action and reaction are equal. But if the reaction be of a different kind to the action we can no longer apply the axiom. If a hot body *melt* a cold one, the latter *cools* the former: here, then, is reaction; but so long as the action and reaction are stated in this form, we cannot assert any equality between them.

D. ON THE PROCESS OF DISCOVERY

Of the Colligation of Facts

APHORISM [I]

Science begins with common observation of facts; but even at this stage, requires that the observations be precise. Hence the sciences which depend upon space and number were the

This article is from **Novum Organon Renovatum** by William Whewell (London, 1858), pp. 59–120 (passim). Reprinted by permission of Longmans, Green & Company, Ltd.

earliest formed. After common observation, come Scientific Observation and Experiment.

APHORISM [II]

The Conceptions by which Facts are bound together, are suggested by the sagacity of discoverers. This sagacity cannot be taught. It commonly succeeds by guessing; and this success seems to consist in framing several tentative hypotheses and selecting the right one. But a supply of appropriate hypotheses cannot be constructed by rule, nor without inventive talent.

APHORISM [III]

The truth of tentative hypotheses must be tested by their application to facts. The discoverer must be ready, carefully to try his hypotheses in this manner, and to reject them if they will not bear the test, in spite of indolence and vanity. . . .

Of Certain Characteristics of Scientific Induction

APHORISM [IV]

The process of scientific discovery is cautious and rigorous, not by abstaining from hypotheses, but by rigorously comparing hypotheses with facts, and by resolutely rejecting all which the comparison does not confirm.

APHORISM [V]

Hypotheses may be useful, though involving much that is superfluous, and even erroneous: for they may supply the true bond of connexion of the facts; and the superfluity and errour may afterwards be pared away.

APHORISM [VI]

It is a test of true theories not only to account for, but to predict phenomena.

APHORISM [VII]

Induction is a term applied to describe the process of a true Colligation of Facts by means of an exact and appropriate Conception. An Induction is also employed to denote the proposition which results from this process.

APHORISM [VIII]

The Consilience of Inductions takes place when an Induction, obtained from one class of facts, coincides with an Induction, obtained from another different class. This Consilience is a test of the Theory in which it occurs.

APHORISM [IX]

An Induction is not the mere sum of the Facts which are

colligated. The Facts are not only brought together, but seen in a new point of view. A new mental Element is super-induced; and a peculiar constitution and discipline of mind are requisite in order to make this Induction.

APHORISM [x]

Although in Every Induction a new conception is super-induced upon the Facts; yet this once effectually done, the novelty of the conception is overlooked, and the conception is considered as a part of the fact. . . .

Of the Logic of Induction

APHORISM [xi]

The Logic of Induction consists in stating the Facts and the Inference in such a manner, that the Evidence of the Inference is manifest; just as the Logic of Deduction consists in stating the Premises and the Conclusion in such a manner that the Evidence of the Conclusion is manifest.

APHORISM [xii]

The Logic of Deduction is exhibited by means of a certain Formula; namely, a Syllogism; and every train of deductive reasoning, to be demonstrative, must be capable of resolution into a series of such Formulæ legitimately constructed. In like manner, the Logic of Induction may be exhibited by means of certain Formulæ; and every train of inductive inference, to be sound, must be capable of resolution into a scheme of such Formulæ, legitimately constructed.

APHORISM [xiii]

The inductive act of thought by which several Facts are colligated into one Proposition, may be expressed by saying: The several Facts are exactly expressed as one Fact, if, and only if, we adopt the Conceptions and the Assertion of the Proposition.

APHORISM [xiv]

The One Fact, thus inductively obtained from several Facts, may be combined with other Facts, and colligated with them by a new act of Induction. This process may be indefinitely repeated: and these successive processes are the Steps of Induction, or of Generalization, from the lowest to the highest.

APHORISM [xv]

The relation of the successive Steps of Induction may be exhibited by means of an Inductive Table, in which the several

Facts are indicated, and tied together by a Bracket, and the Inductive Inference placed on the other side of the Bracket; and this arrangement repeated, so as to form a genealogical Table of each Induction, from the lowest to the highest.

APHORISM [XVI]

The Logic of Induction is the Criterion of Truth inferred from Facts, as the Logic of Deduction is the Criterion of Truth deduced from necessary Principles. The Inductive Table enables us to apply such a Criterion; for we can determine whether each Induction is verified and justified by the Facts which its Bracket includes; and if each induction in particular be sound, the highest, which merely combines them all, must necessarily be sound also.

APHORISM [XVII]

The distinction of Fact and Theory is only relative. Events and phenomena, considered as Particulars which may be colligated by Induction, are Facts; considered as Generalities already obtained by colligation of other Facts, they are Theories. The same event or phenomenon is a Fact or a Theory, according as it is considered as standing on one side or the other of the Inductive Bracket. . . .

Of Laws of Phenomena and of Causes

APHORISM [XVIII]

Inductive truths are of two kinds, Laws of Phenomena, and Theories of Causes. It is necessary to begin in every science with the Laws of Phenomena; but it is impossible that we should be satisfied to stop short of a Theory of Causes. In Physical Astronomy, Physical Optics, Geology, and other sciences, we have instances showing that we can make a great advance in inquiries after true Theories of Causes.

In the first attempts at acquiring an exact and connected knowledge of the appearances and operations which nature presents, men went no further than to learn *what* takes place, not *why* it occurs. They discovered an Order which the phenomena follow, Rules which they obey; but they did not come in sight of the Powers by which these rules are determined, the Causes of which this order is the effect. Thus, for example, they found that many of the celestial motions took place as if the sun and stars were carried round by the revolutions of certain celestial spheres; but what causes kept these spheres in constant motion, they were never able to explain. In like manner in modern times, Kepler discovered that the planets describe ellipses, before Newton explained why they select this particular curve, and describe it in a particular manner. The laws of reflection, refraction, dispersion, and other properties of light have long been known; the causes of these laws are at present under discussion. And the same might be said of many other sciences. The discovery of *the Laws of Phenomena* is, in all cases, the first step in exact knowledge; these Laws may often for a long period constitute the whole of our science; and it is always a matter requiring great talents and great efforts, to advance to a knowledge of the *Causes* of the phenomena.

Hence the larger part of our knowledge of nature, at least of the certain portion of it, consists of the knowledge of the Laws of Phenomena. In Astronomy indeed, besides knowing the rules which guide the appearances, and resolving them into the real motions from which they arise, we can refer these motions to the forces which produce them. In Optics, we have become acquainted with a vast number of laws by which varied and beautiful phenomena are governed; and perhaps we may assume, since the evidence of the Undulatory Theory has been so fully developed, that we know also the Causes of the Phenomena. But in a large class of sciences, while we have learnt many Laws of Phenomena, the causes by which these are produced are still unknown or disputed. Are we to ascribe to the operation of a fluid or fluids, and if so, in what manner, the facts of heat, magnetism, electricity, galvanism? What are the forces by which the elements of chemical compounds are held together? What are the forces, of a higher order, as we cannot help believing, by which the course of vital action in organized bodies is kept up? In these and other cases, we have extensive departments of science; but we are as yet unable to trace the effects to their causes; and our science, so far as it is positive and certain, consists entirely of the laws of phenomena.

In those cases in which we have a division of the science which teaches us the doctrine of the causes, as well as one which states the rules which the effects follow, I have, in the *History*, distinguished the two portions of the science by certain terms. I have thus spoken of *Formal* Astronomy and *Physical* Astronomy. The latter phrase has long been commonly employed to describe that department of Astronomy which deals with those forces by which the heavenly bodies are guided in their motions; the former adjective appears well suited to describe a collection of rules depending on those ideas of space, time, position, number, which are, as we have already said, the *forms* of our apprehension of phenomena. The laws of phenomena may be considered as *formulæ*, expressing results in terms of those ideas. In like manner, I have spoken of Formal Optics and Physical Optics; the latter division including all speculations concerning the machinery by which the effects are produced. Formal Acoustics and Physical Acoustics may be distinguished in like manner, although these two portions of science have been a good deal mixed together by most of those who have treated of them. Formal Thermotics, the knowledge of the laws of the phenomena of heat, ought in like manner to lead to Physical Thermotics, or the Theory of Heat with reference to the cause by which its effects are produced;—a branch of science which as yet can hardly be said to exist.

What *kinds of cause* are we to admit in science? This is an important, and by no means an easy question. In order to answer it, we must consider in what manner our progress in the knowledge of causes has hitherto been made. By far the most conspicuous instance of success in such researches, is the discovery of the causes of the motions of the heavenly bodies. In this case, after the formal laws of the motions,—their conditions as to space and time,—had become known, men were enabled to go a step further; to reduce them to the familiar and general cause of motion—mechanical force; and to determine the laws which this force follows. That this was a step in addition to the knowledge previously possessed, and that it was a real and peculiar truth, will not be contested. And a step in any other subject which should be analogous to this in astronomy;—a discovery of causes and forces as certain and clear as the discovery of universal gravitation;—would undoubtedly be a vast advance upon a body of science consisting only of the laws of phenomena.

Chapter 4
JOHN STUART MILL
(1806–1873)

INTRODUCTION

*J*OHN STUART MILL, son of the historian, economist, and philosopher James Mill (1773–1836), was born in London on May 20, 1806. His early education was carefully conducted by his father. When the younger Mill was three years old he began to learn Greek; this subject remained his main area of study for five years to which English and Arithmetic were later added. Mill undertook Latin, geometry, and algebra when he was eight and thereafter began to teach the younger children of the family. At twelve he started his study of logic and two years later delved into psychology and political economy. For a period of about eleven years Mill spent most of his time in his father's study and also often accompanied him on his walks. Through their countless conversations and discussions the elder Mill gradually contributed to the development of his son's mind, trying to impart to him clarity and exactness of thought and confidence in thinking.

Mill went to France in 1820 and stayed there for a year with the family of Sir Samuel Bentham; he studied the French language, French literature, politics, chemistry, botany, and advanced mathematics. When he returned from France his father introduced him to Jeremy Bentham's thinking, which deeply impressed him. He studied Roman Law under John Austin and continued his studies in psychology at that time, too.

In 1823 the East India Company appointed him to the post of Assistant Examiner, under his father, in the Office of the Examiner of India Correspondence, a position he held until he became chief of the office in 1856.

The overstrain of his early education and the hard work which followed brought a breakdown in Mill's health in 1828, poignantly recounted in his autobiography. He struggled with his various problems in solitude and eventually emerged a "new man" with broader and more personal insights, with deeper and more intense sympathies. Between 1830 and 1843 Mill wrote many essays and articles on a variety of subjects, including literary, philosophical, political, social, and economical, but it was not until 1843 that he was able to finish his first great intellectual work, *A System of Logic, Ratiocinative and Inductive*, which he himself considered later to be one of his best works. During the thirty years which followed the appearance of this book he wrote many constructive and critical works on philosophy, economics, and politics. He was active in philosophy, science, politics, economy, and social practice over the years and remained vital in all these spheres until his death on May 8, 1873.

Mill's *System of Logic* was meant to supply an unimpeachable method of proof for conclusions in moral and social science. There is no doubt that his striving toward this ideal was strongly influenced by August Comte, his father, Bentham, and Hume. Mill considered Newton's physics as a model of scientific exposition whôse inductive logic was to function as a paradigm case for the moral and social sciences. In his view, however, this new logic could not be conceived of as antagonistic to Aristotle's classical logic; inductive logic was to supplement, not supersede, the old logic. Mill searched for several years to find a way of concatenating these two forms of logic. Under the combined influence of Herschel's *Preliminary Discourse on the Study of Natural Philosophy* and Whewell's *Philosophy of the Inductive Sciences* he was finally able to formulate the methods of scientific investigation and to join the new inductive logic harmoniously to its old deductive counterpart. His ideas were published in *System of Logic* in 1843; between 1851 and 1872 he made several revisions which appeared from the third to the eighth editions.

At the beginning of the book Mill states that the goal of the treatise on logic is ". . . to attempt a correct analysis of the intellectual process called reasoning and inference, and of such other mental operations as are intended to facilitate this: as well as, on the foundation of this analysis, and *pari passu* with it, to bring together or frame a set of rules or canons for testing the sufficiency of any given evidence to prove any given proposition." He explains this by stating that he includes induction as well as deduction in reasoning, and that by "other mental operations" he means naming, definition, and classification.

Mill attempts to bridge the gap between inductive and classical logic by claiming that deductive inference is a form of apparent, not real, reasoning, whereas ampliative induction is a form of real inference. But to say that deduction is a form of apparent inference is not to say that deduction is meaningless. Although it is apparent inference and circular in nature, deduction considered as a mode of verifying any given argument is of importance and its value even manifold, in Mill's view. Although deductive logic is not a logic of truth, it is and remains a logic of consistency.

With respect to induction, Mill refers with it to a method which enables one to prove general statements, not as Whewell held, merely a method of discovery. Furthermore in Mill's conception of induction, the question of whether or not the proof of individual facts involves induction was left open.

The selections give a substantial idea of Mill's doctrine on induction. They were chosen with emphasis on induction in connection with the inductive sciences, particularly the sciences of nature, rather than on typically logical problems.

SELECTIVE BIBLIOGRAPHY

Mill, J. S., *A System of Logic: Ratiocinative and Inductive* (2 vols., 1843). London, 1868.

Anschutz, R. P., *The Philosophy of John Stuart Mill.* Oxford, 1953.

Cohen M. and Nagel E., *Introduction to Logic and Scientific Method.* New York, 1934. Chap. 13.

Jackson, R., *An Examination of the Deductive Logic of J. S. Mill.* Oxford, 1953.

Ducasse, C. J., "John Stuart Mill's System of Logic," in *Theories of Scientific Method: The Renaissance Through the Nineteenth Century,* ed. E. H. Madden. Seattle, 1960. Pp. 218–232.

Britton, K., *John Stuart Mill.* Harmondsworth, Eng., 1953.

OF INDUCTION

> According to the doctrine now stated, the highest, or rather the only proper object of physics, is to ascertain those established conjunctions of successive events, which constitute the order of the universe; to record the phenomena which it exhibits to our observation or which it discloses to our experiments; and to refer these phenomena to their general laws.
> —D. STEWART, *Elements of the Philosophy of the Human Mind*, vol. ii. chap. iv. sect. 1.

PRELIMINARY OBSERVATIONS ON INDUCTION IN GENERAL

*T*he portion of the present inquiry upon which we are now about to enter may be considered as the principal, both from its surpassing in intricacy all the other branches, and because it relates to a process which has been shown in the preceding Book to be that in which the investigation of nature essentially consists. We have found that all Inference, consequently all Proof, and all discovery of truths not self-evident, consists of inductions, and the interpretation of inductions; that all our knowledge, not intuitive, comes to us exclusively from that source. What Induction is, therefore, and what conditions render it legitimate, cannot but be deemed the main question of the science of logic—the question which includes all others. It is, however, one which professed writers on logic have almost entirely passed over. . . .

For the purposes of the present inquiry, Induction may be defined, the operation of discovering and proving general propositions. It is true that (as already shown) the process of indirectly ascertaining individual facts is as truly inductive as that by which we establish general truths. But it is not a different kind of induction; it is a form

This article is from **A System of Logic: Ratiocinative and Inductive** by John Stuart Mill. (London, 1868), Vol. I, pp. 313–365 **(passim)**. Reprinted by permission of Longmans, Green & Company, Ltd.

of the very same process: since, on the one hand, generals are but collections of particulars, definite in kind but indefinite in number; and on the other hand, whenever the evidence which we derive from observation of known cases justifies us in drawing an inference respecting even one unknown case, we should on the same evidence be justified in drawing a similar inference with respect to a whole class of cases. The inference either does not hold at all, or it holds in all cases of a certain description; in all cases which, in certain definable respects, resemble those we have observed.

If these remarks are just; if the principles and rules of inference are the same whether we infer general propositions or individual facts; it follows that a complete logic of the sciences would be also a complete logic of practical business and common life. Since there is no case of legitimate inference from experience, in which the conclusion may not legitimately be a general proposition, an analysis of the process by which general truths are arrived at is virtually an analysis of all induction whatever. Whether we are inquiring into a scientific principle or into an individual fact, and whether we proceed by experiment or by ratiocination, every step in the train of inferences is essentially inductive, and the legitimacy of the induction depends in both cases on the same conditions. . . .

OF INDUCTIONS IMPROPERLY SO CALLED

Induction, then, is that operation of the mind by which we infer that what we know to be true in a particular case or cases, will be true in all cases which resemble the former in certain assignable respects. In other words, Induction is the process by which we conclude that what is true of certain individuals of a class is true of the whole class, or that what is true at certain times will be true in similar circumstances at all times.

This definition excludes from the meaning of the term Induction, various logical operations, to which it is not unusual to apply that name.

Induction, as above defined, is a process of inference; it proceeds from the known to the unknown; and any operation involving no inference, any process in which what seems the conclusion is no wider than the premises from which it is drawn, does not fall within the meaning of the term. Yet in the common books of Logic we find this laid down as the most perfect, indeed the only quite perfect, form of induction. . . .

There are several processes used in mathematics which require to be distinguished from Induction, being not unfrequently called by that name, and being so far similar to Induction properly so called, that the propositions they lead to are really general propositions. For example, when we have proved with respect to the circle that a straight line cannot meet it in more than two points, and when the same thing has been successively proved of the ellipse, the parabola, and the hyperbola, it may be laid down as an universal property of the sections of the cone. The distinction drawn in the two previous examples can have no place here, there being no difference between all *known* sections of the cone and *all* sections, since a cone demonstrably cannot be intersected by a plane except in one of these four lines. It would be difficult, therefore, to refuse to the proposition arrived at the name of a generalisation, since there is no room for any generalisation beyond it. But there is no induction, because there is no inference: the conclusion is a mere summing up of what was asserted in the various propositions from which it is drawn.

There are, nevertheless, in mathematics, some examples of so-called Induction, in which the conclusion does bear the appearance of a generalisation grounded on some of the particular cases included in it. A mathematician, when he has calculated a sufficient number of the terms of an algebraical or arithmetical series to have ascertained what is called the *law* of the series, does not hesitate to fill up any number of the succeeding terms without repeating the calculations. But I apprehend he only does so when it is apparent from *à priori* consideration (which might be exhibited in the form of demonstration) that the mode of formation of the subsequent terms, each from that which preceded it, must be similar to the formation of the terms which have been already calculated. And when the attempt has been hazarded without the sanction of such general considerations, there are instances on record in which it has led to false results. . . .

There remains a third improper use of the term Induction, which it is of real importance to clear up, because the theory of Induction has been, in no ordinary degree, confused by it, and because the confusion is exemplified in the most recent and elaborate treatise on the inductive philosophy which exists in our language. The error in question is that of confounding a mere description, by general terms, of a set of observed phenomena, with an induction from them.

Suppose that a phenomenon consists of parts, and that these parts are only capable of being observed separately, and as it were piecemeal. When the observations have been made, there is a convenience (amounting for many purposes to a necessity) in obtaining a representation of the phenomenon as a whole, by combining, or, as we may say, piecing these detached fragments together. A navigator sailing in the midst of the ocean discovers land: he cannot at first, or by any one observation, determine whether it is a continent or an island; but he coasts along it, and after a few days finds himself to have sailed completely round it: he then pronounces it an island. Now there was no particular time or place of observation at which he could perceive that this land was entirely surrounded by water; he ascertained the fact by a succession of partial observations, and then selected a general expression which summed up in two or three words the whole of what he so observed. But is there anything of the nature of an induction in this process? Did he infer anything that had not been observed, from something else which had? Certainly not. He had observed the whole of what the proposition asserts. That the land in question is an island, is not an inference from the partial facts which the navigator saw in the course of his circumnavigation; it is the facts themselves; it is a summary of those facts; the description of a complex fact, to which those simpler ones are as the parts of a whole. . . .

OF THE GROUND OF INDUCTION

Induction, properly so called, as distinguished from those mental operations, sometimes though improperly designated by the name, which I have attempted in the preceding chapter to characterise, may, then, be summarily defined as Generalisation from Experience. It consists in inferring from some individual instances in which a phenomenon is observed to occur, that it occurs in all instances of a certain class; namely, in all which *resemble* the former, in what are regarded as the material circumstances.

In what way the material circumstances are to be distinguished from those which are immaterial, or why some of the circumstances are material and others not so, we are not yet ready to point out. We must first observe that there is a principle implied in the very statement of what Induction is; an assumption with regard to the course of nature and the order of the universe; namely, that there are such things in nature as parallel cases; that what happens once will, under a sufficient degree of similarity of circumstances, happen again, and not only again, but as often as the same circumstances recur. This, I say, is an assumption involved in every case of induction. And if we consult the actual course of nature, we find that the assumption is warranted. The universe, so far as known to us, is so constituted, that whatever is true in any one case, is true in all cases of a certain description; the only difficulty is, to find what description.

This universal fact, which is our warrant for all inferences from experience, has been described by different philosophers in different forms of language; that the course of nature is uniform; that the universe is governed by general laws; and the like. One of the most usual of those modes of expression, but also one of the most inadequate, is that which has been brought into familiar use by the metaphysicians of the school of Reid and Stewart. The disposition of the human mind to generalise from experience,—a propensity considered by these philosophers as an instinct of our nature,—they usually describe under some such name as "our intuitive conviction that the future will resemble the past." Now it has been well pointed out by Mr. Bailey,[1] that (whether the tendency be or not an original and ultimate element of our nature) Time, in its modifications of past, present, and future has no concern either with the belief itself, or with the grounds of it. We believe that fire will burn to-morrow, because it burned to-day and yesterday; but we believe, on precisely the same grounds, that it burned before we were born, and that it burns this very day in Cochin-China. It is not from the past to the future, as past and future, that we infer, but from the known to the unknown; from facts observed to facts unobserved; from what we have perceived, or been directly conscious of, to what has come within our experience. In this last predicament is the whole region of the future; but also the vastly greater portion of the present and of the past.

Whatever be the most proper mode of expressing it, the proposition that the course of nature is uniform is the fundamental principle, or general axiom, of Induction. It would yet be a great error to offer this large generalisation as any explanation of the inductive process. On the contrary, I hold it to be itself an instance of induction, and induction by no means of the most obvious kind. Far from being the first induction we make, it is one of the last, or at all events one of those which are latest in attaining strict philosophical accuracy. As a general maxim, indeed, it has scarcely entered into the minds of any but philosophers; nor even by them, as we shall have many opportunities of remarking, have its extent and limits been always very justly conceived. The truth is, that this great generalisation is itself

---

[1] *Essays on the Pursuit of Truth.*

founded on prior generalisations. The obscurer laws of nature were discovered by means of it, but the more obvious ones must have been understood and assented to as general truths before it was ever heard of. We should never have thought of affirming that all phenomena take place according to general laws, if we had not first arrived, in the case of a great multitude of phenomena, at some knowledge of the laws themselves; which could be done no otherwise than by induction. In what sense, then, can a principle, which is so far from being our earliest induction, be regarded as our warrant for all the others? In the only sense in which (as we have already seen) the general propositions which we place at the head of our reasonings when we throw them into syllogisms ever really contribute to their validity. As Archbishop Whately remarks, every induction is a syllogism with the major premise suppressed; or (as I prefer expressing it) every induction may be thrown into the form of a syllogism by supplying a major premise. If this be actually done, the principle which we are now considering, that of the uniformity of the course of nature, will appear as the ultimate major premise of all inductions, and will, therefore, stand to all inductions in the relation in which, as has been shown at so much length, the major proposition of a syllogism always stands to the conclusion; not contributing at all to prove it, but being a necessary condition of its being proved; since no conclusion is proved for which there cannot be found a true major premise.

The statement that the uniformity of the course of nature is the ultimate major premise in all cases of induction may be thought to require some explanation. The immediate major premise in every inductive argument it certainly is not. Of that Archbishop Whately's must be held to be the correct account. The induction, "John, Peter, &c., are mortal, therefore all mankind are mortal," may, as he justly says, be thrown into a syllogism by prefixing as a major premise, (what is at any rate a necessary condition of the validity of the argument,) namely, that what is true of John, Peter, &c., is true of all mankind. But how came we by this major premise? It is not self-evident; nay, in all cases of unwarranted generalisation it is not true. How, then, is it arrived at? Necessarily either by induction or ratiocination; and if by induction, the process, like all other inductive arguments, may be thrown into the form of a syllogism. This previous syllogism it is, therefore, necessary to construct. There is, in the long-run, only one possible construction. The real proof that what is true of John, Peter, &c., is true of all mankind, can only be, that a different supposition would be inconsistent with the uniformity which we know to exist in the course of nature. Whether there would be this inconsistency or not, may be a matter of long and delicate inquiry; but unless there would, we have no sufficient ground for the major of the inductive syllogism. It hence appears, that if we throw the whole course of any inductive argument into a series of syllogisms, we shall arrive by more or fewer steps at an ultimate syllogism, which will have for its major premise the principle or axiom of the uniformity of the course of nature.

It was not to be expected that in the case of this axiom, any more than of other axioms, there should be unanimity among thinkers with respect to the ground on which it is to be received as true. I have already stated that I regard it as itself a generalisation from experience. Others hold it to be a principle which, antecedently to any verification by experience, we are compelled by the constitution of our thinking faculty to assume as true. . . .

OF LAWS OF NATURE

In the contemplation of that uniformity in the course of nature which is assumed in every inference from experience, one of the first observations that present themselves is, that the uniformity in question is not properly uniformity, but uniformities. . . .

The first point, therefore, to be noted in regard to what is called the uniformity of the course of nature is, that it is itself a complex fact, compounded of all the separate uniformities which exist in respect to single phenomena. These various uniformities, when ascertained by what is regarded as a sufficient induction, we call in common parlance, Laws of Nature. Scientifically speaking, that title is employed in a more restricted sense to designate the uniformities when reduced to their most simple expression. . . .

[T]he following are three uniformities, or call them laws of nature: the law that air has weight, the law that pressure on a fluid is propagated equally in all directions, and the law that pressure in one direction, not opposed by equal pressure in the contrary direction, produces motion, which does not cease until equilibrium is restored. From these three uniformities we should be able to predict another uniformity, namely, the rise of the mercury in the Torricellian tube. This, in the stricter use of the phrase, is not a law of nature. It is the result of laws of nature. It is a *case* of each and every one of the three laws; and is the only occurrence by which they could all be fulfilled. If the mercury were not sustained in the barometer, and sustained at such a height that the column of mercury were equal in weight to a column of the atmosphere of the same diameter; here would be a case, either of the air not pressing upon the surface of the mercury with the force which is called its weight, or of the downward pressure on the mercury not being propagated equally in an upper direction, or of a body pressed in one direction and not in the direction opposite, either not moving in the direction in which it is pressed, or stopping before it had attained equilibrium. If we knew, therefore, the three simple laws, but had never tried the Torricellian experiment, we might *deduce* its result from those laws. The known weight of the air, combined with the position of the apparatus, would .bring the mercury within the first of the three inductions; the first induction would bring it within the second, and the second within the third, in the manner which we characterised in treating of Ratiocination. We should thus come to know the more complex uniformity, independently of specific experience, through our knowledge of the simpler ones from which it results; though, for reasons which will appear hereafter, *verification* by specific experience would still be desirable, and might possibly be indispensable.

Complex uniformities which, like this, are mere cases of simpler ones, and have, therefore, been virtually affirmed in affirming those, may with propriety be called *laws*, but can scarcely, in the strictness of scientific speech, be termed Laws of Nature. It is the custom in science, wherever regularity of any kind can be traced, to call the general proposition which expresses the nature of that regularity a law; as when, in mathematics, we speak of the law of decrease of the successive terms of a converging series. . . . According to one mode of expression, the question, What are the laws of nature? may be stated thus: What are the fewest and simplest assumptions, which being granted, the whole existing order of nature would result? Another mode of stating it would be thus: What are the fewest general propositions from which all the uniformities which exist in the universe might be deductively inferred? . . .

According to this language, every well-grounded inductive generalisation is either a law of nature or a result of laws of nature, capable, if those laws are known, of being predicted from them. And the problem of Inductive Logic may be summed up in two questions: how to ascertain the laws of nature; and how, after having ascertained them, to follow them into their results. On the other hand, we must not suffer ourselves to imagine that this mode of statement amounts to a real analysis, or to anything but a mere verbal transformation of the problem; for the expression, Laws of Nature, *means* nothing but the uniformities which exist among natural phenomena (or, in other words, the results of induction) when reduced to their simplest expression. It is, however, something to have advanced so far as to see that the study of nature is the study of laws, not *a* law; of uniformities in the plural number; that the different natural phenomena have their separate rules or modes of taking place, which, though much intermixed and entangled with one another, may, to a certain extent, be studied apart; that (to resume our former metaphor) the regularity which exists in nature is a web composed of distinct threads, and only to be understood by tracing each of the threads separately; for which purpose it is often necessary to unravel some portion of the web, and exhibit the fibres apart. The rules of experimental inquiry are the contrivances for unravelling the web.

In thus attempting to ascertain the general order of nature by ascertaining the particular order of the occurrence of each one of the phenomena of nature, the most scientific proceeding can be no more than an improved form of that which was primitively pursued by the human understanding while undirected by science. When mankind first formed the idea of studying phenomena according to a stricter and surer method than that which they had in the first instance spontaneously adopted, they did not, conformably to the well-meant but impracticable precept of Descartes, set out from the supposition that nothing had been already ascertained. Many of the uniformities existing among phenomena are so constant, and so open to observation, as to force themselves upon involuntary recognition. Some facts are so perpetually and familiarly accompanied by certain others, that mankind learnt, as children learn, to expect the one where they found the other, long before they knew how to put their expectation into words by asserting, in a proposition, the existence of a connection between those phenomena. No science was needed to teach that food nourishes, that water drowns, or quenches thirst, that the sun gives light and heat, that bodies fall to the ground. The first scientific inquirers assumed these and the like as known truths, and set out from them to discover others which were unknown: nor were they wrong in so doing, subject, however, as they afterwards began to see, to an ulterior revision of these spontaneous generalisations themselves, when the progress of knowledge pointed out limits to them, or showed their truth to be contingent on some circumstance not originally attended to. It will appear, I think, from the subsequent part of our inquiry, that there is no logical fallacy in this mode of proceeding; but we may see already that any other mode is rigorously impracticable: since it is impossible to frame any scientific method of induction, or test of the correctness of inductions, unless on the hypothesis that some inductions deserving of reliance have been already made.

Let us revert, for instance, to one of our former illustrations, and consider why it is that, with exactly the same amount of evidence, both negative and positive we did not reject the assertion that there are black swans, while we should refuse credence to any testimony which asserted that there were men wearing their heads underneath their shoulders. The first assertion was more credible than the latter. But why more credible? So long as neither phenomenon had been actually witnessed, what reason was there for finding the one harder to be believed than the other? Apparently because there is less constancy in the colours of animals than in the general structure of their anatomy. But how do we know this? Doubtless, from experience. It appears, then, that we need experience to inform us in what degree, and in what cases, or sorts of cases, experience is to be relied on. Experience must be consulted in order to learn from it under what circumstances arguments from it will be valid. We have no ulterior test to which we subject experience in general; but we make experience its own test. Experience testifies that among the uniformities which it exhibits or seems to exhibit, some are more to be relied on than others; and uniformity, therefore, may be presumed, from any given number of instances, with a greater degree of assurance, in proportion as the case belongs to a class in which the uniformities have hitherto been found more uniform.

This mode of correcting one generalisation by means of another, a narrower generalisation by a wider, which common sense suggests and adopts in practice, is the real type of scientific Induction. All that art can do is but to give accuracy and precision to this process, and adapt it to all varieties of cases, without any essential alteration in its principle.

There are of course no means of applying such a test as that above described, unless we already possess a general knowledge of the prevalent character of the uniformities existing throughout nature. The indispensable foundation, therefore, of a scientific formula of induction must be a survey of the inductions to which mankind have been conducted in unscientific practice, with the special purpose of ascertaining what kinds of uniformities have been found perfectly invariable, pervading all nature, and what are those which have been found to vary with difference of time, place, or other changeable circumstances.

The necessity of such a survey is confirmed by the consideration that the stronger inductions are the touchstone to which we always endeavour to bring the weaker. If we find any means of deducing one of the less strong inductions from stronger ones, it acquires, at once, all the strength of those from which it is deduced; and even adds to that strength; since the independent experience on which the weaker induction previously rested becomes additional evidence of the truth of the better established law in which it is now found to be included. . . .

On the other hand, if an induction conflicts with stronger inductions, or with conclusions capable of being correctly deduced from them, then, unless on reconsideration it should appear that some of the stronger inductions have been expressed with greater universality than their evidence warrants, the weaker one must give way. . . .

It may be affirmed as a general principle, that all inductions, whether strong or weak, which can be connected by ratiocination, are confirmatory of one another; while any which lead deductively to consequences that are incompatible become mutually each other's test, showing that one or other must be given up, or at least more guardedly expressed. In the case of inductions which confirm each other, the one which becomes a conclusion from ratiocination rises to at least the level of certainty of the weakest of those from which it is deduced; while in general all are more or less increased in certainty. Thus the Torricellian experiment, though a mere case of three more general laws, not only strengthened greatly the evidence on which those laws rested, but converted one of them (the weight of the atmosphere) from a still doubtful generalisation into a completely established doctrine.

If, then, a survey of the uniformities which have been ascertained to exist in nature should point out some which, as far as any human purpose requires certainty, may be considered quite certain and quite universal, then by means of these uniformities we may be able to raise multitudes of other inductions to the same point in the scale. For if we can show, with respect to any inductive inference, that either it must be true, or one of these certain and universal inductions must admit of an exception, the former generalisation will attain the same certainty, and indefeasibleness within the bounds assigned to it, which are the attributes of the latter. It will be proved to be a law; and if not a result of other and simpler laws, it will be a law of nature.

There are such certain and universal inductions; and it is because there are such, that a *Logic of Induction* is possible.

OF THE LAW OF UNIVERSAL CAUSATION

The phenomena of nature exist in two distinct relations to one another; that of simultaneity, and that of succession. Every phenomenon is related, in an uniform manner, to some phenomena that co-exist with it, and to some that have preceded and will follow it.

Of the uniformities which exist among synchronous phenomena, the most important, on every account, are the laws of number; and next to them those of space, or, in other words of extension and figure. The laws of number are common to synchronous and successive phenomena. That two and two make four, is equally true whether the second two follow the first two or accompany them. It is as true of days and years as of feet and inches. The laws of extension and figure (in other words, the theorems of geometry, from its lowest to its highest branches) are, on the contrary, laws of simultaneous phenomena only. The various parts of space, and of the objects which are said to fill space, co-exist; and the unvarying laws which are the subject of the science of geometry are an expression of the mode of their coexistence.

This is a class of laws, or, in other words, of uniformities, for the comprehension and proof of which it is not necessary to suppose any lapse of time, and variety of facts or events succeeding one another. The propositions of geometry are independent of the succession of events. All things which possess extension, or, in other words, which fill space, are subject to geometrical laws. Possessing extension, they possess figure; possessing figure, they must possess some figure in particular, and have all the properties which geometry assigns to that figure. If one body be a sphere and another a cylinder, of equal height and diameter, the one will be exactly two-thirds of the other, let the nature and quality of the material be what it will. . . .

In the laws of number, then, and in those of space, we recognise in the most unqualified manner the rigorous universality of which we are in quest. Those laws have been in all ages the type of certainty, the standard of comparison for all inferior degrees of evidence. Their invariability is so perfect, that it renders us unable even to conceive any exception to them; and philosophers have been led, though (as I have endeavoured to show) erroneously, to consider their evidence as lying not in experience, but in the original constitution of the intellect. If, therefore, from the laws of space and number we were able to deduce uniformities of any other description, this would be conclusive evidence to us that those other uniformities possessed the same rigorous certainty. But this we cannot do. From laws of space and number alone, nothing can be deduced but laws of space and number.

Of all truths relating to phenomena, the most valuable to us are those which relate to the order of their succession. On a knowledge of these is founded every reasonable anticipation of future facts, and whatever power we possess of influencing those facts to our advantage. Even the laws of geometry are chiefly of practical importance to us as being a portion of the premises from which the order of the succession of phenomena may be inferred. Inasmuch as the motion of bodies, the action of forces, and the propagation of influences of all sorts, take place in certain lines and over definite spaces, the properties of those lines and spaces are an important part of the laws to which those phenomena are themselves subject. Again, motions, forces, or other influences, and times are numerable quantities; and the properties of number are applicable to them as to all other things. But though the laws of number and space are important elements in the ascertainment of uniformities of succession, they can do nothing towards it when taken by themselves. They can only be made instrumental to that purpose when we combine with them additional premises, expressive of uniformities of succession already known. . . .

It is not, therefore, enough for us that the laws of space, which are only laws of simultaneous phenomena, and the laws of number, which though true of successive phenomena do not relate to their succession, possess the rigorous certainty and universality of which we are in search. We must endeavour to find some law of succession which has those same attributes, and is therefore fit to be made the foundation of processes for discovering, and of a test for verifying, all other uniformities of succession. This fundamental law must resemble the truths of geometry in their most remarkable peculiarity, that of never being, in any instance whatever, defeated or suspended by any change of circumstances.

Now among all those uniformities in the succession of phenomena which common observation is sufficient to bring to light, there are very few which have any, even apparent, pretension to this rigorous indefeasibility; and of those few, one only has been found capable of completely sustaining it. In that one, however, we recognize a law which is universal also in another sense; it is co-extensive with the entire field of successive phenomena, all instances whatever of succession being examples of it. This law is the Law of Causation. The truth that every fact which has a beginning has a cause, is co-extensive with human experience.

This generalisation may appear to some minds not to amount to much, since after all it asserts only this: "it is a law that every event depends on some law"; "it is a law that there is a law for everything." We must not, however, conclude that the generality of the principle is merely verbal; it will be found on inspection to be no vague or unmeaning assertion, but a most important and really fundamental truth.

The notion of Cause being the root of the whole theory of Induction, it is indispensable that this idea should, at the very outset of our inquiry, be, with the utmost practicable degree of precision, fixed and determined. If, indeed, it were necessary for the purpose of inductive logic that the strife should be quelled which has so long raged among the different schools of metaphysicians respecting the origin and analysis of our idea of causation, the promulgation, or at least the general reception, of a true theory of induction might be considered desperate for a long time to come. But the science of the Investigation of Truth by means of Evidence is happily independent of many of the controversies which perplex the science of the ultimate constitution of the human mind, and is under no necessity of pushing the analysis of mental phenomena to that extreme limit which alone ought to satisfy a metaphysician.

I premise, then, that when in the course of this inquiry I speak of the cause of any phenomenon, I do not mean a cause which is not itself a phenomenon; I make no research into the ultimate or ontological cause of anything. To adopt a distinction familiar in the writings of the Scotch metaphysicians, and especially of Reid, the causes with which I concern myself are not *efficient*, but *physical* causes. They are causes in that sense alone in which one physical fact is said to be the cause of another. Of the efficient causes of phenomena, or whether any such causes exist at all, I am not called upon to give an opinion. The notion of causation is deemed by the schools of metaphysics most in vogue at the present moment to imply a mysterious and most powerful tie, such as cannot, or at least does not, exist between any physical fact and that other physical fact on which it is invariably consequent, and which is popularly termed its cause: and thence is deduced the supposed necessity of ascending higher, into the essences and inherent constitution of things, to find the true cause, the cause which is not only followed by, but actually produces, the effect. No such necessity exists for the purposes of the present inquiry, nor will any such doctrine be found in the following pages. The only notion of a cause which the theory of induction requires is such a notion as can be gained from experience. The Law of Causation, the recognition of which is the main pillar of inductive science, is but the familiar truth that invariability of succession is found by observation to obtain between every fact in nature and some other fact which has preceded it, independently of all considerations respecting the ultimate mode of production of phenomena, and of every other question regarding the nature of "Things in themselves."

Between the phenomena, then, which exist at any instant, and the phenomena which exist at the succeeding instant, there is an invariable order of succession; and, as we said in speaking of the general uniformity of the course of nature, this web is composed of separate fibres; this collective order is made up of particular sequences, obtaining invariably among the separate parts. To certain facts, certain facts always do, and, as we believe, will continue to, succeed. The invariable antecedent is termed the cause; the invariable consequent, the effect. And the universality of the law of causation consists in this, that every consequent is connected in this manner with some particular antecedent or set of antecedents. Let the fact be what it may, if it has begun to exist, it was preceded by some fact or facts with which it is invariably connected. For every event there exists some combination of objects or events, some given concurrence of circumstances, positive and negative, the occurrence of which is always followed by that phenomenon. We may not have found out what this concurrence of circumstances may be; but we never doubt that there is such a one, and that it never occurs without having the phenomenon in question as its effect or consequence. On the universality of this truth depends the possibility of reducing the inductive process to rules. The undoubted assurance we have that there is a law to be found if we only knew how to find it, will be seen presently to be the source from which the canons of the Inductive Logic derive their validity.

OF THE EVIDENCE OF THE LAW OF
UNIVERSAL CAUSATION

We have now completed our review of the logical processes by which the laws, or uniformities, of the sequence of phenomena, and those uniformities in their co-existence which depend on the laws of their sequence, are ascertained or tested. As we recognised in the commencement, and have been enabled to see more clearly in the progress of the investigation, the basis of all these logical operations is the law of causation. The validity of all the Inductive Methods depends on the assumption that every event, or the beginning of every phenomenon, must have some cause, some antecedent, on the existence of which it is invariably and unconditionally consequent. . . .

But is this assumption warranted? Doubtless (it may be said) *most* phenomena are connected as effects with some antecedent or cause, that is, are never produced unless some assignable fact has preceded them; but the very circumstance that complicated processes of induction are sometimes necessary, shows that cases exist in which this regular order of succession is not apparent to our unaided apprehension. If, then, the processes which bring these cases within the same category with the rest require that we should assume the universality of the very law which they do not at first sight appear to exemplify, is not this a *petitio principii?* Can we prove a proposition by an argument which takes it for granted? And if not so proved, on what evidence does it rest?

For this difficulty, which I have purposely stated in the strongest terms it will admit of, the school of metaphysicians who have long predominated in this country find a ready salvo. They affirm that the universality of causation is a truth which we cannot help believing; that the belief in it is an instinct, one of the laws of our believing faculty. As the proof of this, they say, and they have nothing else to say, that everybody does believe it; and they number it among the propositions, rather numerous in their catalogue, which may be logically argued against, and perhaps cannot be logically proved, but which are of higher authority than logic, and so essentially inherent in the human mind, that even he who denies them in speculation

This article is from **A System of Logic: Ratiocinative and Inductive** by John Stuart Mill. (London, 1868), Vol. II, pp. 95–105 (passim). Reprinted by permission of Longmans, Green & Company, Ltd.

shows by his habitual practice that his arguments make no impression upon himself.

Into the merits of this question, considered as one of psychology, it would be foreign to my purpose to enter here; but I must protest against adducing, as evidence of the truth of a fact in external nature, the disposition, however strong or however general, of the human mind to believe it. Belief is not proof, and does not dispense with the necessity of proof. I am aware that to ask for evidence of a proposition which we are supposed to believe instinctively is to expose oneself to the charge of rejecting the authority of the human faculties; which of course no one can consistently do, since the human faculties are all which any one has to judge by; and inasmuch as the meaning of the word evidence is supposed to be something which, when laid before the mind, induces it to believe; to demand evidence when the belief is ensured by the mind's own laws is supposed to be appealing to the intellect against the intellect. But this, I apprehend, is a misunderstanding of the nature of evidence. By evidence is not meant anything and everything which produces belief. There are many things which generate belief besides evidence. A mere strong association of ideas often causes a belief so intense as to be un-shakeable by experience or argument. Evidence is not that which the mind does or must yield to, but that which it ought to yield to, namely, that, by yielding to which, its belief is kept conformable to fact. There is no appeal from the human faculties generally, but there is an appeal from one human faculty to another; from the judging faculty to those which take cognisance of fact, the faculties of sense and consciousness. The legitimacy of this appeal is admitted whenever it is allowed that our judgments ought to be conformable to fact. To say that belief suffices for its own justification is making opinion the test of opinion; it is denying the existence of any outward standard, the conformity of an opinion to which constitutes its truth. We call one mode of forming opinions right and another wrong, because the one does and the other does not tend to make the opinion agree with the fact—to make people believe what really is, and expect what really will be. Now a mere disposition to believe, even if supposed instinctive, is no guarantee for the truth of the thing believed. If, indeed, the belief ever amounted to an irresistible necessity, there would then be no *use* in appealing from it, because there would be no possibility of altering it. . . .

As was observed in a former place, the belief we entertain in the universality, throughout nature, of the law of cause and effect, is itself an instance of induction, and by no means one of the earliest which any of us, or which mankind in general, can have made. We arrive at this universal law by generalisation from many laws of inferior generality. We should never have had the notion of causation (in the philosophical meaning of the term) as a condition of all phenomena, unless many cases of causation, or, in other words, many partial uniformities of sequence, had previously become familiar. The more obvious of the particular uniformities suggest, and give evidence of, the general uniformity, and the general uniformity, once established, enables us to prove the remainder of the particular uniformities of which it is made up. As, however, all rigorous processes of induction presuppose the general uniformity, our knowledge of the particular uniformities from which it was first inferred was not, of course, derived from rigorous induction, but from the loose and uncertain mode of induction *per enumerationem simplicem;* and the law of universal causation, being collected from results so obtained, cannot itself rest on any better foundation.

It would seem, therefore, that induction *per enumerationem simplicem* not only is not necessarily an illicit logical process, but is in reality the only kind of induction possible; since the more elaborate process depends for its validity on a law, itself obtained in that inartificial mode. Is there not then an inconsistency in contrasting the looseness of one method with the rigidity of another, when that other is indebted to the looser method for its own foundation?

The inconsistency, however, is only apparent. Assuredly, if induction by simple enumeration were an invalid process, no process grounded on it could be valid; just as no reliance could be placed on telescopes if we could not trust our eyes. But though a valid process, it is a fallible one, and fallible in very different degrees: if therefore we can substitute for the more fallible forms of the process an operation grounded on the same process in a less fallible form, we shall have effected a very material improvement. And this is what scientific induction does.

A mode of concluding from experience must be pronounced untrustworthy when subsequent experience refuses to confirm it. According to this criterion, induction by simple enumeration—in other words, generalisation of an observed fact from the mere absence of any known instance to the contrary—affords in general a precarious and unsafe ground of assurance; for such generalisations are incessantly discovered, on further experience, to be false. Still, however, it affords some assurance, sufficient, in many cases, for the ordinary guidance of conduct. It would be absurd to say that the generalisations arrived at by mankind in the outset of their experience, such as these, Food nourishes, Fire burns, Water drowns, were unworthy of reliance. There is a scale of trustworthiness in the results of the original unscientific Induction; and on this diversity . . . depend the rules for the improvement of the process. The improvement consists in correcting one of these inartificial generalisations by means of another. As has been already pointed out this is all that art can do. To test a generalisation, by showing that it either follows from, or conflicts with, some stronger induction, some generalisation resting on a broader foundation of experience is the beginning and end of the logic of Induction.

Now the precariousness of the method of simple enumeration is in an inverse ratio to the largeness of the generalisation. The process is delusive and insufficient, exactly in proportion as the subject-matter of the observation is special and limited in extent. As the sphere widens, this unscientific method becomes less and less liable to mislead; and the most universal class of truths, the law of causation for instance, and the principles of number and of geometry, are duly and satisfactorily proved by that method alone, nor are they susceptible of any other proof.

With respect to the whole class of generalisations of which we have recently treated, the uniformities which depend on causation, the truth of the remark just made follows by obvious inference from the principles laid down in the preceding chapters. When a fact has been observed a certain number of times to be true, and is not in any instance known to be false; if we at once affirm that fact as an universal truth or law of nature, without either testing it by any of the four methods of induction, or deducing it from other known laws, we shall in general err grossly; but we are perfectly justified in affirming it as an empirical law, true within certain limits of time, place, and circumstance, provided the number of coincidences be greater than can with any probability be ascribed to chance. The reason for not extending it beyond those limits is, that the fact of its holding true within them may be a consequence of collocations, which cannot be concluded to exist in one place because they exist in another; or may be dependent on the accidental absence of counteracting agencies, which any variation of time, or the smallest change of circumstances, may possibly bring into play. If we suppose, then, the subject-matter of any generalisation to be so widely diffused that there is no time, no place, and no combination of circumstances, but must afford an example either of its truth or of its falsity, and if it be never found otherwise than true, its truth cannot be contingent on any collocations, unless such as exist at all times and places; nor can it be frustrated by any counteracting agencies, unless by such as never actually occur. It is, therefore, an empirical law co-extensive with all human experience, at which point the distinction between empirical laws and laws of nature vanishes, and the proposition takes its place among the most firmly established as well as largest truths accessible to science.

Now, the most extensive in its subject-matter of all generalisations which experience warrants, respecting the sequences and co-existences of phenomena, is the law of causation. It stands at the head of all observed uniformities in point of universality, and therefore (if the preceding observations are correct) in point of certainty. And if we consider, not what mankind would have been justified in believing in the infancy of their knowledge, but what may rationally be believed in its present more advanced state, we shall find ourselves warranted in considering this fundamental law, though itself obtained by induction from particular laws of causation, as not less certain, but, on the contrary, more so, than any of those from which it was drawn. It adds to them as much proof as it receives from them. For there is probably no one even of the best established laws of causation which is not sometimes counteracted, and to which, therefore, apparent exceptions do not present themselves, which would have necessarily and justly shaken the confidence of mankind in the universality of those laws, if inductive processes founded on the universal law had not enabled us to refer those exceptions to the agency of counteracting causes, and thereby reconcile them with the law with which they apparently conflict. Errors, moreover, may have slipped into the statement of any one of the special laws, through inattention to some material circumstance; and instead of the true proposition, another may have been enunciated, false as an universal law, though leading, in all cases hitherto observed, to the same result. To the law of causation, on the contrary, we not only do not know of any exception, but the exceptions which limit or apparently invalidate the special laws, are so far from contradicting the universal one, that they confirm it; since in all cases which are sufficiently open to our observation, we are able to trace the difference of result, either to the absence of a cause which had been present in ordinary cases, or to the presence of one which had been absent.

The law of cause and effect, being thus certain, is capable of imparting its certainty to all other inductive propositions which can be deduced from it; and the narrower inductions may be regarded as receiving their ultimate sanction from that law, since there is no one of them which is not rendered more certain than it was before, when we are able to connect it with that larger induction, and to show that it cannot be denied, consistently with the law that everything which begins to exist has a cause. And hence we are justified in the seeming inconsistency of holding induction by simple enumeration to be good for proving this general truth, the foundation of scientific induction, and yet refusing to rely on it for any of the narrower inductions. I fully admit that if the law of causation were unknown, generalisation in the more obvious cases of uniformity in phenomena would nevertheless be possible, and though in all cases more or less precarious, and in some extremely so, would suffice to constitute a certain measure of probability; but what the amount of this probability might be we are dispensed from estimating, since it never could amount to the degree of assurance which the proposition acquires, when, by the application to it of the Four Methods, the supposition of its falsity is shown to be inconsistent with the Law of Causation. We are therefore logically entitled, and, by the necessities of scientific induction, required to disregard the probabilities derived from the early rude method of generalising, and to consider no minor generalisation as proved except so far as the law of causation confirms it, nor probable except so far as it may reasonably be expected to be so confirmed.

Part II
THE LATTER PART
OF THE NINETEENTH
CENTURY:
1870-1899

Chapter 5
HERMANN LUDWIG FERDINAND VON HELMHOLTZ (1821–1894)

INTRODUCTION

*H*ERMANN LUDWIG FERDINAND VON HELMHOLTZ was born in Potsdam, Germany, not far from Berlin, on August 31, 1821. He was a frail child who was given an ordinary education; even in these simple studies he did not show promise. During the years of his secondary education he read extensively. From his choice of books it is clear that he had neither interest in nor the gift for languages, but that his real bent was in the mathematical and physical sciences. When he was seventeen he decided to become a physicist. Financial considerations, however, dictated his entry into a medicosurgical institute in Berlin, where his plan was to become a surgeon in the Prussian army. He studied there from 1838 to 1842, and then served as an army surgeon in Berlin for seven years. Whatever spare time he had was devoted exclusively to the study of mathematics and physics. Although he was never a student at the university, he did become acquainted with such leading scientists as the physicist Magnus and the physiologist Müller, who offered to guide him in his investigations.

In 1842 Helmholtz published his first paper (his dissertation) on the relationship between the nerve cells of ganglia; the view which he defended in this work foreshadowed the later neuron theory. In 1847, while still a surgeon, he read his famous paper on the conservation of energy (*Über die Erhaltung der Kraft*) before the *Physikalische Gesellschaft* in Berlin. In this paper, which aroused a vigorous discussion among the leading physicists, Helmholtz brought together much of the work done in this area after Newton, and gave the theory its first mathematical formulation.

Its success was responsible for Helmholtz' being called to Königsberg in 1848 as professor of physiology and general pathology, a post which he held for seven years. This was a fruitful period for him.

He wrote a paper on the measurement of the rate of conduction of the nervous impulse, invented some very important optical instruments, and also completed his *Handbuch der physiologischen Optik*. It was during these years, too, that Helmholtz became interested in empiricism; a careful study of the great English empiricists influenced his development gradually away from Kantian philosophy in the direction of Locke, Hume, and Mill.

In 1855 Helmholtz went to Bonn as professor of physiology, and then three years later to Heidelberg in a similar capacity; finally in 1871, following Magnus' death, he went to Berlin as professor of physics, realizing at last the desire of his youth. In 1887 he assumed duties in addition to his professorship by accepting the post of director of the physicotechnical institute at Charlottenburg, near Berlin; he held both these posts simultaneously until his death on September 8, 1894.

During his Heidelberg years Helmholtz completed his *Optik* (1867) and published his famous *Tonempfindungen* (1863). These works are still heralded as classics in the experimental psychology of sight and hearing. The later years of his life were devoted primarily to research in physics, as a result of which he made substantial contributions in the following areas: conservation of energy, hydrodynamics, theory of electricity, meteorological physics, optics, and fundamental principles of mechanics.

Helmholtz was unquestionably a neo-Kantian. With Friedrich Albert Lange (1828–1875) he stands as the most important representative of what became known as the "Physiological School" within this general movement. Both held with Kant that all our theoretical and practical knowledge involves empirical and nonempirical data. The physiological school, however, interpreted the creative and modifying function of our mind in terms of the organization of external stimuli by the nervous system, resulting in signs (not merely pictures) of external objects. The coherence of these signs is also explained physiologically. Proceeding in this way Helmholtz claimed that he was able to maintain the fundamental insights of the Kantian philosophy, only adapting them to the insight of modern science. He admitted, however, that the philosophy of space, time, and mathematics was to be modified in accordance with the discovery of non-Euclidean geometries.

That interest in empiricism which Helmholtz had begun to show in the 1850's, particularly in the wake of his investigations in the realm of perception, which gradually led him to the insight that nativism is to be rejected and a certain form of genetism to be accepted, culminated in the expression of his final view in *Die Tatsachen in der Wahrnemung* in 1878. Here he says about our perception: "The empiricist theory tries to prove that at least no other forces are necessary for their origin beyond the known faculties of our minds. . . . As it is in general a useful rule for scientific inquiry to make no new hypotheses as long as the known facts appear to be sufficient for explanation, I have thought it necessary to prefer the empiricist view in its essentials. The nativistic theory gives even less explanation of the origin of our perceptual pictures for it simply plunges into the midst of the matter by assuming that certain perceptual images of space would be produced directly by an inborn mechanism provided certain nerve fibers were stimulated. . . . Thus in this theory, not only is Kant's assertion adopted, that the general perception of space is an original form of our intuition, but there are laid down as innate certain special space perceptions." Helmholtz was maintaining that the nativistic theory cannot be disproved, but with Lotze, he claimed that it was no theory at all, for it said nothing about space except that the latter is not given in experience and must, therefore, be sensed innately. He was convinced that the development of perceptions in experience can be demonstrated to a certain extent at least and that, for this reason, there is no need for additional hypotheses.

According to Kant's view geometrical axioms are examples of a priori intuitions; hence Helmholtz tried to show that they, too, are products of experience, and in fact between 1866 and 1894 published seven papers on the subject. In his view the fundamental proof of all geometrical theorems lies in the demonstration of the congruence of figures, and this proof supposes superposition of the one figure upon the other; still superposition supposes movement, and this can be known only by experience. The most important parts of these papers, however, are those presenting his pictures of non-Euclidean geometries.

SELECTIVE BIBLIOGRAPHY

Helmholtz, H. L. F. von, *Über die Erhaltung der Kraft* (1847). Berlin, 1915.

———, *On the Sensations of Tone* (1863). London, 1912.

———, *A Treatise on Physiological Optics* (1867). New York, 1924.

————, *Populärwissenschaftliche Vorträge.* Berlin, 1870.

————, *Gesammelte wissenschaftliche Abhandlungen* (3 vols.). Berlin, 1882–1895.

————, *Popular Lectures on Scientific Subjects* (2 vols.), trans. E. Atkinson. New York, 1881.

————, *Schriften zur Erkenntnistheorie,* ed. P. Hertz and M. Schlick. Berlin, 1921.

Königsberger, L., *Hermann von Helmholtz,* trans. F. A. Welby. Oxford, 1906.

————, *Hermann von Helmholtz's Untersuchungen über die Grundlagen der Mathematik und Mechanik.* Leipzig, 1896.

Schwertschlager, J., *Kant und Helmholtz erkenntnis-theoretisch verglichen.* Freiburg im Breisgau, 1883.

Erdmann, B., *Die Axiome der Geometrie.* Leipzig, 1877.

Hall, G. S., *Founders of Modern Psychology.* New York, 1812. Pp. 247–308.

McKendrick, J. H., *Hermann Ludwig Ferdinand von Helmholtz.* London, 1899.

ON THE ORIGIN
AND SIGNIFICANCE
OF GEOMETRICAL AXIOMS

*T*HE fact that a science can exist and can be developed as has been the case with geometry, has always attracted the closest attention among those who are interested in questions relating to the bases of the theory of cognition. . . .

[This fact] has always been used as a prominent example in the discussion on that question, which forms, as it were, the centre of all antitheses of philosophical systems, that there can be a cognition of principles destitute of any bases drawn from experience. In the answer to Kant's celebrated question, "How are synthetical principles *a priori* possible?" geometrical axioms are certainly those examples which appear to show most decisively that synthetical principles are *a priori* possible at all. The circumstance that such principles exist, and force themselves on our conviction, is regarded as a proof that space is an *a priori* mode of all external perception. It appears thereby to postulate, for this *a priori* form, not only the character of a purely formal scheme of itself quite unsubstantial, in which any given result experience would fit; but also to include certain peculiarities of the scheme, which bring it about that only a certain content, and one which, as it were, is strictly defined, could occupy it and be apprehended by us.[1]

[1] In his book, *On the Limits of Philosophy*, Mr. W. Tobias maintains that axioms of a kind which I formerly enunciated are a misunderstanding of Kant's opinion. But Kant specially adduces the axioms, that the straight line is the shortest (*Kritik der reinen Vernunft*, Introduction, v. 2nd ed. p. 16); that space has three dimensions (*Ibid.* part i. sect. i. § 3, p. 41); that only one straight line is possible between two points (*Ibid.* part ii. sect. i. "On the Axioms of Intuition"), as axioms which express *a priori* the conditions of intuition by the senses. It is not

This article is from **Popular Lectures on Scientific Subjects** by H. von Helmholtz, trans. E. Atkinson, Second Series (New York, 1881), pp. 223–247. Reprinted by permission of Appleton-Century-Crofts, Publishers.

It is precisely this relation of geometry to the theory of cognition which emboldens me to speak to you on geometrical subjects in an assembly of those who for the most part have limited their mathematical studies to the ordinary instruction in schools. Fortunately, the amount of geometry taught in our gymnasia will help you to follow, at any rate the tendency, of the principles I am about to discuss.

I intend to give you an account of a series of recent and closely connected mathematical researches which are concerned with the geometrical axioms, their relations to experience, with the question whether it is logically possible to replace them by others.

Seeing that the researches in question are more immediately designed to furnish proofs for experts in a region which, more than almost any other, requires a higher power of abstraction, and that they are virtually inaccessible to the non-mathematician, I will endeavour to explain to such a one the question at issue. I need scarcely remark that my explanation will give no proof of the correctness of the new views. He who seeks this proof must take the trouble to study the original researches.

Anyone who has entered the gates of the first elementary axioms of geometry, that is, the mathematical doctrine of space, finds on his path that unbroken chain of conclusions of which I just spoke, by which the ever more varied and more complicated figures are brought within the domain of law. But even in their first elements certain principles are laid down, with respect to which geometry confesses that she cannot prove them, and can only assume that anyone who understands the essence of these principles will at once admit their correctness. These are the so-called axioms.

For example, the proposition that if the shortest line drawn between two points is called a *straight* line, there can be only one such straight line. Again, it is an axiom that through any three points in space, not lying in a straight line, a plane may be drawn, i.e. a surface which will wholly include every straight line joining any two of its points. Another axiom, about which there has been much discussion, affirms that through a point lying without a straight line only one straight line can be drawn parallel to the first; two straight lines that lie in the same plane and never meet, however far they may be produced, being called parallel. There are also axioms that determine

---

here the question, whether these axioms were originally given as intuition of space, or whether they are only the starting-points from which the understanding can develop such axioms *a priori* on which my critic insists.

the number of dimensions of space and its surfaces, lines and points, showing how they are continuous; as in the propositions, that a solid is bounded by a surface, a surface by a line and a line by a point, that the point is indivisible, that by the movement of a point a line is described, by that of a line a line or a surface, by that of a surface a surface or a solid, but by the movement of a solid a solid and nothing else is described.

Now what is the origin of such propositions, unquestionably true yet incapable of proof in a science where everything else is reasoned conclusion? Are they inherited from the divine source of our reason as the idealistic philosophers think, or is it only that the ingenuity of mathematicians has hitherto not been penetrating enough to find the proof? Every new votary, coming with fresh zeal to geometry, naturally strives to succeed where all before him have failed. And it is quite right that each should make the trial afresh; for, as the question has hitherto stood, it is only by the fruitlessness of one's own efforts that one can be convinced of the impossibility of finding a proof. Meanwhile solitary inquirers are always from time to time appearing who become so deeply entangled in complicated trains of reasoning that they can no longer discover their mistakes and believe they have solved the problem. The axiom of parallels especially has called forth a great number of seeming demonstrations.

The main difficulty in these inquiries is, and always has been, the readiness with which results of everyday experience become mixed up as apparent necessities of thought with the logical processes, so long as Euclid's method of constructive intuition is exclusively followed in geometry. It is in particular extremely difficult, on this method, to be quite sure that in the steps prescribed for the demonstration, we have not involuntarily and unconsciously drawn in some most general results of experience, which the power of executing certain parts of the operation has already taught us practically. In drawing any subsidiary line for the sake of his demonstration, the well-trained geometer always asks if it is possible to draw such a line. It is well known that problems of construction play an essential part in the system of geometry. At first sight, these appear to be practical operations, introduced for the training of learners; but in reality they establish the existence of definite figures. They show that points, straight lines, or circles such as the problem requires to be constructed are possible under all conditions, or they determine any exceptions that there may be. The point on which the investigations turn, that we are about to consider, is essentially of this nature. The foundation of all proof by Euclid's method consists in establishing the congruence of lines, angles, plane figures, solids, &c. To make the congruence evident, the geometrical figures are supposed to be applied to one another, of course without changing their form and dimensions. That this is in fact possible we have all experienced from our earliest youth. But, if we proceed to build necessities of thought upon this assumption of the free translation of fixed figures, with unchanged form, to every part of space, we must see whether the assumption does not involve some presupposition of which no logical proof is given. We shall see later on that it does indeed contain one of the most serious import. But if so, every proof by congruence rests upon a fact which is obtained from experience only.

I offer these remarks, at first only to show what difficulties attend the complete analysis of the presuppositions we make, in employing the common constructive method. We evade them when we apply, to the investigation of principles, the analytical method of modern algebraical geometry. The whole process of algebraical calculation is a purely logical operation; it can yield no relation between the quantities submitted to it that is not already contained in the equations which give occasion for its being applied. The recent investigations in question have accordingly been conducted almost exclusively by means of the purely abstract methods of analytical geometry.

However, after discovering by the abstract method what are the points in question, we shall best get a distinct view of them by taking a region of narrower limits than our own world of space. Let us, as we logically may, suppose reasoning beings of only two dimensions to live and move on the surface of some solid body. We will assume that they have not the power of perceiving anything outside this surface, but that upon it they have perceptions similar to ours. If such beings worked out a geometry, they would of course assign only two dimensions to their space. They would ascertain that a point in moving describes a line, and that a line in moving describes a surface. But they could as little represent to themselves what further spatial construction would be generated by a surface moving out of itself, as we can represent what would be generated by a solid moving out of the space we know. By the much-abused expression "to represent" or "to be able to think how something happens" I understand— and I do not see how anything else can be understood by it without loss of all meaning—the power of imagining the whole series of sensible impressions that would be had in such a case. Now as no sensible impression is known relating to such an unheard-of event, as the movement to a fourth dimension would be to us, or as a movement to our third dimension would be to the inhabitants of a surface, such a "representation" is as impossible as the "representation" of colours would be to one born blind, if a description of them in general terms could be given to him.

Our surface-beings would also be able to draw shortest lines in their superficial space. These would not necessarily be straight lines in our sense, but what are technically called *geodetic lines* of the surface on which they live; lines such as are described by a *tense* thread laid along the surface, and which can slide upon it freely. I will henceforth speak of such lines as the *straightest* lines of any particular surface or given space, so as to bring out their analogy with the straight line in a plane. I hope by this expression to make the conception more easy for the apprehension of my non-mathematical hearers without giving rise to misconception.

Now if beings of this kind lived on an infinite plane, their geometry would be exactly the same as our planimetry. They would affirm that only one straight line is possible between two points; that through a third point lying without this line only one line can be drawn parallel to it; that the ends of a straight line never meet though it is produced to infinity, and so on. Their space might be infinitely extended, but even if there were limits to the movements and perception, they would be able to represent to themselves a continuation beyond these limits; and thus their space would appear to them infinitely extended, just as ours does to us, although our bodies cannot leave the earth, and our sight only reaches as far as the visible fixed stars.

But intelligent beings of the kind supposed might also live on the surface of a sphere. Their shortest or straightest line between two points would then be an arc of the great circle passing through them. Every circle, passing through two points, is by these divided into two parts; and if they are unequal, the shorter is certainly the shortest line on the sphere between the two points, but also the other or larger arc of the same great circle is a geodetic or straightest line, i.e., every smaller part of it is the shortest line between its ends. Thus the notion of the geodetic or straightest line is not quite identical with that of the shortest line. If the two given points are the ends of a diameter of the sphere, every plane passing through this diameter cuts semi-circles, on the surface of the sphere, all of which are shortest lines between the given points. Accordingly, the axiom of there being only one shortest line between two points would not hold without a certain exception for the dwellers on a sphere.

Of parallel lines the sphere-dwellers would know nothing. They would maintain that any two straightest lines, sufficiently produced, must finally cut not in one only but in two points. The sum of the angles of a triangle would be always greater than two right angles, increasing as the surface of the triangle grew greater. They could thus have no conception of geometrical similarity between greater and smaller figures of the same kind, for with them a greater triangle must have different angles from a smaller one. Their space would be unlimited, but would be found to be finite or at least represented as such.

It is clear, then, that such beings must set up a very different system of geometrical axioms from that of the inhabitants of a plane, or from ours with our space of three dimensions, though the logical powers of all were the same; nor are more examples necessary to show that geometrical axioms must vary according to the kind of space inhabited by beings whose powers of reason are quite in conformity with ours. But let us proceed still farther.

Let us think of reasoning beings existing on the surface of an egg-shaped body. Shortest lines could be drawn between three points of such a surface and a triangle constructed. But if the attempt were made to construct congruent triangles at different parts of the surface, it would be found that two triangles, with three pairs of equal sides, would not have their angles equal. The sum of the angles of a triangle drawn at the sharper pole of the body would depart farther from two right angles than if the triangle were drawn at the blunter pole or at the equator. Hence it appears that not even such a simple figure as a triangle can be moved on such a surface without change of form. It would also be found that if circles of equal radii were constructed at different parts of such a surface (the length of the radii being always measured by shortest lines along the surface) the periphery would be greater at the blunter than at the sharper end.

We see accordingly that, if a surface admits of the figures lying on it being freely moved without change of any of their lines and angles as measured along it, the property is a special one and does not belong to every kind of surface. The condition under which a surface possesses this important property was pointed out by Gauss in his celebrated treatise on the curvature of surfaces.[2] The "measure of curvature," as he called it, i.e. reciprocal of the product of the greatest

---

[2] Gauss, *Werke*, Bd. IV. p. 215, first published in *Commentationes Soc. Reg Scientt. Gottengensis recentiores*, vol. VI., 1828.

and least radii of curvature, must be everywhere equal over the whole extent of the surface.

Gauss showed at the same time that this measure of curvature is not changed if the surface is bent without distension or contraction of any part of it. Thus we can roll up a flat sheet of paper into the form of a cylinder, or of a cone, without any change in the dimensions of the figures taken along the surface of the sheet. Or the hemispherical fundus of a bladder may be rolled into a spindle-shape without altering the dimensions on the surface. Geometry on a plane will therefore be the same as on a cylindrical surface; only in the latter case we must imagine that any number of layers of this surface, like the layers of a rolled sheet of paper, lie one upon another, and that after each entire revolution round the cylinder a new layer is reached different from the previous ones.

These observations are necessary to give the reader a notion of a kind of surface the geometry of which is on the whole similar to that of the plane, but in which the axiom of parallels does not hold good. This is a kind of curved surface which is, as it were, geometrically the counterpart of a sphere, and which has therefore been called the *pseudospherical surface* by the distinguished Italian mathematician E. Beltrami, who has investigated its properties.[3] It is a saddle-shaped surface of which only limited pieces or strips can be connectedly represented in our space, but which may yet be thought of as infinitely continued in all directions, since each piece lying at the limit of the part constructed can be conceived as drawn back to the middle of it and then continued. The piece displaced must in the process change its flexure but not its dimensions, just as happens with a sheet of paper moved about a cone formed out of a plane rolled up. Such a sheet fits the conical surface in every part, but must be more bent near the vertex and cannot be so moved over the vertex as to be at the same time adapted to the existing cone and to its imaginary continuation beyond.

Like the plane and the sphere, pseudospherical surfaces have their measure of curvature constant, so that every piece of them can be exactly applied to every other piece, and therefore all figures constructed at one place on the surface can be transferred to any other

[3] *Saggio di Interpretazione della Geometria Non-Euclidea*, Napoli, 1868.—*Teoria fondamentale degli Spazii di Curvatura costante*, Annali di Matematica, Ser. II. Tom. II pp. 232–55. Both have been translated into French by J. Hoüel, *Annales Scientifiques de l'Ecole Normale*, Tom. V., 1869.

place with perfect congruity of form, and perfect equality of all dimensions lying in the surface itself. The measure of curvature as laid down by Gauss, which is positive for the sphere and zero for the plane, would have a constant negative value for pseudospherical surfaces, because the two principal curvatures of a saddle-shaped surface have their concavity turned opposite ways.

A strip of a pseudospherical surface may, for example, be represented by the inner surface (turned towards the axis) of a solid anchor-ring. If the plane figure *aabb* (Fig. 1) is made to revolve on its axis of symmetry AB, the two arcs *ab* will describe a pseudospherical concave-convex surface like that of the ring. Above and below, towards *aa* and *bb*, the surface will turn outwards with ever-increasing flexure, till it becomes perpendicular to the axis, and ends at the edge with one curvature infinite. Or, again, half of a pseudospherical surface may be rolled up into the shape of a champagne-glass (Fig. 2), with tapering stem infinitely prolonged. But the surface is always necessarily bounded by a sharp edge beyond which it cannot be directly continued. Only by supposing each single piece of the edge cut loose and drawn along the surface of the ring or glass, can it be brought to places of different flexure, at which farther continuation of the piece is possible.

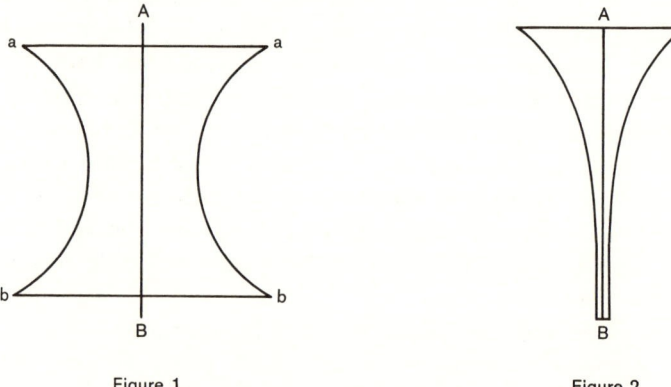

Figure 1   Figure 2

In this way too the straightest lines of the pseudospherical surface may be infinitely produced. They do not, like those on the sphere, return upon themselves, but, as on a plane, only one shortest line is possible between the two given points. The axiom of parallels does not, however, hold good. If a straightest line is given on the surface and a point without it, a whole pencil of straightest lines may pass through the point, no one of which, though infinitely produced, cuts the first line; the pencil itself being limited by two straightest lines, one of which intersects one of the ends of the given line at an infinite distance, the other the other end.

Such a system of geometry, which excluded the axiom of parallels, was devised on Euclid's synthetic method, as far back as the year 1829, by N. J. Lobatchewsky, professor of mathematics at Kasan,[4] and it was proved that this system could be carried out as consistently as Euclid's. It agrees exactly with the geometry of the pseudospherical surfaces worked out recently by Beltrami.

Thus we see that in the geometry of two dimensions a surface is marked out as a plane, or a sphere, or a pseudospherical surface, by the assumption that any figure may be moved about in all directions without change of dimensions. The axiom, that there is only one shortest line between any two points, distinguishes the plane and the pseudospherical surface from the sphere, and the axiom of parallels marks off the plane from the pseudosphere. These three axioms are in fact necessary and sufficient, to define as a plane the surface to which Euclid's planimetry has reference, as distinguished from all other modes of space in two dimensions.

The difference between plane and spherical geometry has been long evident, but the meaning of the axiom of parallels could not be understood till Gauss had developed the notion of surfaces flexible without dilatation, and consequently that of the possibly infinite continuation of pseudospherical surfaces. Inhabiting, as we do, a space of three dimensions and endowed with organs of sense for their perception, we can represent to ourselves the various cases in which beings on a surface might have to develop their perception of space; for we have only to limit our own perceptions to a narrower field. It is easy to think away perceptions that we have; but it is very difficult to imagine perceptions to which there is nothing analogous in our experience. When, therefore, we pass to space of more than three dimensions, we are stopped in our power of representation, by the structure of our organs and the experiences got through them which correspond only to the space in which we live.

There is however another way of treating geometry scientifically.

[4] *Principien der Geometrie,* Kasan, 1829–30.

All known space-relations are measurable, that is, they may be brought to determination of magnitudes (lines, angles, surfaces, volumes). Problems in geometry can therefore be solved, by finding methods of calculation for arriving at unknown magnitudes from known ones. This is done in *analytical geometry*, where all forms of space are treated only as quantities and determined by means of other quantities. Even the axioms themselves make reference to magnitudes. The straight line is defined as the *shortest* between two points, which is a determination of quantity The axiom of parallels declares that if two straight lines in a plane do not intersect (are parallel), the alternate angles, or the corresponding angles, made by a third line intersecting them, are equal; or it may be laid down instead that the sum of the angles of any triangle is equal to two right angles. These, also, are determinations of quantity.

Now we may start with this view of space, according to which the position of a point may be determined by measurement in relation to any given figure (system of co-ordinates), taken as fixed, and then inquire what are the special characteristics of our space as manifested in the measurements that have to be made, and how it differs from other extended quantities of like variety. This path was first entered by one too early lost to science, B. Riemann of Göttingen.[5] It has the peculiar advantage that all its operations consist in pure calculation of quantities, which quite obviates the danger of habitual perceptions being taken for necessities of thought.

The number of measurements necessary to give the position of a point, is equal to the number of dimensions of the space in question. In a line the distance from one fixed point is sufficient, that is to say, one quantity; in a surface the distances from two fixed points must be given; in space, the distances from three; or we require, as on the earth, longitude, latitude, and height above the sea, or, as is usual in analytical geometry, the distances from three co-ordinate planes. Riemann calls a system of differences in which one thing can be determined by n measurements an "$n$fold extended aggregate" or an "aggregate of n dimensions." Thus the space in which we live is a threefold, a surface is a twofold, and a line is a simple extended aggregate of points. Time also is an aggregate of one dimension. The system of colours is an aggregate of three dimensions, inasmuch as

[5] Über die Hypothesen welche der Geometrie zu Grunde liegen, Habilitationsschrift vom 10 Juni 1854. (*Abhandl. der königl. Gesellsch. zu Göttingen*, Bd. XIII.)

each colour, according to the investigations of Thomas Young and of Clerk Maxwell may be represented as a mixture of three primary colours, taken in definite quantities. The particular mixtures can be actually made with the colour-top.

In the same way we may consider the system of simple tone as an aggregate of two dimensions, if we distinguish only pitch and intensity, and leave out of account differences of timbre. This generalisation of the idea is well suited to bring out the distinction between space of three dimensions and other aggregates. We can, as we know from daily experience, compare the vertical distance of two points with the horizontal distance of two others, because we can apply a measure first to the one pair and then to the other. But we cannot compare the difference between two tones of equal pitch and different intensity, with that between two tones of equal intensity and different pitch. Riemann showed, by considerations of this kind, that the essential foundation of any system of geometry, is the expression that it gives for the distance between two points lying in any direction towards one another, beginning with the infinitesimal interval. He took from analytical geometry the most general form for this expression, that, namely, which leaves altogether open the kind of measurements by which the position of any point is given.[6] Then he showed that the kind of free mobility without change of form which belongs to bodies in our space can only exist when certain quantities yielded by the calculation[7]—quantities that coincide with Gauss's measure of surface-curvature when they are expressed for surfaces—have everywhere an equal value. For this reason Riemann calls these quantities, when they have the same value in all directions for a particular spot, the measure of curvature of the space at this spot. To prevent misunderstanding,[8] I will once more observe that this so-called measure of space-curvature is a quantity obtained by purely analytical calculation, and that its introduction involves no suggestion of relations that would have a meaning only for sense-perception. The name is merely taken, as a short expression for a complex relation, from the one case in which the quantity designated admits of sensible representation.

[6] For the square of the distance of two infinitely near points the expression is a homogeneous quadric function of the differentials of their co-ordinates.
[7] They are algebraical expressions compounded from the coefficients of the various terms in the expression for the square of the distance of two contiguous points and from their differential quotients.
[8] As occurs, for instance, in the above-mentioned work of Tobias, pp. 70, etc.

Now whenever the value of this measure of curvature in any space is everywhere zero, that space everywhere conforms to the axioms of Euclid; and it may be called a *flat* (*homaloid*) space in contradistinction to other spaces, analytically constructible, that may be called *curved*, because their measure of curvature has a value other than zero. Analytical geometry may be as completely and consistently worked out for such spaces as ordinary geometry can for our actually existing homaloid space.

If the measure of curvature is positive we have *spherical* space, in which straightest lines return upon themselves and there are no parallels. Such a space would, like the surface of a sphere, be unlimited but not infinitely great. A constant negative measure of curvature on the other hand gives *pseudospherical* space, in which straightest lines run out to infinity, and a pencil of straightest lines may be drawn, in any flattest surface, through any point which does not intersect another given straightest line in that surface.

Beltrami[9] has rendered these last relations imaginable by showing that the points, lines, and surfaces of a pseudospherical space of three dimensions, can be so portrayed in the interior of a sphere in Euclid's homaloid space, that every straightest line or flattest surface of the pseudospherical space is represented by a straight line or a plane, respectively, in the sphere. The surface itself of the sphere corresponds to the infinitely distant points of the pseudospherical space; and the different parts of this space, as represented in the sphere, become smaller, the nearer they lie to the spherical surface, diminishing more rapidly in the direction of the radii than in that perpendicular to them. Straight lines in the sphere, which only intersect beyond its surface, correspond to straightest lines of the pseudospherical space which never intersect.

Thus it appeared that space, considered as a region of measurable quantities, does not at all correspond with the most general conception of an aggregate of three dimensions, but involves also special conditions, depending on the perfectly free mobility of solid bodies without change of form to all parts of it and with all possible changes of direction; and, further, on the special value of the measure of curvature which for our actual space equals, or at least is not distinguishable from, zero. This latter definition is given in the axioms of straight lines and parallels.

Whilst Riemann entered upon this new field from the side of the

[9] *Teoria fondamentale, &c., ut sup.*

most general and fundamental questions of analytical geometry, I myself arrived at similar conclusions,[10] partly from seeking to represent in space the system of colours, involving the comparison of one threefold extended aggregate with another, and partly from inquiries on the origin of our ocular measure for distances in the field of vision. Riemann starts by assuming the above-mentioned algebraical expression which represents in the most general form the distance between two infinitely near points, and deduces therefrom the conditions of mobility of rigid figures. I, on the other hand, starting from the observed fact that the movement of rigid figures is possible in our space, with the degree of freedom that we know, deduce the necessity of the algebraic expression taken by Riemann as an axiom. The assumptions that I had to make as the basis of the calculation were the following.

First, to make algebraical treatment at all possible, it must be assumed that the position of any point A can be determined, in relation to certain given figures taken as fixed bases, by measurement of some kind of magnitudes, as lines, angles between lines, angles between surfaces, and so forth. The measurements necessary for determining the position of A are known as its co-ordinates. In general, the number of co-ordinates necessary for the complete determination of the position of a point, marks the number of the dimensions of the space in question. It is further assumed that with the movement of the point A, the magnitudes used as co-ordinates vary continuously.

Secondly, the definition of a solid body, or rigid system of points, must be made in such a way as to admit of magnitudes being compared by congruence. As we must not, at this stage, assume any special methods for the measurement of magnitudes, our definition can, in the first instance, run only as follows: Between the co-ordinates of any two points belonging to a solid body, there must be an equation which, however the body is moved, expresses a constant spatial relation (proving at last to be the distance) between the two points, and which is the same for congruent pairs of points, that is to say, such pairs as can be made successively to coincide in space with the same fixed pair of points.

However indeterminate in appearance, this definition involves most important consequences, because with increase in the number of points, the number of equations increases much more quickly than

[10] Über die Thatsachen die der Geometrie zum Grunde liegen (*Nachrichten von der königl. Ges. d. Wiss. zu Göttingen*, Juni 3, 1868).

the number of co-ordinates which they determine. Five points, A, B, C, D, E, give ten different pairs of points

$$AB, AC, AD, AE,$$
$$BC, BD, BE,$$
$$CD, CE,$$
$$DE,$$

and therefore ten equations, involving in space of three dimensions fifteen variable co-ordinates. But of these fifteen, six must remain arbitrary, if the system of five points is to admit of free movement and rotation, and thus the ten equations can determine only nine co-ordinates as functions of the six variables. With six points we obtain fifteen equations for twelve quantities, with seven points twenty-one equations for fifteen, and so on. Now from n independent equations we can determine n contained quantities, and if we have more than n equations, the superfluous ones must be deducible from the first n. Hence it follows that the equations which subsist between the co-ordinates of each pair of points of a solid body must have a special character, seeing that, when in space of three dimensions they are satisfied for nine pairs of points as formed out of any five points, the equation for the tenth pair follows by logical consequence. Thus our assumption for the definition of solidity becomes quite sufficient to determine the kind of equations holding between the co-ordinates of two points rigidly connected.

Thirdly, the calculation must further be based on the fact of a peculiar circumstance in the movement of solid bodies, a fact so familiar to us that but for this inquiry it might never have been thought of as something that need not be. When in our space of three dimensions two points of a solid body are kept fixed, its movements are limited to rotations round the straight line connecting them. If we turn it completely round once, it again occupies exactly the position it had at first. This fact, that rotation in one direction always brings a solid body back into its original position, needs special mention. A system of geometry is possible without it. This is most easily seen in the geometry of a plane. Suppose that with every rotation of a plane figure its linear dimensions increased in proportion to the angle of rotation, the figure after one whole rotation through 360 degress would no longer coincide with itself as it was originally. But any second figure that was congruent with the first in its original position might be made to coincide with it in its second position by being also turned through 360 degrees. A consistent system of geometry would be possible upon this supposition, which does not come under Riemann's formula.

On the other hand I have shown that the three assumptions taken together form a sufficient basis for the starting-point of Riemann's investigation, and thence for all his further results relating to the distinction of different spaces according to their measure of curvature.

It still remained to be seen whether the laws of motion, as dependent on moving forces, could also be consistently transferred to spherical or pseudospherical space. This investigation has been carried out by Professor Lipschitz of Bonn.[11] It is found that the comprehensive expression for all the laws of dynamics, Hamilton's principle, may be directly transferred to spaces of which the measure of curvature is other than zero. Accordingly, in this respect also, the disparate systems of geometry lead to no contradiction.

We have now to seek an explanation of the special characteristics of our own flat space, since it appears that they are not implied in the general notion of an extended quanity of three dimensions and of the free mobility of bounded figures therein. *Necessities of thought,* such as are involved in the conception of such a variety, and its measurability, or from the most general of all ideas of a solid figure contained in it, and of its free mobility, they undoubtedly are not. Let us then examine the opposite assumption as to their origin being empirical, and see if they can be inferred from facts of experience and so established, or if, when tested by experience they are perhaps to be rejected. If they are of empirical origin, we must be able to represent to ourselves connected series of facts, indicating a different value for the measure of curvature from that of Euclid's flat space. But if we can imagine such spaces of other sorts, it cannot be maintained that the axioms of geometry are necessary consequences of an *a priori* transcendental form of intuition, as Kant thought.

The distinction between spherical, pseudospherical, and Euclid's geometry depends, as was above observed, on the value of a certain constant called, by Riemann, the measure of curvature of the space in question. The value must be zero for Euclid's axioms to hold good. If it were not zero, the sum of the angles of a large triangle would differ from that of the angles of a small one, being larger in spherical, smaller in pseudospherical, space. Again, the geometrical similarity

---

[11]"Untersuchungen über die ganzen homogenen Functionen von *n* Differentialen" (Borchardt's *Journal für Mathematik*, Bd. lxx. 3, 71; lxxiii. 3, 1); "Untersuchung eines Problems der Variationsrechnung" (*Ibid.* Bd. lxxiv.).

of large and small solids or figures is possible only in Euclid's space. All systems of practical mensuration that have been used for the angles of large rectilinear triangles, and especially all systems of astronomical measurement which make the parallax of the immeasurably distant fixed stars equal to zero (in pseudospherical space the parallax even of infinitely distant points would be positive), confirm empirically the axiom of parallels, and show the measure of curvature of our space thus far to be indistinguishable from zero. It remains, however, a question, as Riemann observed, whether the result might not be different if we could use other than our limited base-lines, the greatest of which is the major axis of the earth's orbit.

Meanwhile, we must not forget that all geometrical measurements rest ultimately upon the principle of congruence. We measure the distance between points by applying to them the compass, rule, or chain. We measure angles by bringing the divided circle or theodolite to the vertex of the angle. We also determine straight lines by the path of rays of light which in our experience is rectilinear; but that light travels in shortest lines as long as it continues in a medium of constant refraction would be equally true in space of a different measure of curvature. Thus all our geometrical measurements depend on our instruments being really, as we consider them, invariable in form, or at least on their undergoing no other than the small changes we know of, as arising from variation of temperature, or from gravity acting differently at different places.

In measuring, we only employ the best and surest means we know of to determine, what we otherwise are in the habit of making out by sight and touch or by pacing. Here our own body with its organs is the instrument we carry about in space. Now it is the hand, now the leg, that serves for a compass, or the eye turning in all directions is our theodolite for measuring arcs and angles in the visual field.

Every comparative estimate of magnitudes or measurement of their spatial relations proceeds therefore upon a supposition as to the behaviour of certain physical things, either the human body or other instruments employed. The supposition may be in the highest degree probable and in closest harmony with all other physical relations known to us, but yet it passes beyond the scope of pure space-intuition.

It is in fact possible to imagine conditions for bodies apparently solid such that the measurements in Euclid's space become what they would be in spherical or pseudospherical space. Let me first remind the reader that if all the linear dimensions of other bodies, and our own, at the same time were diminished or increased in like proportion, as for instance to half or double their size, we should with our means of space-perception be utterly unaware of the change. This would also be the case if the distension or contraction were different in different directions, provided that our own body changed in the same manner, and further that a body in rotating assumed at every moment, without suffering or exerting mechanical resistance, the amount of dilatation in its different dimensions corresponding to its position at the time. Think of the image of the world in a convex mirror. The common silvered globes set up in gardens give the essential features, only distorted by some optical irregularities. A well-made convex mirror of moderate aperture represents the objects in front of it as apparently solid and in fixed positions behind its surface. But the images of the distant horizon and of the sun in the sky lie behind the mirror at a limited distance, equal to its focal length. Between these and the surface of the mirror are found the images of all the other objects before it, but the images are diminished and flattened in proportion to the distance of their objects from the mirror. The flattening, or decrease in the third dimension, is relatively greater than the decrease of the surface-dimensions. Yet every straight line or every plane in the outer world is represented by a straight line or a plane in the image. The image of a man measuring with a rule a straight line from the mirror would contract more and more the farther he went, but with his shrunken rule the man in the image would count out exactly the same number of centimetres as the real man. And, in general, all geometrical measurements of lines or angles made with regularly varying images of real instruments would yield exactly the same results as in the outer world, all congruent bodies would coincide on being applied to one another in the mirror as in the outer world, all lines of sight in the outer world would be represented by straight lines of sight in the mirror. In short I do not see how men in the mirror are to discover that their bodies are not rigid solids and their experiences good examples of the correctness of Euclid's axioms. But if they could look out upon our world as we can look into theirs, without overstepping the boundary, they must declare it to be a picture in a spherical mirror, and would speak of us just as we speak of them; and if two inhabitants of the different worlds could communicate with one another, neither, so far as I can see, would be able to convince the other that he had the true, the other the distorted, relations. Indeed I cannot see that such a question would have any meaning at all, so long as mechanical considerations are not mixed up with it.

Now Beltrami's representation of pseudospherical space in a sphere of Euclid's space, is quite similar, except that the background is not a plane as in the convex mirror, but the surface of a sphere, and that the proportion in which the images as they approach the spherical surface contract, has a different mathematical expression. If we imagine then, conversely, that in the sphere, for the interior of which Euclid's axioms hold good, moving bodies contract as they depart from the centre like the images in a convex mirror, and in such a way that their representatives in pseudospherical space retain their dimensions unchanged—observers whose bodies were regularly subjected to the same change would obtain the same results from the geometrical measurements they could make as if they lived in pseudospherical space.

We can even go a step further, and infer how the objects in a pseudospherical world, were it possible to enter one, would appear to an observer, whose eye-measure and experiences of space had been gained like ours in Euclid's space. Such an observer would continue to look upon rays of light or the lines of vision as straight lines, such as are met with in flat space, and as they really are in the spherical representation of pseudospherical space. The visual image of the objects in pseudospherical space would thus make the same impression upon him as if he were at the centre of Beltrami's sphere. He would think he saw the most remote objects round about him at a finite distance,[12] let us suppose a hundred feet off. But as he approached these distant objects, they would dilate before him, though more in the third dimension than superficially, while behind him they would contract. He would know that his eye judged wrongly. If he saw two straight lines which in his estimate run parallel for the hundred feet to his world's end, he would find on following them that the farther he advanced the more they diverged, because of the dilatation of all the objects to which he approached. On the other hand, behind him, their distance would seem to diminish, so that as he advanced they would appear always to diverge more and more. But two straight lines which from his first position seemed to converge to one and the same point of the background a hundred feet

[12] The reciprocal of the square of this distance, expressed in negative quantity, would be the measure of curvature of the pseudospherical space.

distant, would continue to do this however far he went, and he would never reach their point of intersection.

Now we can obtain exactly similar images of our real world, if we look through a large convex lens of corresponding negative focal length, or even through a pair of convex spectacles if ground somewhat prismatically to resemble pieces of one continuous larger lens. With these, like the convex mirror, we see remote objects as if near to us, the most remote appearing no farther distant than the focus of the lens. In going about with this lens before the eyes, we find that the objects we approach dilate exactly in the manner I have described for pseudospherical space. Now any one using a lens, were it even so strong as to have a focal length of only sixty inches, to say nothing of a hundred feet, would perhaps observe for the first moment that he saw objects brought nearer. But after going about a little the illusion would vanish, and in spite of the false images he would judge of the distances rightly. We have every reason to suppose that what happens in a few hours to any one beginning to wear spectacles would soon enough be experienced in pseudospherical space. In short, pseudospherical space would not seem to us very strange, comparatively speaking; we should only at first be subject to illusions in measuring by eye the size and distance of the more remote objects.

There would be illusions of an opposite description, if, with eyes practised to measure in Euclid's space, we entered a spherical space of three dimensions. We should suppose the more distant objects to be more remote and larger than they are, and should find on approaching them that we reached them more quickly than we expected from their appearance. But we should also see before us objects that we can fixate only with diverging lines of sight, namely, all those at a greater distance from us than the quadrant of a great circle. Such an aspect of things would hardly strike us as very extraordinary, for we can have it even as things are if we place before the eye a slightly prismatic glass with the thicker side towards the nose: the eyes must then become divergent to take in distant objects. This excites a certain feeling of unwonted strain in the eyes, but does not perceptibly change the appearance of the objects thus seen. The strangest sight, however, in the spherical world would be the back of our own head, in which all visual lines not stopped by other objects would meet again, and which must fill the extreme background of the whole perspective picture.

At the same time it must be noted that as a small elastic flat disk, say of india-rubber, can only be fitted to a slightly curved spherical surface with relative contraction of its border and distension of its centre, so our bodies, developed in Euclid's flat space, could not pass into curved space without undergoing similar distensions and contractions of their parts, their coherence being of course maintained only in as far as their elasticity permitted their bending without breaking. The kind of distension must be the same as in passing from a small body imagined at the centre of Beltrami's sphere to its pseudospherical or spherical representation. For such passage to appear possible, it will always have to be assumed that the body is sufficiently elastic and small in comparison with the real or imaginary radius of curvature of the curved space into which it is to pass.

These remarks will suffice to show the way in which we can infer from the known laws of our sensible perceptions the series of sensible impressions which a spherical or pseudospherical world would give us, if it existed. In doing so, we nowhere meet with inconsistency or impossibility any more than in the calculation of its metrical proportions. We can represent to ourselves the look of a pseudospherical world in all directions just as we can develop the conception of it. Therefore it cannot be allowed that the axioms of our geometry depend on the native form of our perceptive faculty, or are in any way connected with it.

It is different with the three dimensions of space. As all our means of sense-perception extend only to space of three dimensions, and a fourth is not merely a modification of what we have, but something perfectly new, we find ourselves by reason of our bodily organisation quite unable to represent a fourth dimension.

In conclusion, I would again urge that the axioms of geometry are not propositions pertaining only to the pure doctrine of space. As I said before, they are concerned with quantity. We can speak of quantities only when we know of some way by which we can compare, divide, and measure them. All space-measurements, and therefore in general all ideas of quantities applied to space, assume the possibility of figures moving without change of form or size. It is true we are accustomed in geometry to call such figures purely geometrical solids, surfaces, angles, and lines, because we abstract from all the other distinctions, physical and chemical, of natural bodies; but yet one physical quality, rigidity, is retained. Now we have no other mark of rigidity of bodies or figures but congruence, whenever they are applied to one another at any time or place, and after any revolution. We cannot, however, decide by pure geometry, and without mechanical considerations, whether the coinciding bodies may not both have varied in the same sense.

If it were useful for any purpose, we might with perfect consistency look upon the space in which we live as the apparent space behind a convex mirror with its shortened and contracted background; or we might consider a bounded sphere of our space, beyond the limits of which we perceive nothing further, as infinite pseudospherical space. Only then we should have to ascribe to the bodies which appear to us to be solid, and to our own body at the same time, corresponding distensions and contractions, and we should have to change our system of mechanical principles entirely; for even the proposition that every point in motion, if acted upon by no force, continues to move with unchanged velocity in a straight line, is not adapted to the image of the world in the convex-mirror. The path would indeed be straight, but the velocity would depend upon the place.

Thus the axioms of geometry are not concerned with space-relations only but also at the same time with the mechanical deportment of solidest bodies in motion. The notion of rigid geometrical figure might indeed be conceived as transcendental in Kant's sense, namely, as formed independently of actual experience, which need not exactly correspond therewith, any more than natural bodies do ever in fact correspond exactly to the abstract notion we have obtained of them by induction. Taking the notion of rigidity thus as a mere ideal, a strict Kantian might certainly look upon the geometrical axioms as propositions given, *a priori*, by transcendental intuition, which no experience could either confirm or refute, because it must first be decided by them whether any natural bodies can be considered as rigid. But then we should have to maintain that the axioms of geometry are not synthetic propositions, as Kant held them; they would merely define what qualities and deportment a body must have to be recognised as rigid.

But if to the geometrical axioms we add propositions relating to the mechanical properties of natural bodies, were it only the axiom of inertia, or the single proposition, that the mechanical and physical properties of bodies and their mutual reactions are, other circumstances remaining the same, independent of place, such a system of propositions has a real import which can be confirmed or refuted by experience, but just for the same reason can also be gained by experience. The mechanical axiom, just cited, is in fact of the utmost importance for the whole system of our mechanical and physical conceptions. That rigid solids, as we call them, which are really nothing else than elastic solids of great resistance, retain the same form in every part of space if no external force affects them, is a single case falling under the general principle.

In conclusion, I do not, of course, maintain that mankind first arrived at space-intuitions, in agreement with the axioms of Euclid, by any carefully executed systems of exact measurement. It is rather a succession of everyday experiences, especially the perception of the geometrical similarity of great and small bodies, only possible in flat space, that led to the rejection, as impossible, of every geometrical representation at variance with this fact. For this no knowledge of the necessary logical connection between the observed fact of geometrical similarity and the axioms was needed; but only an intuitive apprehension of the typical relations between lines, planes, angles, &c., obtained by numerous and attentive observations—an intuition of the kind the artist possesses of the objects he is to represent, and by means of which he decides with certainty and accuracy whether a new combination, which he tries, will correspond or not with their nature. It is true that we have no word but *intuition* to mark this; but it is knowledge empirically gained by the aggregation and reinforcement of similar recurrent impressions in memory, and not a transcendental form given before experience. That other such empirical intuitions of fixed typical relations, when not clearly comprehended, have frequently enough been taken by metaphysicians for *a priori* principles, is a point on which I need not insist.

To sum up, the final outcome of the whole inquiry may be thus expressed:

(1) The axioms of geometry, taken by themselves out of all connection with mechanical propositions, represent no relations of real things. When thus isolated, if we regard them with Kant as forms of intuition transcendentally given, they constitute a form into which any empirical content whatever will fit, and which therefore does not in any way limit or determine beforehand the nature of the content. This is true, however, not only of Euclid's axioms, but also of the axioms of spherical and pseudospherical geometry.

(2) As soon as certain principles of mechanics are conjoined with the axioms of geometry, we obtain a system of propositions which has real import, and which can be verified or overturned by empircal observations, just as it can be inferred from experience. If such a system were to be taken as a transcendental form of intuition and thought, there must be assumed a preestablished harmony between form and reality.

Chapter 6
WILLIAM STANLEY JEVONS
(1835–1882)

INTRODUCTION

WILLIAM STANLEY JEVONS was born in Liverpool, England, on September 1, 1835. He received his education partly at University College School, and later at University College in London. In 1854 his studies of the natural sciences were interrupted by a call to accept a position in a new mint in Sydney, Australia. During the five years he held this post Jevons devoted himself to the study of political economy and sociology. Returning to England he wrote some very important papers and a book on political economy, as a result of which he was appointed to a chair in political economy at Owens College in Manchester in 1866. Five years later he published his most significant work, *Theory of Political Economy*. In 1876 he accepted a professorship in political economy at University College in London, where he remained until his death. He died near Hastings on August 13, 1882.

Jevons made important contributions in the realm of logic and philosophy of science. His most valuable publications in these areas are the following: *Pure Logic: The Logic of Quality apart from Quantity* (1864), *The Substitution of Similars: The Principle of Reasoning* (1869), *Elementary Lessons in Logic* (1870), *The Principles of Science: A Treatise on Logic and Scientific Method* (1873), *Studies and Exercises in Deductive Logic* (1880). He also constructed a "logical machine" which he exhibited before the Royal Society in 1870.

Jevons was a pioneer in the mechanization of logic. He believed himself to be the discoverer of a new kind of logic to which he referred as "combinational logic." He had in mind something similar to the truth-table method, but saw in it not a means of testing formulas but rather a device for drawing inferences. He did not see that in this capacity its value was limited. The work which Jevons did in this area has been largely neglected primarily because logic has taken a different path from the one he envisioned. Unlike Peirce and Frege, Jevons gave little attention to the logic of relations and quantification theory. Instead he took his starting point from Boole's system, and then divested it from its mathematical dress and developed it into a certain logic of terms.

Jevons' general philosophical outlook, more indicated than developed in his works, is close to the basic motives of English empiricism, especially utilitarianism. However, he was vigorously opposed to any form of scientism or evolutionism. In the Conclusion to *The Principles of Science* Jevons wrote that his philosophy aimed at being a positive and affirmative philosophy, and not a negative dogmatism in the sense of Comte's philosophy.

Jevons' most characteristic contribution to philosophy of science consists in the way in which he attempted systematically to relate his views on the nature of induction, probability, and hypothesis. By *deduction* he understood the process through which we pass from more general to less general truths; thus in deduction we are engaged in developing the consequences of a law. *Induction,* however, is exactly the inverse process, which leads us from less to more general truths; given certain results or consequences, we are required to discover the general law from which they flow (*The Principles of Science*, p. 11).

Jevons makes a distinction between a perfect and imperfect induction, speaking of *perfect induction* when all the objects or events which can possibly come under the class dealt with have been examined. In the majority of cases, however, such a procedure is impossible for various reasons (the number of objects or events would often be practically infinite, the greater part of them may be beyond reach, and so on). In all these cases induction is *imperfect* and is affected by more or less uncertainty (*Ib.*, p. 146). Jevons did not agree with the often-held view that perfect induction is useless because it does not increase our knowledge and merely summarizes the knowledge we have already; on the contrary he believed that mere abbreviation of mental labor is one of the most important aids in the acquisition of knowledge, *and* that perfect induction is an absolute prerequisite in the performance of imperfect induction (*Ib.*, pp. 146-147). In explaining the second reason for his view on the importance and necessity of perfect induction Jevons stated that, if one could draw any inference at all concerning objects not examined, it must be on the basis of the data afforded by the objects which have been examined with the help of a perfect induction of this limited class.

It follows then that imperfect induction is and remains always uncertain because two assumptions are made which, although they are postulates of inductive inquiry, nonetheless can never be proved. These are (1) that our past observations give complete knowledge of what exists and (2) that the conditions of things which did exist will continue to be the conditions of things which will exist (assumption of the uniformity of nature) (*Ib.*, pp. 149–150). In this connection Mill's attempt to reduce the first assumption to the second and to justify the latter by means of an analysis of causality is useless because a cause is not the invariable and necessary condition of an event, but only the group of positive and negative conditions which, with more or less probability, precede an event. It is this topic which is dealt with in the Jevons selection to follow.

There it becomes clear that Jevons believed that only perfect knowledge leads to certainty. He says, however, that perfect knowledge of nature would be infinite knowledge, which is beyond our capacities. Thus it is that we must content ourselves with partial knowledge mingled with ignorance and producing doubt. That is to say, all our inferences concerning future events are merely probable, and an accurate appreciation of the degree of probability depends upon a right understanding of the principles of the theory of probability. Here, following Laplace's a priori definition of probability, Jevons understands by the probability of an event the number of favorable cases divided by the total number of "equipossible" cases. After a brief explanation of Laplace's theory of probability, Jevons continues to develop the idea that the inverse application of the classical theory of probability is the very essence of imperfect induction. He justifies this inverse application of the rules of probability with the following principle of Laplace: "If an event can be produced by any one of a certain number of different causes, all equally probable *a priori*, the probabilities of the existence of these causes as inferred from the event, are proportional to the probabilities of the event as derived from these causes" (*Ib.*, pp. 242–243).

SELECTIVE BIBLIOGRAPHY

Jevons, W. S., *Pure Logic*. London 1864.

———, *Substitution of Similars: The Principle of Reasoning*. London, 1869.

————, *The Principles of Science: A Treatise on Logic and Scientific Method.* London, 1873.

————, *Studies and Exercises in Deductive Logic.* London, 1880.

The Letters and Journal of W. Stanley Jevons, ed. Mrs. W. S. Jevons. London, 1886.

Madden, E. H., "W. S. Jevons on Induction and Probability," in *Theories of Scientific Method: The Renaissance Through the Nineteenth Century,* ed. E. H. Madden. Seattle, 1960. Pp. 233–247.

PHILOSOPHY OF INDUCTIVE INFERENCE

INTRODUCTION

We have inquired into the nature of perfect induction, whereby we pass backwards from certain observed combinations of events, to the logical conditions governing such combinations. We have also investigated the grounds of that theory of probability, which must be our guide when we leave certainty behind, and dilute knowledge with ignorance. There is now before us the difficult task of endeavouring to decide how, by the aid of that theory, we can ascend from the facts to the laws of nature; and may then with more or less success anticipate the future course of events. All our knowledge of natural objects must be ultimately derived from observation, and the difficult question arises—How can we ever know anything which we have not directly observed through one of our senses, the apertures of the mind? The utility of reasoning is to assure ourselves that, at a determinate time and place, or under specified conditions, a certain phenomenon will be observed. When we can use our senses and perceive that the phenomenon does occur, reasoning is superfluous. If the senses cannot be used, because the event is in the future, or out of reach, how can reasoning take their place? Apparently, at least, we must infer the unknown from the known, and the mind must itself create an addition to the sum of knowledge. But I hold that it is quite impossible to make any real additions to the contents of our knowledge, except through new impressions upon the senses, or upon some seat of feeling. I shall attempt to show that inference, whether inductive or deductive, is never more than an unfolding of the contents of our experience, and that it always proceeds upon the assumption that the future and the unperceived will be governed by the same conditions as the past and the perceived, an assumption which will often prove to be mistaken.

This article is from **The Principles of Science: A Treatise on Logic and Scientific Method** (New York, 1924), Chap. XI. Reprinted by permission of The Macmillan Company and Dover Publications, Inc.

In inductive as in deductive reasoning the conclusion never passes beyond the premises. Reasoning adds no more to the implicit contents of our knowledge, than the arrangement of the specimens in a museum adds to the number of those specimens. Arrangement adds to our knowledge in a certain sense: it allows us to perceive the similarities and peculiarities of the specimens, and on the assumption that the museum is an adequate representation of nature, it enables us to judge of the prevailing forms of natural objects. Bacon's first aphorism holds perfectly true, that man knows nothing but what he has observed, provided that we include his whole sources of experience, and the whole implicit contents of his knowledge. Inference but unfolds the hidden meaning of our observations, and *the theory of probability shows how far we go beyond our data in assuming that new specimens will resemble the old ones,* or that the future may be regarded as proceeding uniformly with the past.

VARIOUS CLASSES OF INDUCTIVE TRUTHS

It will be desirable, in the first place, to distinguish between the several kinds of truths which we endeavour to establish by induction. Although there is a certain common and universal element in all our processes of reasoning, yet diversity arises in their application. Similarity of condition between the events from which we argue, and those to which we argue, must always be the ground of inference; but this similarity may have regard either to time or place, or the simple logical combination of events, or to any conceivable junction of circumstances involving quality, time, and place. Having met with many pieces of substance possessing ductility and a bright yellow colour, and having discovered, by perfect induction, that they all possess a high specific gravity, and a freedom from the corrosive action of acids, we are led to expect that every piece of substance, possessing like ductility and a similar yellow colour, will have an equally high specific gravity, and a like freedom from corrosion by acids. This is a case of the coexistence of qualities; for the character of the specimens examined alters not with time nor place.

In a second class of cases, time will enter as a principal ground of similarity. When we hear a clock pendulum beat time after time, at equal intervals, and with a uniform sound, we confidently expect that the stroke will continue to be repeated uniformly. A comet having appeared several times at nearly equal intervals, we infer that it will probably appear again at the end of another like interval. A man who has returned home evening after evening for many years, and found his house standing, may, on like grounds, expect that it will be standing the next evening, and on many succeeding evenings. Even the continuous existence of an object in an unaltered state, or the finding again of that which we have hidden, is but a matter of inference depending on experience.

A still larger and more complex class of cases involves the relations of space, in addition to those of time and quality. Having observed that every triangle drawn upon the diameter of a circle, with its apex upon the circumference, apparently contains a right angle, we may ascertain that all triangles in similar circumstances will contain right angles. This is a case of pure space reasoning, apart from circumstances of time or quality, and it seems to be governed by different principles of reasoning. I shall endeavour to show, however, that geometrical reasoning differs but in degree from that which applies to other natural relations.

THE RELATION OF CAUSE AND EFFECT

In a very large part of the scientific investigations which must be considered, we deal with events which follow from previous events, or with existences which succeed existences. Science, indeed, might arise even were material nature a fixed and changeless whole. Endow mind with the power to travel about, and compare part with part, and it could certainly draw inferences concerning the similarity of forms, the coexistence of qualities, or the preponderance of a particular kind of matter in a changeless world. A solid universe, in at least approximate equilibrium, is not inconceivable, and then the relation of cause and effect would evidently be no more than the relation of before and after. As nature exists, however, it is a progressive existence, ever moving and changing as time, the great independent variable, proceeds. Hence it arises that we must continually compare what is happening now with what happened a moment before, and a moment before that moment, and so on, until we reach indefinite periods of past time. A comet is seen moving in the sky, or its constituent particles illumine the heavens with their tails of fire. We cannot explain the present movements of such a body without supposing its prior existence, with a definite amount of energy and a definite direction of motion; nor can we validly suppose that our task is concluded when we find that it came wandering to our solar system through the unmeasured vastness of surrounding space. Every event must have a cause, and that cause again a cause, until we are lost in the obscurity of the past, and are driven to the belief in one First Cause, by whom the course of nature was determined.

FALLACIOUS USE OF THE TERM CAUSE

The words Cause and Causation have given rise to infinite trouble and obscurity, and have in no slight degree retarded the progress of science. From the time of Aristotle, the work of philosophy has been described as the discovery of the causes of things, and Francis Bacon adopted the notion when he said *"vere scire esse per causas scire."* Even now it is not uncommonly supposed that the knowledge of causes is something different from other knowledge, and consists, as it were, in getting possession of the keys of nature. A single word may thus act as a spell, and throw the clearest intellect into confusion, as I have often thought that Locke was thrown into confusion when endeavouring to find a meaning for the word *power*.[1] In Mill's *System of Logic* the term *cause* seems to have re-asserted its old noxious power. Not only does Mill treat the Laws of Causation as almost coextensive with science, but he so uses the expression as to imply that when once we pass within the circle of causation we deal with certainties.

The philosophical danger which attaches to the use of this word may be thus described. A cause is defined as the necessary or invariable antecedent of an event, so that when the cause exists the effect will also exist or soon follow. If then we know the cause of an event, we know what will certainly happen; and as it is implied that science, by a proper experimental method, may attain to a knowledge of causes, it follows that experience may give us a certain knowledge of future events. But nothing is more unquestionable than that finite experience can never give us certain knowledge of the future, so that either a cause is not an invariable antecedent, or else we can never gain certain knowledge of causes. The first horn of this dilemma is hardly to be accepted. Doubtless there is in nature some invariably acting mechanism, such that from certain fixed conditions an invariable result always emerges. But we, with our finite minds and short experience, can never penetrate the mystery of those existences which embody the Will of the Creator, and evolve it throughout time. We are in the position of spectators who witness

---

[1] *Essay concerning Human Understanding*, bk. ii. chap. xxi.

the productions of a complicated machine, but are not allowed to examine its intimate structure. We learn what does happen and what does appear, but if we ask for the reason, the answer would involve an infinite depth of mystery. The simplest bit of matter, or the most trivial incident, such as the stroke of two billiard balls, offers infinitely more to learn than ever the human intellect can fathom. The word cause covers just as much untold meaning as any of the words *substance, matter, thought, existence.*

CONFUSION OF TWO QUESTIONS

The subject is much complicated, too, by the confusion of two distinct questions. An event having happened, we may ask—

(1) Is there any cause for the event?
(2) Of what kind is that cause?

No one would assert that the mind possesses any faculty capable of inferring, prior to experience, that the occurrence of a sudden noise with flame and smoke indicates the combustion of a black powder, formed by the mixture of black, white, and yellow powders. The greatest upholder of *à priori* doctrines will allow that the particular aspect, shape, size, colour, texture, and other qualities of a cause must be gathered through the senses.

The question whether there is any cause at all for an event, is of a totally different kind. If an explosion could happen without any prior existing conditions, it must be a new creation—a distinct addition to the universe. It may be plausibly held that we can imagine neither the creation nor annihilation of anything. As regards matter, this has long been held true; as regards force, it is now almost universally assumed as an axiom that energy can neither come into nor go out of existence without distinct acts of Creative Will. That there exists any instinctive belief to this effect, indeed, seems doubtful. We find Lucretius, a philosopher of the utmost intellectual power and cultivation, gravely assuming that his raining atoms could turn aside from their straight paths in a self-determining manner, and by this spontaneous origination of energy determine the form of the universe.[2] Sir George Airy, too, seriously discussed the mathematical conditions under which a perpetual motion, that is, a perpetual source of self-

2 *De Rerum Natura*, bk. ii. ll. 216–293.

created energy, might exist.[3] The larger part of the philosophic world has long held that in mental acts there is free will—in short, self-causation. It is in vain to attempt to reconcile this doctrine with that of an intuitive belief in causation, as Sir W. Hamilton candidly allowed.

It is obvious, moreover, that to assert the existence of a cause for every event cannot do more than remove into the indefinite past the inconceivable fact and mystery of creation. At any given moment matter and energy were equal to what they are at present, or they were not; if equal, we may make the same inquiry concerning any other moment, however long prior, and we are thus obliged to accept one horn of the dilemma—existence from infinity, or creation at some moment. This is but one of the many cases in which we are compelled to believe in one or other of two alternatives, both inconceivable. My present purpose, however, is to point out that we must not confuse this supremely difficult question with that into which inductive science inquires on the foundation of facts. By induction we gain no certain knowledge; but by observation, and the inverse use of deductive reasoning, we estimate the probability that an event which has occurred was preceded by conditions of specified character, or that such conditions will be followed by the event.

DEFINITION OF THE TERM CAUSE

Clear definitions of the word cause have been given by several philosophers. Hobbes has said, "A cause is the sum or aggregate of all such accidents, both in the agents and the patients, as concur in the producing of the effect propounded; all which existing together, it cannot be understood but that the effect existeth with them; or that it can possibly exist if any of them be absent." Brown, in his *Essay on Causation,* gave a nearly corresponding statement. "A cause," he says,[4] "may be defined to be the object or event which immediately precedes any change, and which existing again in similar circumstances will be always immediately followed by a similar change." Of the kindred word *power,* he likewise says:[5] "Power is nothing more than that invariableness of antecedence which is implied in the belief of causation."

[3] *Cambridge Philosophical Transactions* (1830), vol. iii. pp. 369–372.
[4] *Observations on the Nature and Tendency of the Doctrine of Mr. Hume, concerning the Relation of Cause and Effect.* Second ed. p. 44.
[5] Ibid. p. 97.

These definitions may be accepted with the qualification that our knowledge of causes in such a sense can be probable only. The work of science consists in ascertaining the combinations in which phenomena present themselves. Concerning every event we shall have to determine its probable conditions, or the group of antecedents from which it probably follows. An antecedent is anything which exists prior to an event; a consequent is anything which exists subsequently to an antecedent. It will not usually happen that there is any probable connection between an antecedent and consequent. Thus nitrogen is an antecedent to the lighting of a common fire; but it is so far from being a cause of the lighting, that it renders the combustion less active. Daylight is an antecedent to all fires lighted during the day, but it probably has no appreciable effect upon their burning. But in the case of any given event it is usually possible to discover a certain number of antecedents which seem to be always present, and with more or less probability we conclude that when they exist the event will follow.

Let it be observed that the utmost latitude is at present enjoyed in the use of the term *cause*. Not only may a cause be an existent thing endowed with powers, as oxygen is the cause of combustion, gunpowder the cause of explosion, but the very absence or removal of a thing may also be a cause. It is quite correct to speak of the dryness of the Egyptian atmosphere, or the absence of moisture, as being the cause of the preservation of mummies, and other remains of antiquity. The cause of a mountain elevation, Ingleborough for instance, is the excavation of the surrounding valleys by denudation. It is not so usual to speak of the existence of a thing at one moment as the cause of its existence at the next, but to me it seems the commonest case of causation which can occur. The cause of motion of a billiard ball may be the stroke of another ball; and recent philosophy leads us to look upon all motions and changes, as but so many manifestations of prior existing energy. In all probability there is no creation of energy and no destruction, so that as regards both mechanical and molecular changes, the cause is really the manifestation of existing energy. In the same way I see not why the prior existence of matter is not also a cause as regards its subsequent existence. All science tends to show us that the existence of the universe in a particular state at one moment, is the condition of its existence at the next moment, in an apparently different state. When we analyse the meaning which we can attribute to the word *cause*, it amounts to the existence of suitable portions of matter endowed with suitable quantities of energy.

If we may accept Horne Tooke's assertion, *cause* has etymologically the meaning of *thing before*. Though, indeed, the origin of the word is very obscure, its derivatives, the Italian *cosa*, and the French *chose*, means simply *thing*. In the German equivalent *Ursache*, we have plainly the original meaning of *thing before*, the *sache* denoting "interesting or important object," the English *sake*, and *ur* being the equivalent of the English *ere, before*. We abandon, then, both etymology and philosophy, when we attribute to the *laws of causation* any meaning beyond that of the *conditions* under which an event may be expected to happen, according to our observation of the previous course of nature.

I have no objection to use the words cause and causation, provided they are never allowed to lead us to imagine that our knowledge of nature can attain to certainty. I repeat that if a cause is an invariable and necessary condition of an event, we can never know certainly whether the cause exists or not. To us, then, a cause is not to be distinguished from the group of positive or negative conditions which, with more or less probability, precede an event. In this sense, there is no particular difference between knowledge of causes and our general knowledge of the succession of combinations, in which the phenomena of nature are presented to us, or found to occur in experimental inquiry.

DISTINCTION OF INDUCTIVE
AND DEDUCTIVE RESULTS

We must carefully avoid confusing together inductive investigations which terminate in the establishment of general laws, and those which seem to lead directly to the knowledge of future particular events. That process only can be called induction which gives general laws, and it is by the subsequent employment of deduction that we anticipate particular events. If the observation of a number of cases shows that alloys of metals fuse at lower temperatures than their constituent metals, I may with more or less probability draw a general inference to that effect, and may thence deductively ascertain the probability that the next alloy examined will fuse at a lower temperature than its constituents. It has been asserted, indeed, by Mill,[6] and partially admitted by Mr. Fowler,[7] that we can argue directly from case to case, so that what is true of some alloys will be true of the next.

[6] *System of Logic*, bk. II. chap. iii.
[7] *Inductive Logic*, pp. 13, 14.

Professor Bain has adopted the same view of reasoning. He thinks that Mill has extricated us from the deadlock of the syllogism and effected a total revolution in logic. He holds that reasoning from particulars to particulars is not only the usual, the most obvious and the most ready method, but that it is the type of reasoning which best discloses the real process.[8] Doubtless, this is the usual result of our reasoning, regard being had to degrees of probability; but these logicians fail entirely to give any explanation of the process by which we get from case to case.

It may be allowed that the knowledge of future particular events is the main purpose of our investigations, and if there were any process of thought by which we could pass directly from event to event witout ascending into general truths, this method would be sufficient, and certainly the briefest. It is true, also, that the laws of mental association lead the mind always to expect the like again in apparently like circumstances, and even animals of very low intelligence must have some trace of such powers of association, serving to guide them more or less correctly, in the absence of true reasoning faculties. But it is the purpose of logic, according to Mill, to ascertain whether inferences have been correctly drawn, rather than to discover them.[9] Even if we can, then, by habit, association, or any rude process of inference, infer the future directly from the past, it is the work of logic to analyse the conditions on which the correctness of this inference depends. Even Mill would admit that such analysis involves the consideration of general truths,[10] and in this, as in several other important points, we might controvert Mill's own view by his own statements. It seems to me undesirable in a systematic work like this to enter into controversy at any length, or to attempt to refute the views of other logicians. But I shall feel bound to state, in a separate publication, my very deliberate opinion that many of Mill's innovations in logical science, and especially his doctrine of reasoning from particulars to particulars, are entirely groundless and false.

THE GROUNDS OF INDUCTIVE INFERENCE

I hold that in all cases of inductive inference we must invent hypotheses, until we fall upon some hypothesis which yields deductive results in accordance with experience. Such accordance renders the

[8] Bain, *Deductive Logic*, pp. 208, 209.
[9] *System of Logic*. Introduction, § 4. Fifth ed. pp. 8, 9.
[10] Ibid. bk. II. chap. iii. § 5, pp. 225, &c.

chosen hypothesis more or less probable, and we may then deduce, with some degree of likelihood, the nature of our future experience, on the assumption that no arbitrary change takes place in the conditions of nature. We can only argue from the past to the future, on the general principle set forth in this work, that what is true of a thing will be true of the like. So far then as one object or event differs from another, all inference is impossible, particulars as particulars can no more make an inference than grains of sand can make a rope. We must always rise to something which is general or same in the cases, and assuming that sameness to be extended to new cases we learn their nature. Hearing a clock tick five thousand times without exception or variation, we adopt the very probable hypothesis that there is some invariably acting machine which produces those uniform sounds, and which will, in the absence of change, go on producing them. Meeting twenty times with a bright yellow ductile substance, and finding it always to be very heavy and incorrodible, I infer that there was some natural condition which tended in the creation of things to associate these properties together, and I expect to find them associated in the next instance. But there always is the possibility that some unknown change may take place between past and future cases. The clock may run down, or be subject to a hundred accidents altering its condition. There is no reason in the nature of things, so far as known to us, why yellow colour, ductility, high specific gravity, and incorrodibility, should always be associated together, and in other cases, if not in this, men's expectations have been deceived. Our inferences, therefore, always retain more or less of a hypothetical character, and are so far open to doubt. Only in proportion as our induction approximates to the character of perfect induction, does it approximate to certainty. The amount of uncertainty corresponds to the probability that other objects than those examined may exist and falsify our inferences; the amount of probability corresponds to the amount of information yielded by our examination; and the theory of probability will be needed to prevent us from over-estimating or under-estimating the knowledge we possess.

Chapter 7
JOHANN BERNARD STALLO
(1823–1900)

INTRODUCTION

*J*OHANN BERNARD STALLO was born on March 16, 1823, in Damme (Oldenburg), Germany. The schoolteaching profession was quite prominent on both sides of his family; the first instruction he received, in German, English, French, and arithmetic, came from his grandfather and father and was later followed by Latin and Greek studies. At the age of thirteen Stallo entered the teacher's college at Vechta while studying simultaneously at the nearby Gymnasium. Two years later he was ready to enter a university, but his father lacked the financial means to support further education. Stallo did not find the idea of becoming a country schoolteacher particularly attractive and so decided to set out for America to join his uncle, Franz Joseph Stallo, who after some initial misfortune had settled in Cincinnati in the early 1830's.

In 1839 Stallo arrived in Cincinnati ready to take his chances. It was difficult for him to find a suitable occupation, however, in that he was too well educated to become a laborer, but not prepared to engage in a trade. Eventually he found a position as a German teacher in a local parish school. Finding that there was no suitable primer for the teaching of German, Stallo undertook immediately the task of preparing one; it was published anonymously in 1840 under the title of *A.B.C: Buchstabir und Lesebuch für die deutschen Schulen Amerikas.* The book went through many editions and was widely used by schools in all parts of the country.

By the fall of 1841 Stallo had secured a teaching position at St. Xavier's College in Cincinnati. He remained there until 1844 teaching German and continuing his studies in Greek and mathematics. During his last two years at St. Xavier's he found himself teaching more mathematics courses than German. In his spare time he studied physics and chemistry and was so successful in these areas that in 1844 St. John's College at Fordham called him as professor of physics, chemistry, and mathematics, a position he held until 1848. During his four-year stay he found ample opportunity to resume the study of philosophy which he had had to forego in 1839. He concentrated on Leibniz, Kant, Fichte, Schelling, Oken, and Hegel, and as a result of these studies wrote his first serious work, *General Principles of the Philosophy of Nature,* which appeared in Boston in 1848.

Stallo's purpose in writing the book was mainly to acquaint the American public better with modern German philosophy. The first part of the book is an attempt to describe the fundamental principles of a philosophy of the natural sciences. The second part locates this "programmatic part" in a broader historical context, and an attempt is made "at a delineation of German 'philosophy of nature' (*Naturphilosophie*) in some of its most notable phases, embracing principally the philosophical system of Schelling (with Oken's System of Nature) and Hegel" (*General Principles of the Philosophy of Nature,* p. *vi*).

The book as a whole is written within the general perspective of Hegel's philosophical system, although in many aspects it bears a highly personal mark. Typical of the book is its severe criticism of all forms of empiricism and naturalism in philosophy of science, as for example: "Unfortunately, the materialistic, utilitarian tendencies, which at present pervade every branch of science under color of a misconceived Baconism, have revoked every alliance between philosophical pursuits and the investigation of nature. Speculation is perfectly disavowed (sometimes justly, perhaps), and the very worst passport which a naturalist could carry about him is that of a metaphysician" (*Ib.,* p. *vii*). The book was a flat failure. At first Stallo attributed this failure to an incapacity of the English-speaking public to take a genuine interest in things of the "Spirit," but later even he repudiated his own book as an immature, juvenile work, saying: "That book was written while I was under the spell of Hegel's ontological reveries—at a time when I was barely of age and still seriously affected with the metaphysical malady which seems to be one of the unavoidable disorders of intellectual infancy" (*Concepts,* p. 11).

Stallo returned to Cincinnati and began to study law. He passed the bar examination at the end of 1849 and started practicing immediately thereafter. He was successful not only in private legal practice, but in politics as well. In 1884 he became ambassador to Italy. At the end of this assignment four years later, he decided to remain in Italy, settling in Florence, where he devoted himself to the exclusive pursuit of scientific and cultural interests. He died there on January 6, 1900.

Between 1849 and 1881 Stallo was steadily involved in the practice of law, but from all appearances must have spent a substantial part of his free time pursuing personal investigations in the realm of philosophy of science, for he published regularly on the subject. In these publications it became clear that he had moved from his original Hegelian view to an antimetaphysical conception of science. His first investigations led to the publishing of his most important book, *The Concepts and Theories of Modern Physics*, in 1882. The main theme of the book is Stallo's criticism of mechanism as accepted by most of the physicists of his time. In the second chapter, Stallo defined this mechanical theory of the universe by means of the following theses: (1) the primary elements of all natural phenomena, and, therefore the ultimates of scientific analysis, are mass and motion; (2) mass and motion are disparate (mass is indifferent to motion which may be imparted to it "from the outside"); (3) both mass and motion are constant. In Stallo's estimation most of the physicists and chemists of his time combined this general outlook with the assumption of the atomic constitution of matter, which in turn leads to four other theses: (1) the elementary units of mass are simple and in all respects equal; (2) they are absolutely hard and inelastic; (3) they are absolutely inert and passive; (4) all potential energy is in reality kinetic energy.

It was not too difficult for Stallo to show that this view was based on metaphysical prejudices clearly discernible in the history of philosophy from Democritus on. Stallo, however, went too far in his criticism, rejecting the reality of atoms completely. His view is similar to that of Mach in many respects and certainly had some influence on Ostwald's criticism of mechanism.

The selection which follows is a part of Stallo's critique of mechanism in the particular case of the kinetic theory of gases. This selection was chosen specifically because it contains Stallo's theory on the meaning of scientific hypotheses.

SELECTIVE BIBLIOGRAPHY

Stallo, J. B., *General Principles of Philosophy of Nature*. Boston, 1848.

———, *The Concepts and Theories of Modern Physics* (1882), ed. Percy W. Bridgman. Cambridge, Mass., 1960.

———, *Reden, Abhandlungen und Briefe*. New York, 1893.

Bridgman, P. W., *Introduction* to Stallo's *The Concepts, op. cit.*

Drake, S., "J. B. Stallo and the Critique of Classical Physics," in *Men and Moments in the History of Science,* ed. M. Evans. Seattle, 1959.

Kleinpeter, H., "J. B. Stallo als Erkenntniskritiker," *Vierteljahrschr. für wissenschaftliche Philosophie* **25** (1901) 401–440.

THE KINETIC THEORY OF GASES—
CONDITIONS OF THE VALIDITY
OF SCIENTIFIC HYPOTHESES

In the fourth chapter[1] I have already given an outline of the doctrine now generally known and accepted as the kinetic theory of gases. The assumptions of this theory are that a gaseous body consists of a great number of minute solid particles—molecules or atoms—in perpetual rectilinear motion, which, as a whole, is conserved by reason of the absolute elasticity of the moving particles, while the directions of the movements of the individual particles are incessantly changed by their mutual encounters or collisions. The colliding particles are supposed to act upon each other only within very small distances and for very short times before and after collision, their motion being free, and consequently rectilinear, in the intervals between such distances and times. The durations of the rectilinear motions in free paths are, moreover, assumed to be indefinitely large as compared with the durations of the encounters and of the mutual actions.

This theory was first advanced by Kroenig,[2] and has since been elaborated by Clausius, Maxwell, Boltzmann, Stefan, Pfaundler, and other physicists of the highest note. As in the case of the atomic hypothesis generally, I propose for the present to discuss, not so much the logical warrant, as the scientific value, of the theory in question. To this end it will be necessary, however, first to ascertain the true nature and function of a scientific hypothesis—not only the criteria of its value, but also the conditions of its validity

---

[1] [*Cf. Concepts*, p. 71.]

[2] *Pogg. Ann.*, XCIX [1856], 315f. As is usual in such cases, preclusions of the theory have since been discovered in the writings of various older physicists—cf. E. Du Bois-Reymond in *Pogg. Ann.*, CVII [1859], 490f.

A scientific hypothesis may be defined in general terms as a provisional or tentative explanation of physical phenomena.[3] But what is an explanation in the true scientific sense? The answers to this question which are given by logicians and men of science, though differing in their phraseology, are essentially of the same import. Phenomena are explained by an exhibition of their partial or total identity with other phenomena. Science is knowledge; and all knowledge, in the language of Sir William Hamilton,[4] is a "unification of the multiple." "The basis of all scientific explanation," says Bain,[5] "consists in assimilating a fact to some other fact or facts. It is identical with the generalizing process." And "generalization is only the apprehension of the One in the Many."[6] Similarly Jevons:[7] "Science arises from the discovery of identity amid diversity," and[8] "every great advance in science consists in a great generalization pointing out deep and subtle resemblances." The same thing is stated by the author just quoted in another place.[9] "Every act of explanation consists in detecting and pointing out a resemblance between facts, or in showing that a greater or less degree of identity exists between apparently diverse phenomena."

All this may be expressed in familiar language thus: When a new phenomenon presents itself to the man of science or to the ordinary observer, the question arises in the mind of either: What is it?—and this question simply means: Of what known, familiar fact is this apparently strange, hitherto unknown fact a new presentation—of what known, familiar fact or facts is it a disguise or complication? Or, inasmuch as the partial or total identity of several phenomena is the basis of classification (a class being a number of objects having one or more properties in common), it may also be said that all explanation, including explanation by hypothesis, is in its nature classification.

Such being the essential nature of a scientific explanation of which

---

[3] Wundt has lately (*Logik*, I, 403) sought to distinguish hypotheses from "anticipations of fact" and to restrict the term "hypothesis" to a sense which, notwithstanding its etymological warrant, is at variance with ordinary as well as scientific usage.

[4] *Lectures on Metaphysics [and Logic]*, Boston ed. [1860], pp. 47–48.

[5] [Alexander Bain] *Logic*, II (Induction) [London, 1870], chap. xii, § 2.

[6] Hamilton, *Lectures on Metaphysics [and Logic]*, p. 48.

[7] *Principles of Science*, I, 1.

[8] *Ibid.*, II, 281.

[9] *Ibid.*, II, 166.

an hypothesis is a probatory form, it follows that no hypothesis can be valid which does not identify the whole or a part of the phenomenon, for the explanation of which it is advanced, with some other phenomenon or phenomena previously observed. This first and fundamental canon of all hypothetical reasoning in science is formally resolvable into two propositions, the first of which is that every valid hypothesis must be an identification of two terms—the fact to be explained and a fact by which it is explained; and the second that the latter fact must be known to experience.

Tested by the first of these propositions, all hypotheses are futile which merely substitute an assumption for a fact, and thus, in the language of the schoolmen, explain *obscurum per obscurius* ["the obscure by means of the more obscure"], or (the assumption being simply the statement of the fact itself in another form—the "fact over again") illustrate *idem per idem*. And the futility of such hypotheses goes to the verge of mischievous puerility when they replace a single fact by a number of arbitrary assumptions, among which is the fact itself. Some of the uses made of the atomic hypothesis, both in physics and chemistry, which have been discussed in the last chapter, afford conspicuous examples of this class of bootless assumptions; and similar instances abound among the mathematical formulæ that are not infrequently paraded as physical theories. These formulæ are in many cases simply results of a series of transformations of an equation which embodies an hypothesis whose elements are neither more nor less than the elements of the phenomenon to be accounted for, the sole merit of the emerging formula being that it is not in conflict with the initial one.[10]

[10] I hope not to be misunderstood as disparaging the services for which physical science is indebted to mathematics. These services—especially those rendered by modern analysis—are incalculable. But there are mathematicians who imagine that they have compassed a solution of all the mysteries involved in a case of physical action when they have reduced it to the form of a differential expression preceded by a group of integral signs. Even when their equations are integrable they should bear in mind that the operations of mathematics are essentially deductive, and, while they may extend, can never deepen a physical theory. Granting that mathematics are much more than καθάρματα ψυχῆς ["purgations of the soul"] and that their office in the investigation of the causes of natural phenomena is far more important than the purely regulative functions of formal logic in science generally—conceding that the application of mathematics to physics has not only brought to light the significance of many experimental results, but has often been a trustworthy guide to successful research—neverthe-

In order to comply with the first condition of its validity, an hypothesis must bring the fact to be explained into relation with some other fact or facts by identifying the whole or a part of the former with the whole or a part of the latter. In this sense it has been well said that a valid hypothesis reduces the number of the uncompre-

---

less some of our prominent physicists and mathematicians might still read with profit the ninety-sixth aphorism in the first book of Bacon's *Novum Organum:* "Naturalis Philosophia adhuc sincera no invenitur, sed infecta et corrupta; in Aristotelis schola per logicam; in Platonis schola per theologiam naturalem; in secunda schola Platonis, Procli et aliorum *per Mathematicam quae philosophiam naturalem terminare, non generare aut procreare debet.*" ["Natural philosophy so far is not found in its pure form but only as infected and corrupted: in the school of Aristotle through logic; in the school of Plato through natural theology; in the second school of Plato, of Proclus and others, through mathematics, which should terminate natural philosophy, not generate or procreate it."] As to the value of the class of formulae referred to in the text it may not be inappropriate to cite the words of Cournot (*De l'Enchainement,* I, 249): "Tant qu'un calcul ne fait que rendre ce que l'on a tiré de l'observation pour dans les éléments du calcul à vrai dire il n'ajoute rien aux données de l'observation." ["In so far as a calculation uses as its elements only that which has been deduced from observation it cannot be validly claimed to add anything to the data of observation."] To the same effect are the admirable reflections of M. [Louis] Poinsot (*Théorie Nouvelle de la Rotation des Corps,* 1851 ed., p. 79): "Ce qui a pu faire illusion à quelques esprits sur cette espèce de force qu'ils supposent aux formules de l'analyse, c'est qu'on en retire, avec assez de facilité, des vérités déjà connues, et qu'on y a, pour ainsi dire, soi-même introduites, et il semble alors que l'analyse nous donne ce qu'elle ne fait que nous rendre dans un autre langage. Quand un théorème est connu, on n'a qu'à l'exprimer par des équations; si le théorème est vrai, chacune d'elles ne peut manquer d'être exacte, aussi bien que les transformées qu'on en peut déduire; et si l'on arrive ainsi à quelque formule évidente, ou bien établie d'ailleurs, on n'a qu'à prendre cette expression comme un point de départ, à revenir sur ses pas, et le calcul seul paraît avoir conduit comme de lui-même au théorème dont il s'agit. Mais c'est en cela que le lecteur est trompé," ["Some people are deceived with regard to the potency which they ascribe to the formulas of analysis by the fact that they can so easily extract from the formulas truths already known and which they have, so to speak, introduced themselves. It appears then as though the analysis were giving something new instead of merely repeating in different language. When a theorem is understood it can be expressed in equations. If the theorem is true, each one of the equations cannot fail to be true, as are also the equations that can be deduced from them by mathematical transformation. If in this way one arrives at a formula which is obviously true, or which can be established by some independent method, then one has only to take this formula as the new point of departure from which one can retrace his steps, and the calculation alone will appear to have automatically led to the theorem from which one started. But in this the reader is deceived."]

hended elements of a phenomenon by at least one.[11] In the same sense it is sometimes said that every true theory or hypothesis is in effect a simplification of the data of experience—an assertion which must be understood, however, with due regard to the second proposition to be discussed presently, i.e., with the proviso that the theory be not a mere *asylum ignorantae*, of the kind denoted by the schoolmen as a *principium expressivum*, such as the explanation of the phenomena of life by reference to a *vital principle*, or of certain chemical processes by *catalytic action*. True scientific explanations are generally complicated in form, not only because most phenomena, on proper analysis, prove to be complex, but because the simplest fact is not the effect of a single cause, but the product of a great and often indeterminate multiplicity of agencies—the outcome of the concurrence of numerous conditions. The Newtonian theory of planetary motion is much more intricate than that of Kepler, according to which every planet is conducted along its path by an *angelus rector;* and the account given by modern celestial mechanics of the precession of the equinoxes is far less simple than the announcement that among the great periods originally established by the Author of the universe was the Hipparchian cycle. The old brocard, *simplex veri judicium*, is to be taken with many grains of allowance before it can be trusted as a safe rule in determining the validity or value of scientific doctrines.

I now come to the second requirement of the validity of an

[11] "Der Verstand hat das Beduerfniss jede Erscheinung zu erklaeren d.h. dieselbe als das Resultat bekannter Kraefte oder Erscheinungen begrifflich abzuleiten. . . . Es geht hieraus hervor, dass jede Hypothese bekannte Kraefte oder Erscheinungen annehmen darf, indem die Annahme einer bisher unbekannten Kraft nur die Qualitaet des zu erklaerenden Phaenomen's aendern, aber nicht die Zahl der unerklaerten Momente reduciren kann. Soll eine Hypothese nicht vollkommen unnuetz und demgemaess die Verstandesarbeit, welche sie zur Befriedigung eines Beduerfnisses erzeugte, keine zwecklose sein, *so muss jede Hypothese die Zahl der unbegriffene Momente einer Erscheinung mindestens um eine erniedrigen.*" ["The intellect has a tendency to explain every phenomenon, that is, to derive it conceptually as the result of known forces or phenomena. As a result, every hypothesis can assume forces or phenomena already known, because the assumption of a hitherto unknown force can only alter the character of the phenomenon to be explained without reducing the number of the unexplained factors. If a hypothesis is not to be completely sterile and the intellectual effort which it generated for the satisfaction of a need is not to be useless, it is necessary that it reduce by at least one the number of unexplained factors in a situation."] [John Karl Friedrich] Zoellner, [*Ueber die*] *Natur der Kometen* ["The Nature of the Comets"; Leipzig, 1872], p. 189f.

hypothesis: that the explanatory phenomenon (i.e., that with which the phenomenon to be explained is identified) must be a datum of experience. This proposition is in substance equivalent to that part of Newton's first *regula philosophandi*,[12] in which he insists that the cause assigned for the explanation of natural things must be a *vera causa*—a term which he does not expressly define in the Principia, but whose import may be gathered from the following passage of his Opticks:[13] "To tell us that every species of things is endowed with an occult specific quality by which it acts and produces manifest effects is to tell us nothing. But to derive two or three general principles of motion *from phenomena* and afterward to tell us how the properties and actions of all corporeal things follow from these manifest principles would be a very great step in philosophy, though the causes of those principles were not yet discovered."

The requirement in question has long been the subject of animated discussion by J. S. Mill, Whewell, and others; but it will be found, I think, that, after making due allowance for necessary implications, there is little real disagreement among thinkers. The recent statement of G. H. Lewes[14] that "an explanation to be valid must be expressed in terms of phenomena already observed," and the counter-statement of Jevons[15] that "agreement with fact (i.e., the fact to be explained) is the one sole and sufficient test of a true hypothesis," are both far too broad, and are, indeed, modified by Lewes and Jevons themselves in the progress of the discussion; but the claim of Mr. Lewes is nevertheless true in the sense that no explanation is real unless it is an identification of experiential data. The confusion which, as in so many other cases of scientific controversy, is at the bottom of the seeming disagreement between the contending parties, arises from a disregard of the circumstance that the identification of two phenomena may be both partial and indirect—that it may be effected by showing that the phenomena have some known feature in common on condition that the existence, in one or both of the phenomena, of some other feature not yet directly observed, and perhaps incapable of direct observation, be assumed. The aptest illustration of this is the much-debated undulatory theory of light. This hypothesis identifies light with other

---

[12] *Phil. Nat. Princ. Math.*, Book III.
[13] 4th ed., p. 377.
[14] [George Henry Lewes] *Problems of Life and Mind* [Boston, 1879–1880; 2 vols.], II, 7.
[15] *Principles of Science*, II, 138.

forms of radiance, and even with sound, by showing that all these phenomena have the element of vibration or undulation (which is well known to experience) in common, on the assumption of an all-pervading material medium, of a kind wholly unknown to experience, as the bearer of the luminar undulations. In this case, as in all similar cases, the identity lies, not in the *fictitious* element, the æther, but in the *real* element, the *undulation*. It consists, not in the *agent*, but in *the law of its action*. And it is obvious that every hypothesis which establishes coincidences between phenomena in particulars that are purely fictitious is wholly vain, because it is in no sense an identification of phenomena. It is worse than vain: it is meaningless—a mere collection of words or symbols without comprehensive import. As Jevons expresses it:[16] "No hypothesis can be so much as framed in the mind, unless it be more or less conformable to experience. As the material of our ideas is undoubtedly derived from sensation so we can not figure to ourselves any existence or agent but as endowed with some of the properties of matter. All that the mind can do in the creation of new existences is to alter combinations, or by analogy to alter the intensity of sensuous properties." J. S. Mill is, therefore, clearly wrong when he says[17] that, "an hypothesis being a mere supposition, there can be no other limits to hypotheses than those of the human imagination," and that "we may, if we please, imagine, by way of accounting for an effect, some cause of a kind utterly unknown and acting according to a law altogether fictitious." The unsoundness of the latter part of this proposition is evidently felt by Mill himself, for he adds at the end of the next sentence that "there is *probably* no hypothesis in the history of science in which both the agent itself and the law of its operation were fictitious." There *certainly* is no such hypothesis—at least none which has in any way subserved the interests of science.

An hypothesis may involve not only one but several fictitious assumptions, provided they bring into relief, or point to the probability, or at least possibility, of an agreement between phenomena in a particular that is real and observable. This is especially legitimate when the agreement thus brought to light is not between two, but a greater number of phenomena, and still more so when the agreement is not merely in one but in several real particulars between

---

[16] *Ibid.*, 141.
[17] [*A System of*] *Logic* [*Ratiocinative and Inductive*], 8th ed. [London, 1872], p. 394.

diverse phenomena, so that, in the language of Whewell,[18] "the hypotheses which were assumed for one class of cases are found to explain another of a different nature—a consilience of induction." An instance of this is afforded by the hypothesis just referred to of the luminiferous æther, which was at first believed also to explain the retardation of comets. But, while the probability of the truth of an hypothesis is in direct ratio to the number of phenomena thus brought into relation, it is in the inverse ratio of the number of such fictions, or, more accurately, its improbability increases geometrically while the series of independent fictions expands arithmetically.[19] This finds

[18] [William Whewell] *History of the Inductive Sciences,* American ed. [New York, 1858], II, 186.

[19] "En général," says Cournot (*De l'Enchainement,* I, 103), "une théorie scientifique quelconque, imaginée pour relier un certain nombre de faits donnés par l'observation, peut être assimilée à la courbe que l'on trace d'après une loi géométrique, en s'imposant la condition de la faire passer par un certain nombre de points donnés d'advance. Le jugement que la raison porte sur la valeur intrinsèque de cette théorie est un jugement probable, une induction dont la probabilité tient d'une part à la simplicité de la formule théorique, d'autre part au nombre des faits ou des groupes de faits qu'elle relie, le même groupe devant comprendre tous les faits qui s'expliquent déjà les uns par les autres, indépendamment de l'hypothèse théorique. *S'il faut compliquer la formule à mesure que de nouveaux faits se révèlent à l'observation, elle devient de moins en moins probable en tant que loi de la Nature;* ce n'est bientôt plus qu'un échafaudage artificiel qui croule enfin lorsque, par un surcroit de complication, elle perd même l'utilité d'un système artificiel, celle d'aider le travail de la pensée et de diriger les recherches. Si au contraire les faits acquis à l'observation postérieurement à la construction de l'hypothèse sont reliés par elle aussi bien que les faits qui ont servi à la construire, si surtout des faits prévus comme conséquences de l'hypothèse reçoivent des observations postérieures une confirmation éclatante, la probabilité peut aller jusqu'à ne laisser aucune place au doute dans un esprit éclairé."

["In general, any scientific theory whatever designed to connect a certain number of facts of observation can be respresented by a curve traced according to some geometrical law so as to satisfy the condition that it pass through a certain number of points given in advance. The judgment which reason makes on the intrinsic importance of this theory is a judgment made on a probability basis, the probability depending partly on the simplicity of the theoretical formula and partly on the number of facts or groups of facts which it ties together, a single group of facts including all those which are so related that they explain each other, without reference to the hypothetical theory. *If it becomes necessary to make the formula more complicated in proportion as new facts of observation are discovered, the theory becomes in so far less and less probable as a law of nature;* it quickly becomes merely an artificial scaffolding which finally collapses when, by an excess of complication, it loses even the usefulness of an artificial system in assisting thought or in directing research. If, on the contrary,

illustration again in the undulatory theory of light. The multitude of fictitious assumptions embodied in this hypothesis, in conjunction with the failure of the consiliences by which it appeared at first to be distinguished, can hardly be looked upon otherwise than as a standing impeachment of its validity in its present form. However ready we may be to accede to the demands of the theorist when he asks us to grant that all space is pervaded, and all sensible matter is penetrated, by an adamantine solid exerting at each point in space an elastic force 1,148,000,000,000 times that of air at the earth's surface, and a pressure upon the square inch of 17,000,000,000,000 pounds[20] —a solid which, at the same time, wholly eludes our senses, is utterly impalpable and offers no appreciable resistance to the motions of ordinary bodies—we are appalled when we are told that the alleged existence of this adamantine medium, the æther, does not, after all, explain the observed irregularities in the periods of comets; that, furthermore, not only is the supposed luminiferous æther unavailable as a medium for the origination and propagation of dielectric phenomena, so that for these a distinct all-pervading electriferous æther must be assumed,[21] but that it is very questionable whether the assumption of a single æthereal medium is competent to account for all the known facts in optics (as, for instance, the non-interference of two rays originally polarized in different planes when they have been brought to the same plane of polarization, and certain phenomena of double refraction, in view of which it is necessary to suppose that the rigidity of the medium varies with the direction of the strain—a supposition discountenanced by the facts relating to the intensities of reflected light), and that for the adequate explanation of the phenomena of light it is "necessary to consider what we term the æther as consisting of two media, each possessed of equal and enormous self-repulsion or elasticity, and both existing in equal quantities throughout space, whose vibrations take place in perpendicular planes, the two media being mutually indifferent, neither

---

the facts acquired by observation after the construction of the hypothesis are covered by it as well as the facts used in its construction, and if, above all, facts anticipated as consequences of the hypothesis receive a brilliant confirmation from posterior observation, the probability of the hypothesis becomes so great as to leave no room for doubt in an enlightened mind."]

[20] Cf. [Sir John] Herschel, *Familiar Lectures* [*on Scientific Subjects* (London and New York, 1866)], p. 282; F. De Wrede (President Royal Academy of Sciences in Stockholm), address, *Phil. Mag.*, 4th ser., XLIV [August 1872], 82.
[21] W. A. Norton, "On Molecular Physics," *ibid.*, XXVIII [1864], 193.

attracting nor repelling."[22] In this endless superfetation of æthereal media upon space and ordinary matter, there are omninous suggestions of the three kinds of æthereal substances postulated by Leibnitz and Cartesius alike as a basis for their vortical systems. There is an impulsive whirl in our thoughts, at least, when we are called upon, in the interests of the received form of the undulatory theory, not only to reject all the presumptions arising from our common observation and all the analogies of experience, but to cumulate hypotheses and æthers indefinitely. And we are but partially reassured by the circumstance that the theory in question, besides accounting for the phenomena of optics which had been observed at the time of its promulgation, has the great merit of successful prevision, having led to the prediction of a number of facts subsequently discovered. These predictions, certainly, have not only been numerous, but several of them, such as Hamilton's announcement of conical refraction (afterward verified by Lloyd) and Fresnel's forecast (from the imaginary form of an algebraic formula) of circular polarization after two internal reflections in a rhomb, are very striking. But, although anticipations of this sort just serve to accredit an hypothesis, they are, as Mill has shown,[23] by no means absolute tests of their truth. Using the word "cause" in the sense in which it is commonly

---

[22] Hudson, "On Wave Theories of Light, Heat, and Electricity," *ibid.*, XLIV [September 1872], 210f. In this article the author also points out the crudeness of the subsidiary hypotheses which have been framed to obviate other difficulties of the undulatory theory, among which are those discussed in the last chapter. "Waves of sound," he says, "in our atmosphere are 10,000 times as long as the waves of light, and their velocity of propagation about 850,000 times less, and, even when air has been raised to a temperature at which waves of red light are propagated from matter, the velocity of sound-waves is only increased to about double what it was at zero centigrade. Even their velocity through glass is 55,000 times less than the speed of the aethereal undulations, and the extreme slowness of change of temperature in the conduction of heat (as contrasted with the rapidity with which the vibrations of the aether exhaust themselves, becoming insensible almost instantly when the action of the existing cause ceases) marks distinctly the essential difference between molecular and aethereal vibrations. It appears to me, therefore, a very crude hypothesis to imagine a combination of aethereo-molecular vibrations as accounting for the very minute difference in the retardation of doubly refracted rays in crystals."
[23] *Logic*, p. 356. Long before Mill, Leibnitz observed that success in explaining (or predicting) facts is no proof of the validity of an hypothesis, inasmuch as right conclusions may be drawn from wrong premises—as Leibnitz expresses it, "Comme le vrai peut être tiré de faux" ["As it is posssible to deduce the true from the false"]. Cf. *Nouveaux Essais*, chap. xvii, sec. 5—Leibnitz, *Opp.*, ed. Erdmann, p. 397.

understood, an effect may be due to any one of several causes, and may, therefore, in many cases be accounted for by any one of several conflicting hypotheses, as becomes evident to the most cursory glance at the history of science. When an hypothesis successfully explains a number of phenomena with reference to which it was constructed, it is not strange that it should also explain others connected with them that are subsequently discovered. There are few discarded physical theories that could not boast the prevision of phenomena to which they pointed and which were afterward observed; among them are the one-fluid theory of electricity and the corpuscular theory of light.

There are, of course, other conditions of the validity of an hypothesis to which I have not yet adverted. Among them are those specified by Sir W. Hamilton, Mill, Bain, and others, such as that the hypothesis must not be contradictory of itself or in conflict with the known laws of nature (which latter requirement is, however, somewhat doubtful, inasmuch as the laws in question may be incomplete inductions from past experience to be supplemented by the very elements postulated by the hypothesis); that it must be of a nature to admit of deductive inferences, etc. Upon all these it is not necessary, in view of my present purpose, to dilate. The two conditions which I have sought to enforce and illustrate are, in my judgment, sufficient tests of the validity and merits of the kinetic theory of gases.

The fundamental fact to be accounted for by this theory is that gases are bodies which, at constant temperatures and in the absence of external pressure, expand at even rate. From this fact the two great empirical laws, so called, expressive of those physical properties of a gas which are directly attested by experience, are the necessary and immediate consequences, being, indeed, nothing more than partial and complementary statements of it. The limitation of gaseous volume being produced by pressure alone—the cohibition of the bulk of a gas being due *solely* to pressure—it follows that it must be proportional to it; in other words, that the volume of a gas must be inversely as the pressure; and this is the law of Boyle or Mariotte. Again: temperature is measured by the uniform expansion of a column of gas (in the air-thermometer); hence, if all gases expand equally, temperature is proportional to the volume of a gas and conversely; this is the law of Charles.[24]

24 One of the strangest incidents in the history of physics is the grave discussion of the question respecting the true law of gaseous expansion. "According to Gay-Lussac," Balfour Stewart ([*An Elementary*] *Treatise on Heat* [Oxford, 1866], p. 60), "the augmentation of volume which a gas receives when the temperature

The foregoing real definition (i.e., exhibition of the properties) of a gas applies only to ideal or perfect gases. In actual experience we meet with no gas which, in the absence of pressure, expands with absolute uniformity; and for that reason we do not know experimentally of any gas behaving in strict conformity to the laws of Boyle and Charles. Moreover, we are unable directly to observe a gas which is wholly free from pressure; the datum of experience is simply that gases expand (other things being equal) in proportion to the diminution of the pressure to which they are subjected. But in the case of many gases—those which are either wholly incoercible, or coercible (i.e., reducible to the liquid or solid state) with great difficulty, and of nearly all gases at very high temperatures—the deviation from uniformity of expansion is very slight.

---

increases 1° is a certain fixed proportion of *its initial volume at 0° C.*; while, according to Dalton, a gas at any temperature increases in volume for a rise of 1° by a constant fraction of *its volume at that temperature*. . . . The dilation of gases has since been investigated by Rudberg, Dulong and Petit, Magnus and Regnault, and the result of their labors leaves little doubt that Gay-Lussac's method of expressing the law is much nearer the truth than Dalton's." Inasmuch as the experiments of Rudberg and others were necessarily made on the supposition that the coefficient of expansion was the same for all gases (the question relating, not to the expansion of some particular gas, but of gases generally), and, as the standard temperatures were those of the air thermometer, it would have been surprising, indeed, if the result had been confirmatory of Dalton's view. A thermometer is graduated by dividing a given length of a tube of even bore into equal parts. It is clear, therefore, that the increment of volume resulting from the expansion of the air in such a tube through one degree is a fixed part of a constant volume initially assumed, and not of a constantly increasing volume; and the same thing is, of course, true of any other gas if it expands at the same rate. Dalton's form of the law of expansion would yield the following remarkable series of equal ratios—in which the first represents the rate of expansion of air in the thermometer, and the others stand for the rate (or rather rates) of expansion of the gas under examination (a being the linear expansion of the air in the thermometer, v its initial volume, a' the corresponding expansion in the gas under examination, v' its initial volume):

$$\frac{a}{v} = \frac{a'}{v'} = \frac{a'}{v' + a'} = \frac{a'}{v' + 2a'} = \frac{a'}{v' + 3a'} = \frac{a'}{v' + 4a'} \text{ etc., etc.}$$

The attempts at an experimental solution of the question here referred to are suggestive, by the way, of a doubt as to the correctness of the prevailing systems of thermometry, which are founded on the assumption of equalities of volume-ratios in which one of the terms is constant while the other is variable, i.e., of fractions which have the same numerator, but different denominators. These suggestions are but imperfectly met by the reflection that the bores of our thermometrical tubes are very small.

Now, how does the kinetic theory of gases explain the experiential fact or facts just stated? It professes to explain them on the basis of at least three arbitrary assumptions, not one of which is a datum of experience, viz.:

1. That a gas is composed of solid particles which are indestructible and of constant mass and volume.

2. That these constituent particles are absolutely elastic.

3. That these particles are in perpetual motion, and, except at very small distances, in no wise act upon each other, so that their motions are absolutely free and therefore rectilinear.

I refrain from adding a fourth assumption—that of the absolute equality of the particles, in mass at least—because it is claimed (though unjustifiably) to be a corollary from the other assumptions.

The first of these assumptions has been sufficiently considered in the last chapter. The second assumption asserts the absolute elasticity of the constituent solid particles. What is the import and scope of this assumption? The elasticity of a solid body is that property by means of which it occupies, and tends to occupy, portions of space of determinate volume and figure, and therefore reacts against any force or stress producing, or tending to produce, an alteration of such volume or figure with a counter-force or stress which, in the case of perfect elasticity, is exactly proportional to the altering force. Now, it is seen at once that the property—the *fact*—thus assumed in the constituent solid includes the very fact to be accounted for in the gas. A perfect gas reacts against a stress tending to reduce its volume with a spring proportional to the stress; and for this reason gases are defined as elastic fluids. This resilience of the gas against diminution of volume is obviously a simpler fact than the rebound of a solid against both diminution *and increase* of volume, *in addition to the reaction against a change of figure*. The resistance to *several* kinds of change implies a greater number of forces, and is therefore a more complex phenomenon, than the resistance to *one* kind of change.[25]

---

[25] It may be said that the greater simplicity of the properties of a gas is purely conceptual. The identification of concepts with facts is undoubtedly the great fundamental error of speculation; but we are now dealing with the conceptual elements of the hypothesis under discussion. The opinion that a solid of constant volume (or, more accurately expressed, of variable volume, expanding or contracting to a fixed volume *proprio motu*) is a simpler thing than a uniformly expanding body is certainly not based upon any fact of experience, but is a mere prejudice of the intellect akin to the notion that a body at rest is a simpler

It thus appears that the presupposition of absolute elasticity in the solids, whose aggregate is said to constitute a gas, is a flagrant violation of the first condition of the validity of an hypothesis—the condition which requires a reduction of the number of unrelated elements in the fact to be explained, and therefore forbids a mere reproduction of this fact in the form of an assumption, and *a fortiori* a substitution of several arbitrary assumptions for one fact. Manifestly the explanation offered by the kinetic hypothesis, in so far as its second assumption lands us in the very phenomenon from which it starts, the phenomenon of resilience, is (like the explanation of impenetrability, or of the combination of elements in definite proportions by the atomic theory) simply the illustration of *idem per idem*, and the very reverse of a scientific procedure. It is a mere *versatio in loco* ["turning around in the same place"]—movement without progress. It is utterly vain; or rather, inasmuch as it complicates the phenomenon which it professes to explicate, it is worse than vain—a complete inversion of the order of intelligence, a resolution of identity into difference, a dispersion of the One into the Many, an unraveling of the Simple into the Complex, an interpretation of the Known in terms of the Unknown, an elucidation of the Evident by the Mysterious, a reduction of an ostensible and real fact to a baseless and shadowy phantom.[26]

---

phenomenon than a body in uniform motion, and generally that rest is simpler than motion. This prejudice has its root in our habitual oblivion of the essential relativity of all phenomena, which will be discussed hereafter.

[26] All theorists who attempt to account for a physical fact by a multiplication of arbitrary assumptions in which the fact itself is reproduced are liable to Aristotle's acute animadversion upon the Platonic doctrine of ideas—their endeavors are as nugatory as those of a person who, for the purpose of facilitating the operation of counting, begins by multiplying his numbers—"Those who posit the forms as causes in the first place seeking to lay hold on the cause of these existences have come up with a second set equal in number to the first, as if one who wished to make a count should think that he could not do so while the objects to be counted were too few but that he could do it if they were multiplied." *Met.*, A.9, 990a33–b4. Occam's rule "Entia non sunt multiplicanda praeter necessitatem" "Entities are not to be multiplied beyond what is necessary" has its applications in physics no less than in metaphysics; and there are physical theories of which Michel Montaigne, if he lived today, would say what he said of certain scholastic vagaries, three hundred years ago: "On eschange un mot pour un aultre mot, et souvent plus incogneu. . . . Pour satisfaire à un doubte, ils m'en donnent trois; c'est la teste d'Hydra. . . . Nous communiquons une question; on nous en redonne une ruchée." ["One replaces a word by another,

Waiving the question already discussed, whether or not the assumed absolute solidity and constancy of volume of the supposed constituent particles are consistent (in the light of the mechanical theory generally) with their absolute elasticity, I proceed to consider the third assumption of the kinetic hypothesis. This assumption is an unavoidable supplement to the initial theoretical complication of the phenomenon of elasticity, produced by the arbitrary substitution of the resilience of a solid against increase or diminution of volume and change of figure for the reaction of a gas against diminution of volume alone. To get rid of one gratuitous feature of the hypothesis (the addition of the rebound against dilatation and distortion to that against compression) and to bring it into conformity with the fact to be explained, it becomes necessary to add another arbitrary feature— to endow the parts with incessant rectilinear motion in all directions. In respect to this assumption, which, like other assumptions of the mechanical theory, is based upon a total disregard of the relativity and consequent mutual dependence of natural phenomena, it is to be said, for the present, that it is utterly gratuitous, and not only wholly unwarranted by experience, but out of all analogy with it. Bodies which, except on the very verge of immediate contact, move independently without mutual attraction or repulsion or any sort of mutual action and thus present perfect realizations of the abstract concept of free and ceaseless rectilinear motion, are unheard-of strangers in the wide domain of sensible experience. So complete an abandonment of the analogies of experience is all the more surprising in view of the circumstance that the atomic hypothesis, whereof the kinetic theory of gases is a branch, is confessedly a concretion of suggestions derived from celestial mechanics. There is hardly a treatise on modern physics in which the atoms or molecules are not compared to planetary or stellar systems. "A compound atom," says Jevons,[27] "may perhaps be compared with a stellar system, each star a minor system in itself." But the bodies with which celestial mechanics deal are all subject to the law of attraction; and the import of the very first theorem of Newton's *Principia* is, that these bodies, if their motions are at any

---

often even less known than the first. . . . In order to solve a difficulty, they give me three; it is like Hydra's head. . . . We are asking a question; they give us back a whole hive."] *Essais,* III, p. 13.

[27] *Principles of Science,* I, 453. In Arwed Walter's *Untersuchungen ueber Molekularmechanik* ["Research on Molecular Mechanics"; Berlin, 1873], p. 216, the system of Jupiter and his satellites is called a "planetary molecule."

moment out of the same straight line, can never collide, but must always move in curved orbits at a distance from each other. Oblique impacts between them productive of rotations as well as of deviations from their paths before impact, as they are imagined by Clausius and the other promoters of the kinetic theory are impossible. And this is true not only when the mutual actions of the bodies vary inversely as the squares of their distances but whenever they vary as any higher power of these distances—a proposition to be borne in mind in view of certain speculations of Boltzmann, Stefan, and Maxwell, of which I shall presently speak.

There is another very extraordinary and, in the light of all the teachings of science, unwarrantable feature in the assumption respecting the movements of the alleged solid constituent particles. I allude to the absolute discontinuity between the violent mutual action attributed to these particles during the few instants of time before and after their collisions, and their total freedom from mutual action during the comparatively long periods of their rectilinear motion along "free paths." And this leads me to say a few words in regard to certain subsidiary assumptions made by Maxwell and others in order to account for the anomalies exhibited by gases of different degrees of coercibility in their deviations from Boyle's and Charles's law. Maxwell assumes that the gas-molecules are neither strictly spherical nor absolutely elastic, and that their centers repel each other with a force inversely proportional to the fifth power of their distance;[28] while Stefan[29] endeavors to adjust the hypothesis to the phenomena in question by postulating that the molecules are absolutely proportional to the fourth roots of the absolute temperatures of the gases. These assumptions, which are fatal to all claims of simplicity preferred on behalf of the kinetic hypothesis, are in no sense an outgrowth of its original postulates; both are purely gratuitous as well as without experiential analogy, and the first of them, that of Maxwell, is in direct defiance of all the inductions from the wide range of actual observation. They are both mere stop-gaps of the hypothesis, peace-offerings for its non-congruence with the facts,

[28] Since this was written, Maxwell himself has abandoned this assumption as not conformable to the facts.
[29] "Ueber die dynamische Diffusion der Gase" ["On the Mechanical Diffusion of Gases"], Minutes of the Imperial Academy of Science, Division of Math. and Nat. Science, LXV, 323. Cf. also [Ludwig] Boltzmann, "Über das Wirkungsgesetz der Molekularkraefte" ["On the Law of Effect of Molecular Forces"], ibid., LXVI, 213.

pure inventions to satisfy the emergencies created by the hypothesis itself.

It were work of supererogation to review in detail the logical and mathematical methods by which it is attempted, from an hypothesis resting on such foundations, to deduce formulæ corresponding to the facts of experience. I may be permitted to say, however, that the methods of deduction are only less extraordinary than the premises. To account for the laws of Boyle and Charles resort is had to the calculus of probabilities, or, as Maxwell terms it,[30] the method of statistics. It is alleged that, although the individual molecules move with unequal velocities, either because these velocities were originally unequal, or because they have become unequal in consequence of the encounters between them, nevertheless, there will be an average of all the velocities belonging to the molecules of a system (i.e., of a gaseous body) which Maxwell calls the "velocity of mean square." The pressure, on this supposition, is proportional to a product of the square of this average velocity into the number of molecules multiplied by the mass of each molecule. The product of the number of molecules into the mass of each molecule is then replaced by the density—in other words, the whole molecular assumption is, for the nonce, abandoned—and the velocity is eliminated as representing the temperature; it follows, of course, that the pressure is proportional to the density.

Similar procedures lead to the law of Charles and the "law" of Avogadro (according to which the number of molecules in any two equal volumes of gases of whatever kind is the same at the same temperatures and pressures—a law which is itself a mere hypothesis). It is claimed, on statistical grounds again, that not only the average velocity of a number of molecules in a given gaseous body is the same, but that "if two sets of molecules, whose mass is different, are in motion in the same vessel, they will, by their encounters, exchange energy with each other till the average kinetic energy of *a single molecule of either set* is the same."[31] "This," says Maxwell, "follows from the same investigation which determines the law of distribution of velocities in a single set of molecules." All this being granted, the law of Charles and the law of Avogadro (called by Maxwell the law of Gay-Lussac) are readily derived. And at the end of these devious courses of deduction Maxwell adds a disquisition on the properties

[30] *Theory of Heat,* p. 288.
[31] *Ibid.,* p. 289f.

of molecules, in which he claims to have made it evident that the molecules of the same substance are "unalterable by the processes which go on in the present state of things, and every individual of the same species is of exactly the same magnitude as though they had all been cast in the same mold, like bullets, and not merely selected and grouped according to their size, like small shot," and that, therefore, as he expresses it in another place[32] they are not the products of any sort of evolution, but, in the language of Sir John Herschel, "have the essential character of manufactured articles."

Now, on what logical, mathematical, or other grounds is the statistical method applied to the velocities of the molecules in preference to their weights and volumes? What reason is given, or can be given, why the masses of the molecules should not be subjected to the process of averaging as well as their motions? None whatever. And, in the absence of such reason, the deductions of the kinetic theory, besides being founded on rickety premises, are delusive paralogisms.

Upon these considerations I do not hesitate to declare that the kinetic hypothesis has none of the characteristics of a legitimate physical theory. Its premises are as inadmissible as the reasoning upon them is inconclusive. It postulates what it professes to explain; it is a solution in terms more mysterious than the problem—a solution of an equation by imaginary roots of unknown quantities. It is a pretended explanation, of which it were unmerited praise to say that it leaves the facts where it found them, and is obnoxious to the old Horatian stricture: *nil agit exemplum, litem quod lite resolvit.* ["An example is no good if it resolves one dispute only by raising another."]

Much is said about the support derived by the kinetic theory of gases from the revelations of the spectroscope. The spectra of gases, unlike those of solids and liquids, are not continuous, but consist of distinct colored lines or bands—showing, as is claimed, that in gases the vibrations of molecules do not interfere; that incandescent gases emit distinct kinds of light and not (as Jevons expresses it) luminar noises, because there is no clashing of the molecules disturbing the natural periods of vibration.[33] The spectroscope is, no doubt,

---

[32] "Bradford Lecture on the Theory of Molecules," *Popular Science Monthly* (January 1874).

[33] According to the latest interpretation of spectroscopic phenomena, the continuity or discontinuity of a spectrum is indicative, not so much of the state of aggression, as of the molecular complexity of the body examined. It is said that a body yields a spectrum of lines when its molecules contain but a few

the most important witness yet called on behalf of the kinetic theory; but the testimony of this witness is not all in its favor. "The spectroscope," says Maxwell himself,[34] "shows that some molecules can execute a great many different kinds of vibrations. They must, therefore, be systems of very considerable complexity, having far more than six variables. Now, every additional variable introduces an additional amount of capacity for internal motion without increasing the external pressure. Every additional variable, therefore, increases the specific heat, whether reckoned at constant pressure or constant volume. So does any capacity which the molecule may have for storing up energy in the potential form. But the calculated specific heat is already too great when we suppose the molecule to consist of two atoms only. Hence every additional degree of complexity which we attribute to the molecule can only increase the difficulty of reconciling the observed with the calculated value of the specific heat."

It may seem strange that so many of the leaders of scientific research, who have been trained in the severe schools of exact thought and rigorous analysis, should have wasted their efforts upon a theory so manifestly repugnant to all scientific sobriety—an hypothesis in which the very thing to be explained is but a small part of its explanatory assumptions. But even the intellects of men of science are haunted by pre-scientific survivals, not the least of which is the inveterate fancy that the mystery by which a fact is surrounded may be got rid of by minimizing the fact and banishing it to the regions of a solid atom is in less need of explanation than that of a bulky gaseous body is closely related to the conceit that the chasm between the world of matter and that of mind may be narrowed, if not bridged, by a rarefaction of matter, or by its resolution into "forces." The scientific literature of the day teems with theories in the nature of attempts to convert facts into ideas by a process of dwindling or subtilization. All such attempts are nugatory; the intangible specter proves more troublesome in the end than the tangible presence. Faith in spooks (with due respect be it said for Maxwell's thermodynamical "demons" and for the population of the "Unseen Universe") is unwisdom in physics no less than in pneumatology.

---

atoms each; that, when they contain more, the spectrum presents the appearance of fluted bands; and that the spectrum is continuous when each molecule comprises a great number of atoms.

[34] "On the Dynamical Evidence of the Molecular Constitution of Bodies," *Nature*, Nos. 279, 280 (March 4 and 11, 1875).

Chapter 8
ERNST MACH
(1838–1916)

*E*RNST MACH was born on February 18, 1838 in Turas, Moravia, which at that time belonged to Austria. He received his education in Vienna and in 1861 was engaged to teach physics at the University of Vienna. Between 1864 and 1867 he taught mathematics at Graz and then was appointed to a chair in physics at Prague. He returned to Vienna in 1895 and taught physics and "inductive philosophy" until 1901, the year he was made a member of the Austrian house of peers. He died at Haar, a town near Munich, on Feburary 19, 1916.

Mach is the author of many important studies on the most diverse subjects, but all of these works are in one way or another concerned with the philosophical and logical questions raised by the inductive sciences. This is not to say, however, that there were no other important insights in his writings, for in fact Mach made substantial contributions to mechanics, the theory of electricity, acoustics, and thermodynamics, as well as to the experimental psychology of space, time, and perception. He published his first book, *Die Geschichte und die Wurzel des Satzes von der Erhaltung der Arbeit* ("History and Root of the Principle of the Conservation of Energy") in 1872. His other publications followed in regular succession, among them: *Die Mechanik in ihrer Entwicklung historisch-kritisch dargestellt* (1883) ("Science of Mechanics described in its Historico-critical Development"), *Beiträge zur Analyse der Empfindungen* (1886) ("Contributions to the Analysis of Sensation"), *Die Prinzipien der Wärmelehre* (1896) ("Principles of the Theory of Heat), *Populär-wissenschaftliche Vorlesungen* (1896) ("Popular Lectures"), *Erkenntnis und Irrtum. Skizzen zur Psychologie der Forschung* (1905) ("Knowledge and Error. Sketches toward a Psychology of Scientific Research"), and *Die Prinzipien der physikalischen Optik* (1921) ("Principles of Physical Optics").

Mach was influenced by the philosophies of Kant and Fechner in his early days, but later rejected their distinction between phenomena and "things in themselves." After that he returned to the works of Hume, Avenarius, and leading physicist-philosophers such as Helmholtz, Kirchhoff, Boltzmann, and Clifford. The point of view which he adopted during the 1860's under the influence of these men was modified in later years and developed further, but it was maintained substantially throughout the rest of his life. It may be characterized as sensationalism, conventionalism, and a strongly antimetaphysical attitude.

Although in his treatise *On the Definition of Mass* (1860) Mach was still maintaining that in every scientific theory there is an a priori element of a purely formal character as far as its fundamental principles are concerned, later he gradually came to the insight that one must reject any aprioric element in the constitution of our knowledge of things. He saw science as no more than a conceptual reflection upon facts whose elements are contents of consciousness given to us by sensation; science must explain the reciprocal dependence of these elements and connect them methodically, guided by observation and experiment.

In the realm of the empirical sciences such an attitude necessarily leads to positivism. For Mach no statement could be accepted by science unless it was empirically verifiable. His criterion of verifiability was so exceptionally rigorous, however, that he not only had to reject metaphysical conceptions such as that of absolute space, absolute time, and even the ether, but also to oppose the introduction of atoms and molecules. This positivistic attitude was also the root of Mach's claim that scientific laws have only a descriptive meaning, and that hypotheses and theories are merely conventional whose scientific value is to be determined exclusively on the basis of the principle of the economy of thought.

In philosophy Mach defended a phenomenalistic position, maintaining that all empirical statements including those which occur in scientific theories can and must be reduced to statements about sensations. He tried to justify this point of view in his work *Analysis of Sensations*, but the attempt was not a great success. It is clear that considering the very principles of science to be nothing but abbreviated descriptions of sensations does not account fully for the fact that those principles contain simple mathematical relationships which highly transcend in clarity the vague connections among a great number of hazy concepts which are found in a description of our immediate sensations. In addition, to call the principles of scientific theories "economic descriptions of sensations" is not to do justice to the predominant role of reason and discourse in the actual discovery and presentation of these principles. These considerations deeply influenced the members of the later Vienna Circle who, although impressed by many of Mach's insights, nonetheless felt that particularly in these aspects Mach's view was to be further developed.

The selection which follows is an address delivered by Mach before the anniversary meeting of the Imperial Academy of Sciences at Vienna, on May 25, 1882.

SELECTIVE BIBLIOGRAPHY

Mach, E., *History and Root of the Principle of the Conservation of Energy* (1872), trans. P. E. Jourdain. Chicago, 1962.

――――, *The Science of Mechanics* (1883), trans. T. J. McCormack. La Salle, Ill., 1960.

――――, *The Analysis of Sensations* (1886) trans. C. M. Williams. New York, 1959.

――――, *Popular Scientific Lectures* (1896), trans. T. J. McCormack. Chicago, 1943.

――――, *Space and Geometry* (1901–1903), trans. T. J. McCormack. Chicago, 1960.

――――, *Erkenntnis und Irrtum*. Leipzig, 1905.

――――, *The Principles of Physical Optics* (1921), trans. J. S. Anderson and A. F. Young. New York, 1953.

Weinberg, C. B., *Mach's Empirio-Pragmatism in Physical Science*. New York, 1937.

Frank, P., *Modern Science and Its Philosophy*. New York, 1961. Pp. 13–62 and 69–95.

Mises, R. von, *Positivism. A Study in Human Understanding*. Cambridge, Mass., 1951.

――――, *Ernst Mach und die empiristische Wissenschaftsauffassung*. The Hague, 1938.

Popper, K., "A Note on Berkeley as a Precursor of Mach," *Brit. J. Phil. Sci.* 4 (1953) 26–48.

Einstein, A., "Autobiographical Notes," in *Albert Einstein: Philosopher-Scientist,* ed. P. A. Schilpp (2 vols.). New York, 1959. Vol. I, pp. 1–95 (passim).

Alexander, P., "The Philosophy of Science, 1850–1910," in *A Critical History of Western Philosophy,* ed. D. J. O'Connor. New York, 1964. Pp. 403–409.

Dingler, H. *Die Grundgedanken der Machschen Philosophie.* Leipzig, 1924.

Adler, F., *Ernst Machs Überwindung des mechanischen Materialismus.* Vienna, 1918.

THE ECONOMICAL NATURE OF
PHYSICAL INQUIRY

W̅HEN the human mind, with its limited powers, attempts to mirror in itself the rich life of the world, of which it is itself only a small part, and which it can never hope to exhaust, it has every reason for proceeding economically. . . .

The homely beginnings of science will best reveal to us its simple, unchangeable character. Man acquires his first knowledge of nature half-consciously and automatically, from an instinctive habit of mimicking and forecasting facts in thought, of supplementing sluggish experience with the swift wings of thought, at first only for his material welfare. When he hears a noise in the underbrush he constructs there, just as the animal does, the enemy which he fears; when he sees a certain rind he forms mentally the image of the fruit which he is in search of; just as we mentally associate a certain kind of matter with a certain line in the spectrum or an electric spark with the friction of a piece of glass. A knowledge of causality in this form certainly reaches far below the level of Schopenhauer's pet dog, to whom it was ascribed. It probably exists in the whole animal world, and confirms the great thinker's statement regarding the will which created the intellect for its purposes. These primitive psychical functions are rooted in the economy of our organism not less firmly than are motion and digestion. Who would deny that we feel in them, too, the elemental power of a long practised logical and physiological activity, bequeathed to us as an heirloom from our forefathers?

Such primitive acts of knowledge constitute to-day the solidest foundation of scientific thought. Our instinctive knowledge, as we shall briefly call it, by virtue of the conviction that we have consciously and intentionally contributed nothing to its formation, confronts us with an authority and logical power which consciously acquired knowledge even from familiar sources and of easily tested fallibility can never possess. All so-called axioms are such instinctive knowledge. Not consciously gained knowledge alone, but powerful

This article is from **Popular Scientific Lectures** by Ernst Mach, trans. T. J. McCormack, pp. 188–213. Copyright © 1943 by The Open Court Publishing Company, La Salle, Ill.

intellectual instinct, joined with vast conceptive powers, constitute the great inquirer. The greatest advances of science have always consisted in some successful formulation, in clear, abstract, and communicable terms, of what was instinctively known long before, and of thus making it the permanent property of humanity. By Newton's principle of the equality of pressure and counterpressure, whose truth all before him had felt, but which no predecessor had abstractly formulated, mechanics was placed by a single stroke on a higher level. Our statement might also be historically justified by examples from the scientific labors of Stevinus, S. Carnot, Faraday. J. R. Mayer, and others.

All this, however, is merely the soil from which science starts. The first real beginnings of science appear in society, particularly in the manual arts, where the necessity for the communication of experience arises. Here, where some new discovery is to be described and related, the compulsion is first felt of clearly defining in consciousness the important and essential features of that discovery, as many writers can testify. The aim of instruction is simply the saving of experience; the labor of one man is made to take the place of that of another.

The most wonderful economy of communication is found in language. Words are comparable to type, which spare the repetition of written signs and thus serve a multitude of purposes; or to the few sounds of which our numberless different words are composed. Language, with its helpmate, conceptual thought, by fixing the essential and rejecting the unessential, constructs its rigid pictures of the fluid world on the plan of a mosaic, at a sacrifice of exactness and fidelity but with a saving of tools and labor. Like a piano-player with previously prepared sounds, a speaker excites in his listener thoughts previously prepared, but fitting many cases, which respond to the speaker's summons with alacrity and little effort.

The principles which a prominent political economist, E. Hermann,[1] has formulated for the economy of the industrial arts, are also applicable to the ideas of common life and of science. The economy of language is augmented, of course, in the terminology of science. With respect to the economy of written intercourse there is scarcely a doubt that science itself will realise that grand old dream of the philosophers of a Universal Real Character. That time is not far distant. Our numeral characters, the symbols of mathematical analysis, chemical symbols, and musical notes, which might easily be supplemented by

---

[1] *Principien der Wirthschaftslehre*, Vienna, 1873.

a system of color-signs, together with some phonetic alphabets now in use, are all beginnings in this direction. The logical extension of what we have, joined with a use of the ideas which the Chinese ideography furnishes us, will render the special invention and promulgation of a Universal Character wholly superfluous.

The communication of scientific knowledge always involves description, that is, a mimetic reproduction of facts in thought, the object of which is to replace and save the trouble of new experience. Again, to save the labor of instruction and of acquisition, concise, abridged description is sought. This is really all that natural laws are. Knowing the value of the acceleration of gravity, and Galileo's laws of descent, we possess simple and compendious directions for reproducing in thought all possible motions of falling bodies. A formula of this kind is a complete substitute for a full table of motions of descent, because by means of the formula the data of such a table can be easily constructed at a moment's notice without the least burdening of the memory.

No human mind could comprehend all the individual cases of refraction. But knowing the index of refraction for the two media presented, and the familiar law of the sines, we can easily reproduce or fill out in thought every conceivable case of refraction. The advantage here consists in the disburdening of the memory; an end immensely furthered by the written preservation of the natural constants. More than this comprehensive and condensed report about facts is not contained in a natural law of this sort. In reality, the law always contains less than the fact itself, because it does not reproduce the fact as a whole but only in that aspect of it which is important for us, the rest being either intentionally or from necessity omitted. Natural laws may be likened to intellectual type of a higher order, partly movable, partly stereotyped, which last on new editions of experience may become down-right impediments.

When we look over a province of facts for the first time, it appears to us diversified, irregular, confused, full of contradictions. We first succeed in grasping only single facts, unrelated with the others. The province, as we are wont to say, is not *clear*. By and by we discover the simple, permanent elements of the mosaic, out of which we can mentally construct the whole province. When we have reached a point where we can discover everywhere the same facts, we no longer feel lost in this province; we comprehend it without effort; it is *explained* for us.

Let me illustrate this by an example. As soon as we have grasped the fact of the rectilinear propagation of light, the regular course of our thoughts stumbles at the phenomena of refraction and diffraction. As soon as we have cleared matters up by our index of refraction we discover that a special index is necessary for each color. Soon after we have accustomed ourselves to the fact that light added to light increases its intensity, we suddenly come across a case of total darkness produced by this cause. Ultimately, however, we see everywhere in the overwhelming multifariousness of optical phenomena the fact of the spatial and temporal periodicity of light, with its velocity of propagation dependent on the medium and the period. This tendency of obtaining a survey of a given province with the least expenditure of thought, and of representing all its facts by some one single mental process, may be justly termed an economical one.

The greatest perfection of mental economy is attained in that science which has reached the highest formal development, and which is widely employed in physical inquiry, namely, in mathematics. Strange as it may sound, the power of mathematics rests upon its evasion of all unnecessary thought and on its wonderful saving of mental operations. Even those arrangement-signs which we call numbers are a system of marvellous simplicity and economy. When we employ the multiplication-table in multiplying numbers of several places, and so use the results of old operations of counting instead of performing the whole of each operation anew; when we consult our table of logarithms, replacing and saving thus new calculations by old ones already performed; when we employ determinants instead of always beginning afresh the solution of a system of equations; when we resolve new integral expressions into familiar old integrals; we see in this simply a feeble reflexion of the intellectual activity of a Lagrange or a Cauchy, who, with the keen discernment of a great military commander, substituted for new operations whole hosts of old ones. No one will dispute me when I say that the most elementary as well as the highest mathematics are economically-ordered experiences of counting, put in forms ready for use.

In algebra we perform, as far as possible, all numerical operations which are identical in form once for all, so that only a remnant of work is left for the individual case. The use of the signs of algebra and analysis, which are merely symbols of operations to be performed, is due to the observation that we can materially disburden the mind in this way and spare its powers for more important and more difficult duties, by imposing all mechanical operations upon the hand. One result of this method, which attests its economical character, is the construction of calculating machines. The mathematician Babbage, the inventor of the difference-engine, was probably the first who clearly perceived this fact, and he touched upon it, although only cursorily, in his work, *The Economy of Manufactures and Machinery*.

The student of mathematics often finds it hard to throw off the uncomfortable feeling that his science, in the person of his pencil, surpasses him in intelligence,—an impression which the great Euler confessed he often could not get rid of. This feeling finds a sort of justification when we reflect that the majority of the ideas we deal with were conceived by others, often centuries ago. In great measure it is really the intelligence of other people that confronts us in science. The moment we look at matters in this light, the uncanniness and magical character of our impressions cease, especially when we remember that we can think over again at will any one of those alien thoughts.

Physics is experience, arranged in economical order. By this order not only is a broad and comprehensive view of what we have rendered possible, but also the defects and the needful alterations are made manifest, exactly as in a well-kept household. Physics shares with mathematics the advantages of succinct description and of brief, compendious definition, which precludes confusion, even in ideas where, with no apparent burdening of the brain, hosts of others are contained. Of these ideas the rich contents can be produced at any moment and displayed in their full perceptual light. Think of the swarm of well-ordered notions pent up in the idea of the potential. Is it wonderful that ideas containing so much finished labor should be easy to work with?

Our first knowledge, thus, is a product of the economy of self-preservation. By communication, the experience of *many* persons, individually acquired at first, is collected in *one*. The communication of knowledge and the necessity which every one feels of managing his stock of experience with the least expenditure of thought, compel us to put our knowledge in economical forms. But here we have a clue which strips science of all its mystery, and shows us what its power really is. With respect to specific results it yields us nothing that we could not reach in a sufficiently long time without methods. There is no problem in all mathematics that cannot be solved by direct counting. But with the present implements of mathematics many operations of counting can be performed in a few minutes which without mathematical methods would take a lifetime. Just as a single human being, restricted wholly to the fruits of his own labor, could never amass a fortune, but on the contrary the accumulation of the labor of many men in the hands of one is the foundation of wealth and power, so, also, no knowledge worthy of the name can be gathered up in a single human mind limited to the span of a human life and gifted only with finite powers, except by the most exquisite economy of thought and by the careful amassment of the economically ordered experience of thousands of co-workers. What strikes us here as the fruits of sorcery are simply the rewards of excellent house-keeping, as are the like results in civil life. But the business of science has this advantage over every other enterprise, that from *its* amass-ment of wealth no one suffers the least loss. This, too, is its blessing, its freeing and saving power.

The recognition of the economical character of science will now help us, perhaps, to understand better certain physical notions.

Those elements of an event which we call "cause and effect" are certain salient features of it, which are important for its mental re-production. Their importance wanes and the attention is transferred to fresh characters the moment the event or experience in question becomes familiar. If the connexion of such features strikes us as a necessary one, it is simply because the interpolation of certain inter-mediate links with which we are very familiar, and which possess, therefore, higher authority for us, is often attended with success in our explanations. That *ready* experience fixed in the mosaic of the mind with which we meet new events, Kant calls an innate concept of the understanding (*Verstandesbegriff*).

The grandest principles of physics, resolved into their elements, differ in no wise from the descriptive principles of the natural historian. The question, "Why?" which is always appropriate where the explanation of a contradiction is concerned, like all proper habitudes of thought, can overreach itself and be asked where nothing remains to be understood. Suppose we were to attribute to nature the property of producing like effects in like circumstances; just these like circumstances we should not know how to find. Nature exists once only. Our schematic mental imitation alone produces like events. Only in the mind, therefore, does the mutual dependence of certain features exist.

All our efforts to mirror the world in thought would be futile if we found nothing permanent in the varied changes of things. It is this that impels us to form the notion of substance, the source of which is not different from that of the modern ideas relative to the conservation of energy. The history of physics furnishes numerous examples of this impulse in almost all fields, and pretty examples of it may be traced back to the nursery. "Where does the light go to when it is put out?" asks the child. The sudden shrivelling up of a hydrogen balloon is inexplicable to a child; it looks everywhere for the large body which was just there but is now gone.

Where does heat come from? Where does heat go to? Such childish questions in the mouths of mature men shape the character of a century.

In mentally separating a body from the changeable environment in which it moves, what we really do is to extricate a group of sensations on which our thoughts are fastened and which is of relatively greater stability than the others, from the stream of all our sensations. Absolutely unalterable this group is not. Now this, now that member of it appears and disappears, or is altered. In its full identity it never recurs. Yet the sum of its constant elements as compared with the sum of its changeable ones, especially if we consider the continuous character of the transition, is always so great that for the purpose in hand the former usually appear sufficient to determine the body's identity. But because we can separate from the group every single member without the body's ceasing to be for us the same, we are easily led to believe that after abstracting all the members something additional would remain. It thus comes to pass that we form the notion of a substance distinct from its attributes, of a thing-in-itself, whilst our sensations are regarded merely as symbols or indications of the properties of this thing-in-itself. But it would be much better to say that bodies or things are compendious mental symbols for groups of sensations—symbols that do not exist outside of thought. Thus, the merchant regards the labels of his boxes merely as indexes of their contents, and not the contrary. He invests their contents, not their labels, with real value. The same economy which induces us to analyse a group and to establish special signs for its component parts, parts which also go to make up other groups, may likewise induce us to mark out by some single symbol a whole group.

On the old Egyptian monuments we see objects represented which do not reproduce a single visual impression, but are composed of various impressions. The heads and the legs of the figures appear in profile, the head-dress and the breast are seen from the front, and so on. We have here, so to speak, a mean view of the objects, in forming which the sculptor has retained what he deemed essential, and neglected what he thought indifferent. We have living exemplifications of the processes put into stone on the walls of these old temples, in the drawings of our children, and we also observe a faithful analogue of them in the formation of ideas in our own minds. Only in virtue of some such facility of view as that indicated, are we allowed to speak of *a* body. When we speak of a cube with trimmed corners—a figure which is not a cube—we do so from a natural instinct of economy, which prefers to add to an old familiar conception a correction instead of forming an entirely new one. This is the process of all judgment.

The crude notion of "body" can no more stand the test of analysis than can the art of the Egyptians or that of our little children. The physicist who sees a body flexed, stretched, melted, and vaporised, cuts up this body into smaller permanent parts; the chemist splits it up into elements. Yet even an element is not unalterable. Take sodium. When warmed, the white, silvery mass becomes a liquid, which, when the heat is increased and the air shut out, is transformed into a violet vapor, and on the heat being still more increased glows with a yellow light. If the name sodium is still retained, it is because of the continuous character of the transitions and from a necessary instinct of economy. By condensing the vapor, the white metal may be made to reappear. Indeed, even after the metal is thrown into water and has passed into sodium hydroxide, the vanished properties may by skilful treatment still be made to appear; just as a moving body which has passed behind a column and is lost to view for a moment may make its appearance after a time. It is unquestionably very convenient always to have ready the name and thought for a group of properties wherever that group by any possibility can appear. But more than a compendious economical symbol for these phenomena, that name and thought is not. It would be a mere empty word for one in whom it did not awaken a large group of well-ordered sense-impressions. And the same is true of the molecules and atoms into which the chemical element is still further analysed.

True, it is customary to regard the conservation of weight, or, more precisely, the conservation of mass, as a direct proof of the constancy of matter. But this proof is dissolved, when we go to the bottom of it, into such a multitude of instrumental and intellectual operations, that in a sense it will be found to constitute simply an equation which our ideas in imitating facts have to satisfy. That obscure, mysterious lump which we involuntarily add in thought, we seek for in vain outside the mind.

It is always, thus, the crude notion of substance that is slipping unnoticed into science, proving itself constantly insufficient, and ever under the necessity of being reduced to smaller and smaller world-particles. Here, as elsewhere, the lower stage is not rendered indispensable by the higher which is built upon it, no more than the simplest mode of locomotion, walking, is rendered superfluous by the most elaborate means of transportation. Body, as a compound of light and touch sensations, knit together by sensations of space, must be as familiar to the physicist who seeks it, as to the animal who hunts its prey. But the student of the theory of knowledge, like the geologist and the astronomer, must be permitted to reason back from the forms which are created before his eyes to others which he finds ready made for him.

All physical ideas and principles are succinct directions, frequently involving subordinate directions, for the employment of economically classified experiences, ready for use. Their conciseness, as also the fact that their contents are rarely exhibited in full, often invests them with the semblance of independent existence. Poetical myths regarding such ideas,—for example, that of Time, the producer and devourer of all things,—do not concern us here. We need only remind the reader that even Newton speaks of an *absolute* time independent of all phenomena, and of an absolute space—views which even Kant did not shake off, and which are often seriously entertained to-day. For the natural inquirer, determinations of time are merely abbreviated statements of the dependence of one event upon another, and nothing more. When we say the acceleration of a freely falling body is 9·810 metres per second, we mean the velocity of the body with respect to the centre of the earth is 9·810 metres greater when the earth has performed an additional 86400th part of its rotation—a fact which itself can be determined only by the earth's relation to other heavenly bodies. Again, in velocity is contained simply a relation of the position of a body to the position of the earth.[2] Instead of referring events to the earth we may refer them to a clock, or even to our internal sensation of time. Now, because all are connected, and each may be made the measure of the rest, the illusion easily arises that time has significance independently of all.[3]

[2] It is clear from this that all so-called elementary (differential) laws involve a relation to the Whole.
[3] If it be objected, that in the case of perturbations of the velocity of rotation of the earth, we could be sensible of such perturbations, and being obliged to have some measure of time, we should resort to the period of vibration of the

The aim of research is the discovery of the equations which subsist between the elements of phenomena. The equation of an ellipse expresses the universal *conceivable* relation between its co-ordinates, of which only the real values have *geometrical* significance. Similarly, the equations between the elements of *phenomena* express a universal, mathematically conceivable relation. Here, however, for many values only certain directions of change are *physically* admissible. As in the ellipse only certain *values* satisfying the equation are realised, so in the physical world only certain *changes* of value occur. Bodies are always accelerated towards the earth. Differences of temperature, left to themselves, always grow less; and so on. Similarly, with respect to space, mathematical and physiological researches have shown that the space of experience is simply an *actual* case of many conceivable cases, about whose peculiar properties experience alone can instruct us. The elucidation which this idea diffuses cannot be questioned, despite the absurd uses to which it has been put.

Let us endeavor now to summarise the results of our survey. In the economical schematism of science lie both its strength and its weakness. Facts are always represented at a sacrifice of completeness and never with greater precision than fits the needs of the moment. The incongruence between thought and experience, therefore, will continue to subsist as long as the two pursue their course side by side; but it will be continually diminished.

In reality, the point involved is always the completion of some partial experience; the derivation of one portion of a phenomenon from some other. In this act our ideas must be based directly upon sensations. We call this measuring.[4] The condition of science, both in its origin and in its application, is a *great relative stability* of our environment. What it teaches us is interdependence. Absolute forecasts, consequently, have no significance in science. With great changes in celestial space we should lose our co-ordinate systems of space and time.

When a geometer wishes to understand the form of a curve, he first resolves it into small rectilinear elements. In doing this, however, he is fully aware that these elements are only provisional and arbitrary

---

waves of sodium light,—all that this would show is that for practical reasons we should select that event which best served us as the *simplest* common measure of the others.

[4] Measurement, in fact, is the definition of one phenomenon by another (standard) phenomenon.

devices for comprehending in parts what he cannot comprehend as a whole. When the law of the curve is found he no longer thinks of the elements. Similarly, it would not become physical science to see in its self-created, changeable, economical tools, molecules and atoms, realities behind phenomena, forgetful of the lately acquired sapience of her older sister, philosophy, in substituting a mechanical mythology for the old animistic or metaphysical scheme, and thus creating no end of suppositious problems. The atom must remain a tool for representing phenomena, like the functions of mathematics. Gradually, however, as the intellect, by contact with its subject-matter, grows in discipline, physical science will give up its mosaic play with stones and will seek out the boundaries and forms of the bed in which the living stream of phenomena flows. The goal which it has set itself is the *simplest* and *most economical* abstract expression of facts.

The question now remains, whether the same method of research which till now we have tacitly restricted to physics, is also applicable in the psychical domain. This question will appear superfluous to the physical inquirer. Our physical and psychical views spring in exactly the same manner from instinctive knowledge. We read the thoughts of men in their acts and facial expressions without knowing how. Just as we predict the behavior of a magnetic needle placed near a current by imagining Ampère's swimmer in the current, similarly we predict in thought the acts and behavior of men by assuming sensations, feelings, and wills similar to our own connected with their bodies. What we here instinctively perform would appear to us as one of the subtlest achievements of science, far outstripping in significance and ingenuity Ampère's rule of the swimmer, were it not that every child unconsciously accomplished it. The question simply is, therefore, to grasp scientifically, that is, by conceptional thought, what we are already familiar with from other sources. And here much is to be accomplished. A long sequence of facts is to be disclosed between the physics of expression and movement and feeling and thought.

We hear the question, "But how is it possible to explain feeling by the motions of the atoms of the brain?" Certainly this will never be done, no more than light or heat will ever be deduced from the law of refraction. We need not deplore, therefore, the lack of ingenious solutions of this question. The problem is not a problem. A child looking over the walls of a city or of a fort into the moat below sees with astonishment living people in it, and not knowing of the portal which connects the wall with the moat, cannot understand how they could have got down from the high ramparts. So it is with the notions of physics. We cannot climb up into the province of psychology by the ladder of our abstractions, but we can climb down into it.

Let us look at the matter without bias. The world consists of colors, sounds, temperatures, pressures, spaces, times, and so forth, which now we shall not call sensations, nor phenomena, because in either term an arbitrary, one-sided theory is embodied, but simply *elements*. The fixing of the flux of these elements, whether mediately or immediately, is the real object of physical research. As long as, neglecting our own body, we employ ourselves with the interdependence of those groups of elements which, including men and animals, make up *foreign* bodies, we are physicists. For example, we investigate the change of the red color of a body as produced by a change of illumination. But the moment we consider the special influence on the red of the elements constituting our body, outlined by the well-known perspective with head invisible, we are at work in the domain of physiological psychology. We close our eyes, and the red together with the whole visible world disappears. There exists, thus, in the perspective field of every sense a portion which exercises on all the rest a different and more powerful influence than the rest upon one another. With this, however, all is said. In the light of this remark, we call *all* elements, in so far as we regard them as dependent on this special part (our body), *sensations*. That the world is our sensation, in this sense, cannot be questioned. But to make a system of conduct out of this provisional conception, and to abide its slaves, is as unnecessary for us as would be a similar course for a mathematician who, in varying a series of variables of a function which were previously assumed to be constant, or in interchanging the independent variables, finds his method to be the source of some very surprising ideas for him.[5]

[5] I have represented the point of view here taken for more than thirty years and developed it in various writings (*Erhaltung der Arbeit,* 1872, parts of which are published in the article on *The Conservation of Energy* in this collection; *The Forms of Liquids,* 1872, also published in this collection; and the *Bewegungsempfindungen.* 1875). The idea, though known to philosophers, is unfamiliar to the majority of physicists. It is a matter of deep regret to me, therefore, that the title and author of a small tract which accorded with my views in numerous details and which I remember having caught a glance of in a very busy period (1879–1880), have so completely disappeared from my memory that all efforts to obtain a clue to them have hitherto been fruitless.

If we look at the matter in this unbiassed light it will appear indubitable that the method of physiological psychology is none other than that of physics; what is more, that this science is a part of physics. Its subject-matter is not different from that of physics. It will unquestionably determine the relations the sensations bear to the physics of our body. We have already learned from a member of this academy (Hering) that in all probability a sixfold manifoldness of the chemical processes of the visual substance corresponds to the sixfold manifoldness of color-sensation, and a threefold manifoldness of the physiological processes to the threefold manifoldness of space-sensations. The paths of reflex actions and of the will are followed up and disclosed; it is ascertained what region of the brain subserves the function of speech, what region the function of locomotion, etc. That which still clings to our body, namely, our thoughts, will, when those investigations are finished, present no difficulties new in principle. When experience has once clearly exhibited these facts and science has marshalled them in economic and perspicuous order, there is no doubt that we shall *understand* them. For other "understanding" than a mental mastery of facts never existed. Science does not create facts from facts, but simply *orders* known facts.

Let us look, now, a little more closely into the modes of research of physiological psychology. We have a very clear idea of how a body moves in the space encompassing it. With our optical field of sight we are very familiar. But we are unable to state, as a rule, how we have come by an idea, from what corner of our intellectual field of sight it has entered, or by what region the impulse to a motion is sent forth. Moreover, we shall never get acquainted with this mental field of view from self-observation alone. Self-observation, in conjunction with physiological research, which seeks out physical connexions, can put this field of vision in a clear light before us, and will thus first really reveal to us our inner man.

Primarily, natural science, or physics, in its widest sense, makes us acquainted with only the firmest connexions of groups of elements. Provisorily, we may not bestow too much attention on the single constituents of those groups, if we are desirous of retaining a comprehensible whole. Instead of equations between the primitive variables, physics gives us, as much the easiest course, equations between *functions* of those variables. Physiological psychology teaches us how to separate the visible, the tangible, and the audible from bodies—a labor which is subsequently richly requited, as the division of the subjects of physics well shows. Physiology further analyses the visible into light and space sensations; the first into colors, the last also into their component parts; it resolves noises into sounds, these into tones, and so on. Unquestionably this analysis can be carried much further than it has been. It will be possible in the end to exhibit the common elements at the basis of very abstract but definite logical acts of like form,—elements which the acute jurist and mathematician, as it were, *feels* out, with absolute certainty, where the uninitiated hears only empty words. Physiology, in a word, will reveal to us the true real elements of the world. Physiological psychology bears to physics in its widest sense a relation similar to that which chemistry bears to physics in its narrowest sense. But far greater than the mutual support of physics and chemistry will be that which natural science and psychology will render each other. And the results that shall spring from this union will, in all likelihood, far outstrip those of the modern mechanical physics.

What those ideas are with which we shall comprehend the world when the closed circuit of physical and psychological facts shall lie complete before us, (that circuit of which we see now only two disjoined parts,) cannot be foreseen at the outset of the work. The men will be found who will see what is right and will have the courage, instead of wandering in the intricate paths of logical and historical accident, to enter on the straight ways to the heights from which the mighty stream of facts can be surveyed. Whether the notion which we now call matter will continue to have a scientific significance beyond the crude purposes of common life, we do not know. But we certainly shall wonder how colors and tones which were such innermost parts of us could suddenly get lost in our physical world of atoms; how we could be suddenly surprised that something which outside us simply clicked and beat, in our heads should make light and music; and how we could ask whether matter can feel, that is to say, whether a mental symbol for a group of sensations can feel?

We cannot mark out in hard and fast lines the science of the future, but we can foresee that the rigid walls which now divide man from the world will gradually disappear; that human beings will not only confront each other, but also the entire organic and so-called lifeless world, with less selfishness and with livelier sympathy. Just such a presentiment as this perhaps possessed the great Chinese philosopher Licius some two thousand years ago when, pointing to a heap of mouldering human bones, he said to his scholars in the rigid, lapidary style of his tongue: "These and I alone have the knowledge that we neither live nor are dead."

Chapter 9
KARL PEARSON
(1857-1936)

INTRODUCTION

K ARL PEARSON was born in London on March 27, 1857. He received his education at University College School there, and subsequently, after spending one year with a private tutor, at Kings College, Cambridge. Pearson's main area of concentration was pure mathematics, but he also had great interest in applied mathematics, mechanics, and the theory of elasticity. He regularly came in contact with the mathematicians Roth, Cayley, Clerk-Maxwell, Stokes, and Todhunter. He received his degree in 1879 and a year later was awarded a fellowship at King's College. During the two years which followed he continued to study mathematics, but reserved a considerable amount of time for philosophy and theology. Intermittently during that period he went to Germany, Heidelberg and Berlin in particular, to study law, physics, and philosophy. He was called to the Bar in 1881, and then practiced law for three years before accepting an appointment as professor of applied mathematics and mechanics at University College, London, where he taught until his retirement in 1933. He was a lecturer in geometry in 1891, then head of the department of applied mathematics, and finally professor of eugenics. Pearson was the founder and editor of the review for statistics, *Biometrika,* and for a time editor of the review *Annals of Eugenics.* He died on April 27, 1936.

Pearson was a great scholar who harmoniously combined thoroughness in research with a broad field of interest; in addition he was an excellent teacher and prolific writer. A complete bibliography of his work includes several hundred studies on a variety of subjects: mathematics, statistics, biology, problems of heredity and evolution, philosophy, theology, and social politics.

From a philosophical point of view Pearson's *The Grammar of Science,* the first edition of which appeared in 1892, is undoubtedly the most important of his publications. In the second edition (1900),

Pearson added two new chapters dealing with problems connected with the theory of evolution. The third edition (1911) contained only the first eight chapters of the two earlier editions plus two new ones, the last of which was mainly the work of Professor E. Cunningham. The selection which follows is part of Chapter VI ("Space and Time") of the third edition.

Pearson begins his explanation in *The Grammar of Science* by stating that the goal of science is to ascertain truth in all possible realms of knowledge. He maintains that this goal can be reached only by means of a thorough application of the scientific method; this method comprises a careful and accurate classification of facts and the observation of their correlations and sequences, the discovery of scientific laws by means of the creative imagination and, finally, a severe self-criticism whose ultimate criterion is the equal validity of the insights brought to light for all normally constituted human minds (Chap. I).

With respect to epistemology Pearson defends a quite radical form of empiricism. In his view all knowledge is ultimately based upon sensations. The "familiar side of sensations," that is, our immediate sense impressions, form permanent impresses in the human brain and these psychically constitute our memory. The combination of immediate sense impressions and associated stored impressions then leads to the formation of constructs which again, by association and generalization, furnish the whole range of material to which the scientific method applies. We are accustomed to projecting such constructs "outside ourselves" and calling them phenomena. It is the totality of these phenomena which we call the real world. When an interval of time elapses between sense impressions and exertion filled by cerebral activity, marking the revival and combination of past sense impressions stored as impresses in the brain, we are said to think, to be conscious. Knowledge, therefore, is possible only in regard to what is found in immediate sense impressions and legitimate derivations from them. Applied to things outside the plane of thought, the term knowledge is meaningless. (Chap. II).

Scientific laws, according to Pearson, are of a completely different character from social laws in that the former do not presuppose an intelligent lawgiver, nor do they imply a command and a corresponding duty. A scientific law is a brief description in mental shorthand of as wide a range as possible of the sequences of our immediate sense impressions. The expression "natural law" has two different meanings which must be distinguished: it first means the mere routine of perception, whereas the scientific law is a formula describing the field of natural phenomena. Only when we speak of laws in the latter sense is it possible to refer to "reason" in natural laws; it is of great importance to observe that this "reason" is found in scientific laws only because the human mind has placed it there. Here it becomes clear that scientific laws do not explain why our perceptions have a certain order, nor why this order continually repeats itself. Consequently the laws discovered by science introduce no element of necessity into the sequence of our sense-impressions; they merely give a shorthand statement of how those changes are taking place. That such a sequence has occurred and recurred in the past is a matter of experience which we express in the concept of causation. That the sequence will continue to recur in the future is a matter of mere belief to which we give expression in the concept of probability (Chaps. III, IV).

Here the conclusion is that we cannot look at space as an enormous void in which objects have been placed by an agency in no way conditioned by its own perceptive faculty, for such a void is not given in sense perceptions, nor can it be legitimately derived from them. Space is not a thing, but an order of things. To say that a thing exists in space is to assert that our perceptive faculty has distinguished this thing as a group of sense perceptions from other groups of sense perceptions which actually (or possibly) coexist. In other words, space is not a real thing in the sense of an immediately given phenomenon, but merely our mode of perceiving phenomena, in short a mode of our perceptive faculty. The problem which then presents itself is the question of what is to be understood by "conceptual space"; it is this problem which Pearson takes up in the selection which follows.

SELECTIVE BIBLIOGRAPHY

Pearson, K., *The Grammar of Science* (1892). London, 1911.

Pearson, E. S., *Karl Pearson*. London, 1939.

Peirce, C. S., *Collected Papers*, 8 vols., ed. C. Hartshorne and P. Weiss (vols. 1–6) and A. Burks (vols. 7–8). Cambridge (Mass.), 1931–1958, vol. 8.

PERCEPTUAL AND CONCEPTUAL SPACE

CONCEPTIONS AND PERCEPTIONS

*I*F such be the space of perception, we have next to ask: How do we scientifically describe it? What is conceptual space—the space with which we deal in the science of geometry? We have seen that our perceptive faculty presents sense-impressions to us as separated into groups, and further, that though this separation is most serviceable for practical purposes, it is not very exactly and clearly defined "at the limits." How do we represent in thought, in conception, this separation into groups which results from our mode of perception? The answer is: We *conceive* groups of sense-impressions to be bounded by *surfaces*, to be limited by straight or curved *lines*. Thus our consideration of conceptual space leads us at once to a discussion of surfaces and lines—to a study, in fact, of *Geometry*.

Several important problems at once present themselves for investigation. In the first place, have these surfaces and lines a real existence in the world of perception? Are they phenomena? Or are they ideal modes whereby we analyse the manner in which we perceive phenomena? In the second place, if they should be only ideals of conception, what is the historical process by which they have been reached? What is their ultimate root in perception?

Now there is at this stage an important remark to be made, namely, that *what is imperceptible is not therefore inconceivable*. This remark is all the more necessary, for it seems directly opposed to the healthy scepticism of Hume.[1] Yet unless it be true the whole fabric of exact science falls to the ground, neither the concepts of geometry, nor those of mechanics, would be of service; for example, the circle and

[1] See especially the *Treatise on Human Nature*, part II. *Of the Ideas of Space and Time*. Green and Grose: Hume's Works, vol. I. pp. 334–371.

This article is from **The Grammar of Science** by Karl Pearson (Everyman's Library Edition; New York and London, 1911), pp. 191–208. Reprinted by permission of E. P. Dutton & Company, Inc., and J. M. Dent & Sons, Ltd.

the motion of a point would be absurdities if, being imperceptible, they were really inconceivable. The basis of our conceptions doubtless lies in perceptions, but in imagination we can carry on perceptual processes to a limit which is itself not a perception, we can further associate groups of stored sense-impresses, and form ideas which correspond to nothing in our perceptual experience.

Here a word of caution is, however, very necessary. Because we conceive a thing, we must not argue that it is either possible or probable as a perception. Indeed, the process or association by which we have reached our conception may in itself suffice to exhibit its perceptual impossibility or improbability. The appeal to experience can alone determine whether a conception is possible as a perception. For example, experience shows me that there is a sensible limit to the visible and tangible; hence a point, valid as a conception, can never have a real existence as a perception. I reach this conception of a point by carrying to a limit in my imagination a process which cannot be so carried in perception. Exactly of the same character are my conceptions of infinite distance or infinite number; they are the conceptual limits to processes, which may be *started* in perception, but cannot be carried to a limit except in the imagination. Somewhat different from perceptual impossibility is perceptual improbability. I can conceive Her Majesty Queen Mary walking alone down Regent Street, but, tested by my experience of the past actions of royalty, this association of conceptions is hardly a perceptual probability. These instances may be sufficient to indicate that what is improbable or impossible in perception may be valid in conception. But we must ever be careful to bear in mind that the *reality* of the conception, its existence outside thought, can only be demonstrated by an appeal to perceptual experience. The geometrician even asserts the phenomenal impossibility of his points, lines, and surfaces; the physicist by no means postulates the existence of atoms, molecules, and electrons as possible perceptions. Science is content for the present to look upon these concepts as existing only in the sphere of thought, as purely the product of man's mind. It does not, like metaphysics or theology, demand any existence in or beyond sense-impression for its conceptions until experience has shown that the conceptual limit or association can become a perceptual reality.[2] The validity of scientific

---

[2] Leverrier and Adams *conceived* a planet having a definite orbit as a method of accounting for the irregularities perceived in the motions of Uranus. Their conception might have been valid as a manner of describing these irregularities,

conceptions does not in the first place depend on their reality as perceptions, but on the means they provide of classifying and describing perceptions. If a rectangle and a circle have no real existence, they are still invaluable as enabling me to classify my perceptions of form, to describe, however imperfectly, the difference in shape between the face of a page of this book and of my watch. They are symbols in that shorthand by means of which science describes the universe of phenomena. The atom, if a pure conception, still enables us, by codifying our past experience, to economise thought; it preserves within reasonable limits the material upon which we base our prediction of possible future experience. If any one tells us that the storm-god is to some minds as conceivable as the atom, we must, in the first place, reply that the conceivable is not the real; and further, that the value to man of any ideal of conception depends upon the extent to which it subsumes the future in its *résumé* of the past. The conception storm-god may, after all, be of some value as a striking monument to our meteorological ignorance, and as a useful reminder that we must "be prepared for all weathers."

What we have at this stage to notice is that the mind is not limited to perceptual association, and that it can carry on in conception a process which may be begun but cannot be indefinitely continued in the sphere of perception. The scientific value of such conceptions, whether reached by association or as a limit, must in every case be judged by the extent to which they enable us to classify, describe, and predict phenomena.

SAMENESS AND CONTINUITY

Now there are two ideas reached as conceptual limits to perceptual processes which have important bearings on the geometrical representation of space. Those may be expressed by the work *sameness* and *continuity*. So far as our perceptual experience goes, probably no two groups of sense-impressions are exactly the same. The sameness in each depends upon the degree of our examination and observation. To a casual observer all the sheep in a flock appear the same, but the shepherd individualises each. Two coins from one die, or two engravings from one block, will always be found to possess some

---

if Neptune itself had never been perceived—in other words, if their conception had not become a perceptual reality.

distinguishing marks. We may safely assert that absolute sameness has never occurred in our experience. No "permanent" group of sense-impressions or "object" even is exactly the same at two different times. Various elements in the group have changed slightly with the time, the light, or the observer. Take a polished piece of metal and note two parts of its surface; they appear exactly alike, but the microscope reveals their want of sameness. Thus sameness is never a real limit to our experience of phenomena; the more closely we examine, the less is the sameness. Yet, as a conception, the sameness of two groups of sense-impressions is a very valid idea, and the basis of much of our scientific classification. In the sphere of perceptions sameness denotes the identity for certain practical purposes of two slightly different groups of sense-impressions. In the sphere of conceptions, however, sameness denotes absolute identity of all the members of either group; it is a limit to a process of comparison which cannot be reached in the perceptual world.

The idea of continuity, in the sense in which we are now considering the word, involves that of sameness. If I take a vessel of water, I find a certain permanent group of sense-impressions which leads me to term the contents of the vessel water; if I take a small quantity of the water out of the vessel I find the "same" group, and this still remains true if I take a smaller and smaller quantity, even to a drop. I may continue to divide the drop, but apparently as long as the portion taken remains sensible at all, there is the same group of sense-impressions, and I term the fraction of the drop water. Now the question arises, if this division could be carried on indefinitely, should we at last reach a limit at which the group of sense-impressions would change not only quantitatively, that is in intensity, but also qualitatively? If we could magnify the sense-impressions due to the infinitesimal fraction of a drop of water up to a sensible intensity, would they so differ from those characteristic of the contents of the original vessel that we should not give them the name water? Now we cannot test the effects of an indefinitely continued division in the phenomenal world, for we soon reach a stage at which we fail to get, by the means at our disposal, any sense-impressions at all from the divided substances. Our magnifiers of sense-impression have but a limited range.[3] But although in the sphere of perceptions there is no

---

[3] *E.g.*, the microscope, the microphone, the spectroscope, etc. From the spectroscope we obtain, perhaps, positive indications of a qualitative change in many substances as the quantity is diminished.

possibility of carrying division to its ultimate limit, we can yet in conception repeat the process indefinitely. If after an infinite number of divisions we conceive that the same group of sense-impressions would be found, then we are said to conceive the substance as *continuous*. We have then to ask how far the conception of continuity applies to the real bodies of our perceptual experience. From the finite process of division which is possible in perception, we might easily conclude that continuity was a property of real substances; and there is small doubt that a slight amount of observation is favourable to the notion that many real substances are continuous, although the infinite division necessary to the conception of continuity fails to find any perceptual equivalent. Further observation and wider insight, however, contradict this notion. The physicist and the chemist bring many arguments to show us that the finite process of division which suggests continuity would, if carried to an infinite limit, show bodies to be discontinuous. On a first and untrained inspection we find a continuity and a sameness in perceptions which disappear on closer and more critical examination. The ideas conveyed in these words are found to be no real limits to the actual, but ideal limits to processes which can only be carried out in the field of conception. Bearing this in mind we may now return to the geometrical conceptions of space.

CONCEPTUAL SPACE. GEOMETRICAL BOUNDARIES

It has been remarked that we conceive groups of sense-impressions to be limited by surface and lines. We speak of the surface of the table; the fly-leaf of this book appears to be separated from the air above it by a plane surface and that plane to be bounded at its upper edge by a portion of a straight line. In the first place, we have to ask whether our geometrical notions of line and plane corresponds to the limits of anything we actually find in perception or whether they are purely ideal limits to processes begun in perception, but which it is impossible to carry to a limit in perception. The answer to these questions lies in the conceptions of *sameness* and *continuity*. The geometrical ideas of line and plane involve absolute sameness in all their elements and absolute continuity. Every element of a straight line can in conception be made to fit every other element, and this however it be turned about its terminal points. Every element of a plane can be made to fit every other element, and this without regard to side. Further, every element of a straight line or a plane, however often divided up, is in conception, when magnified up, still an element of straight line or plane.

The geometrical ideas correspond to absolute sameness and continuity, but do we experience anything like these in our perceptions? The fly-leaf of this book appears at first sight a plane surface bounded by a straight line, but a very slight inspection with a magnifying lens shows that the surface has hollows and elevations in it, which quite defy all geometrical definition and scientific treatment. The straight line which seems to bound its edge becomes, under a powerful glass, so torn and jagged that its ups and downs are more like a saw-edge than a straight line. The sameness and continuity are seen to be wanting on more careful investigation. We take a glass cube skilfully cut and polished, and its faces appear at first as true planes. But we find that a small body placed upon one of its faces does not slide off when the cube is slightly tilted. The face of the cube must, after all, be *rough,* there are hollows and projections in it which catch those of the superposed body; our plane again appears delusive. Or we may take one of Whitworth's wonderful metal planes obtained by rubbing the faces of three pieces of metal upon each other. Here again a powerful microscope reveals to us that we are still dealing with a surface having ridges and hollows.

The fact remains, that however great the care we take in the preparation of a plane surface, either a microscope or other means can be found of sufficient power to show that it is not a plane surface. It is precisely the same with a straight line; however accurate it appears at first to be, exact methods of investigation invariably show it to be widely removed from the conceptual straight line of geometry. It is a race between our power of representing a straight line or plane and our power of creating instruments which demonstrate that the sameness and continuity of the geometrical conceptions are wanting. Absolutely perfect instruments could probably only be constructed if we were already in possession of a true geometrical line or plane, but the instruments we can make appear invariably to win the race. *Our experience gives us no reason to suppose that with any amount of care we could obtain a perceptual straight line or plane, the elements of which would on indefinite magnification satisfy the condition of ultimate sameness involved in the geometrical definitions.* We are thus forced to conclude that the geometrical definitions are the results of processes which may be started, but the limits of which can never be reached in perception; they are pure perceptual experience. What we have said of straight lines and planes holds equally of all geometrically defined curves and surfaces. The fundamental conceptions of geometry are only ideal symbols which enable us to form an approximate, but in no sense absolute analysis of our sense-impressions. They are the scientific shorthand by which we describe, classify, and formulate the characteristics of that mode of perception which we term perceptual space. Their validity, like that of all other conceptions, lies in the power they give us of codifying past and predicting future experience.

We speak of a spherical or cubical body, and say that it is of such and such a capacity. But no perceptual body is ever truly spherical or cubical, and the size we attribute to it is at best an approximate one. Further analysis of our sense-impressions leads us in each case to find variations from the geometrical definition and measurement. Yet the conceptions of sphere and cube are frequently sufficient to enable us to classify and identify various bodies and predict the different types of sense-impression to which these bodies correspond.[4] Perhaps no better instance than geometry can be taken to show how science *describes* the world of phenomena by aid of conceptions corresponding to no reality in phenomena themselves. That our geometrical conceptions enable us on the whole to so effectually describe perceptual space is only a striking instance of the practically equal development of our perceptive and reasoning faculties.

SURFACES AS BOUNDARIES

Although perceptual boundaries do not, on ultimate analysis, in any way correspond to any special geometrical definition such as that of plane or sphere, we have still to inquire whether they answer to our conception of surface at all. By surface in this sense we are to consider, not something of which it would be possible to analyse the properties by any of the known processes of geometry, but any *continuous* boundary between two groups of sense-impressions or bodies.[5]

---

[4] Our whole system of measuring size will be found to be based on geometrical conceptions having no actuality in perception.

[5] "*That which has position, length and breadth but not thickness,* is called *surface.*" "The word *surface* in ordinary language conveys the idea of extension in two directions, for instance, we speak of the surface of the earth, the surface of

Is there a continuous boundary between the open page of this book and the air above it? Would it be possible to say at any distinct step of the passage from air to paper, here air ends and paper begins? At this point we reach one of the most important problems of science. Are we to consider the groups of sense-impressions which we term bodies *continuous* or not? If bodies are not continuous, then it is clear that boundaries are only mental symbols of separation, and on deeper analysis correspond to no exact reality in the sphere of sense-impression.

Would every element of the surface of a body still appear to us a continuous boundary, however small the element and however much we magnified it up? If I could take the hundredth part of a square inch of this page and magnify it to a billion times its present size, would there still appear a continuous boundary between air and paper?

Consider the boundary of still water. It furnishes us with the impression of a continuous surface. On the other hand, examine a heap of sand closely, and it appears to have no continuous boundary at all. Are there any reasons which would lead us to suppose that, if we could sufficiently magnify a small element of this page of paper, it would produce in us sense-impressions not of continuity but of discontinuity? Would it look, supposing it were still visible, like the surface of water, or rather like a heap of sand, a pile of small shot, or, better still, like a starry patch of the heavens on a clear night? No group of stars is in perception separated from another by a line or surface. We can *imagine* such boundaries drawn across the heavens, but we do not *perceive* them. We have, then, to ask whether the boundary between paper and air, if immensely magnified, would look sideways, not indeed like a geometrical line, but roughly like the first or second of these figures (Figs. 1a and 1b):—

Now no direct answer can really be given to this question, because bodies cease to impress us sensibly long before we reach the point at

---

the sea, the surface of a sheet of paper. Although in some cases the idea of the thickness or the depth of the thing spoken of may be present in the speaker's mind, yet as a rule no stress is laid on depth or thickness. When we speak of a *geometrical surface,* we put aside the idea of depth and thickness altogether" (H. M. Taylor, *Pitt Press Euclid,* i.–ii. p. 3). It seems to me that in ordinary language there is something more than length and breadth involved—there is an idea of *continuous boundary.* It is difficult to say how far this idea is really involved in the word extension. A veil may have extension in two directions, but it fails to fulfil our idea of surface because it is not a continuous boundary.

Figure 1a

Figure 1b

which the appearance of continuity might be expected to disappear. We cannot predict what our sense-impressions would be if we could magnify a drop of water up to the size of the earth. But we may put the question in a slightly different way. We may ask: Would it enable us to classify and describe phenomena better if we *conceived* bodies to be continuous as in Fig. 1a, or discontinuous as in Fig. 1b? The physicist promptly replies: I can only conceive bodies to be discontinuous. Discontinuity is essential to the methods by which I describe and formulate my sense-impressions of the phenomenal world.

CONCEPTUAL DISCONTINUITY OF BODIES. THE ATOM

Foremost among the physicist's reasons for postulating the discontinuity of bodies is the elasticity which we notice in all of them. Air can be placed under a piston in a cylinder and compressed; a bar of wood can be bent—in other words, a portion of it squeezed and another portion stretched. Even the amounts by which we can squeeze iron or granite are capable of measurement. Now it is very hard, I think impossible, to *conceive* how we can alter the size of bodies if we suppose them continuous. We feel ourselves compelled to assert that, if the parts of a body move closer together, they must have something free of body into which they can move. If a body were continuous and yet compressible, there appears to be no reason why it should not be indefinitely compressible, or indefinitely extensible, both results repugnant to our experience. Further, our sense-impressions of temperature in both gaseous and solid bodies, and of colour in solid bodies, the phenomena of pressure in gases, and those of the absorption and emission of light, are easily analysed and described, if we conceive the ultimate parts of bodies to have a capacity for relative motion; but there is no possibility of conceiving such a motion if all the parts of a body are continuous. A crowd of human beings seen from a great height may look like a turbulent fluid in motion at every point. But we know from experience that this motion is only possible because there is some void in the crowd. It may become so densely packed that motion is no longer practicable. Thus it is with that relative motion of the parts of bodies upon which so much of modern physics depends; absolutely close packing, that is continuity, seems to render it impossible. It is only by reducing in conception the complex groups of sense-impressions, which we term bodies, into simple elements directly depending on the motion of discontinuous systems,—of what we may term granular or starlike systems,—that we have been able to resume phenomena in the wide-reaching laws of physics and chemistry. The relative motion of the ultimate parts of bodies, involving the idea of discontinuity, is one of the fundamental conceptions of modern science. These ultimate parts of bodies we are accustomed to speak of as *atoms;* groups of atoms which apparently repeat themselves over and over again in the same body— something like planetary systems in the starry universe—we term *molecules.* The generally accepted atomic or molecular theory of bodies postulates essentially their discontinuity. Take, for example, a spherical drop of water—to follow Lord Kelvin—suppose it to be as big as a football, then if we could magnify the whole drop up to the size of the earth, the structure, he tells us, would be more coarse-grained than a heap of small shot, but probably less coarse-grained than a heap of footballs.[6]

Now I propose later to return to the atomic hypothesis. At present I will only ask the reader to look upon atom and molecule as *conceptions* which very greatly reduce the complexity of our description of phenomena. But what it is necessary to notice at this stage is: that the conception atom, when applied to our perceptions, is opposed to the conception of surface, as the continuous boundary of a body. We have here an important example of what is not an uncommon oc-

[6] *Popular Lectures and Addresses,* vol. I., "The Size of Atoms," p. 217.

currence in science, namely, two conceptions which cannot both correspond to realities in the perceptual world. Either perceptual bodies have continuous boundaries, and the atomic theory has no perceptual validity; or, conversely, bodies have an atomic structure, and geometrical surfaces are perceptually impossible. At first sight this result might appear to the reader to involve a contradiction between geometry and physics; it might seem that either physical or geometrical conceptions must be false. But the whole difficulty really lies in the habit we have formed of considering bodies as objective realities unconditioned by our perceptive faculty. We cannot too often recall the fact that bodies are for us more or less permanent, more or less clearly defined groups of sense-impressions, and that the relationships and sequences among the sense-impressions are largely conditioned by the perceptive faculty. At the present time we have no sense-impressions corresponding to geometrical surface or to atom; we may legitimately doubt whether our perceptive faculty is of such a nature that it could present impressions in any way corresponding to these conceptions. It is impossible, therefore, to say that one of these conceptions must be real and the other unreal, for neither at present has perceptual validity—that is, exists in the world of real things. As conceptions both are equally valid; both are equally ideals, not involved in our sense-impressions themselves, but which the reasoning faculty has discovered and developed as a means of classifying different types of sense-impressions and of resuming in brief formulæ their relationships and sequences.

Thus geometrical truths apply with absolute accuracy to no group whatever of our sense-impressions; but they enable us to classify very wide ranges of phenomena by aid of the notions of position, size, and shape. Geometry enables us to predict with absolute certainty a variety of relations between sense-impressions, when these impressions do not involve more than a certain keenness in our senses, more than a certain degree of exactness in our measuring instruments. The absolute sameness and continuity demanded by geometrical conceptions do not exist as *limits* in the world of perceptual experience, but only as approximations or average.[7] In precisely the same way the theory of atoms treats of ideal conceptions; it enables us to classify another and different range of sense-impressions, and to formulate their mutual relations to a certain degree of keenness again in our

[7] Geometry might almost be termed a branch of statistics, and the definition of the circle has much the same character as that of Quetelet's *l'homme moyen*.

senses, or of exactness in our scientific apparatus. Should the atom become a perception as well as a conception, this would not invalidate the usefulness of geometry. Very probably, however, if we could magnify a football up to the size of the earth, so that the perceptual atom, if it existed, would have a size between small shot and a football, we should find that the sense-impressions which the atom was conceived to distinguish and resume, had themselves disappeared under the new conditions.[8] In other words, our scientific conceptions are valid for the world as we know it, but we cannot in the least predict how they would be related to a world which is at present beyond perception.

CONCEPTUAL CONTINUITY. ETHER

The reader will now be prepared to appreciate scientific conceptions, which, if they corresponded to realities of the phenomenal world, would contradict each other. Having destroyed the continuity of bodies by the idea of atom, it might at first sight appear as if our conceptual space were fundamentally different from perceptual space. The latter, as we have seen, is our mode of distinguishing groups of sense-impressions, and where there is nothing to distinguish, there is no space. The perceptive faculty rather than nature may be said "to abhor a vacuum." On the other hand, having destroyed the continuity of bodies by the atomic hypothesis, we seem at first sight to be postulating a void in conceptual space. But here the physicist compels us to introduce a new continuity. This new continuity is that of the *ether*, a medium which physicists conceive to fill up the interstices between bodies and between the atoms of bodies. By aid of this concept, the ether (to which we shall return later), we are able to classify and resume other wide groups of sense-impressions. With regard to the perceptual existence of the ether, it now stands, some physicists would assert, on a rather different footing from that of the atom. By the *real* existence of anything we mean that it forms a more or less permanent group of sense-impressions. Now this can hardly be asserted of the ether; we conceive it rather as a conduit for the motions by which we interpret sense-impression. The nerves seem to

---

[8] The visibility and tangibility of bodies may possibly be described by the motion of atoms, but we cannot predict that a *single* atom would be either visible or tangible, still less "bounded by a surface."

us conduits of the like kind, but then the nerves also appear to us as permanent groups of sense-impressions apart from their function of conductivity. There are no sense-impressions which we class together and term ether, and on that account it seems far better to consider the ether as a conception rather than a perception. It is true that to some minds the ether may appear as real a perception as the air, and the matter is, perhaps, largely one of definition. Still even wireless telegraphy, for example, does not seem to me to have logically demonstrated the perceptual existence of the ether, but to have immensely increased the validity of the scientific concept, ether, by showing that a wider range of perceptual experience may be described in terms of it, than had hitherto been demonstrated before Hertz's experiment.[9] Further, many of the properties which we associate with the ether are not such as our past experience shows us are likely to become matter for direct sense-impression. I shall therefore continue to speak of the ether as a scientific concept on the same footing as geometrical surface and atom.

ON THE GENERAL NATURE OF SCIENTIFIC CONCEPTIONS

Our discussion of these spatial conceptions will the better have enabled the reader to appreciate the nature of scientific conceptions in general. Geometrical surface, atom, ether, exist only in the human mind, and they are "shorthand" methods of distinguishing, classifying, and resuming phases of sense-impression. They do not exist in or beyond the world of sense-impressions, but are the pure product of our reasoning faculty. The universe is not to be thought of as a real complex of atoms floating in ether, both atom and ether being to us unknowable "things-in-themselves," producing or enforcing upon us the world of sense-impressions. This would indeed be for science to repeat the dogmas of the metaphysicians, the crassest paradoxes of a short-sighted materialism. On the contrary, the scientist postulates nothing of the world beyond sense; for him the atom and the ether are —like the geometrical surface—models by aid of which he resumes the world of sense. The ghostly world of "things-in-themselves" be-

---

[9] Nay, in the nineteen years that have elapsed since the first edition of this book appeared, a perceptual ether has grown less and less possible. Little remains of the "ether" to-day but the conceptions involved in a set of differential equations!

hind sense he leaves as a playground to the metaphysician and the materialist. There these gymnasts, released from the dreary bondage of space and time, can play all sorts of tricks with the unknowable, and explain to the few who can comprehend them how the universe is "created" out of will, or out of atom and ether, and how a knowledge of things beyond perception, *i.e.*, beyond the knowable, may be attained by the favoured few. The scientist bravely asserts that it is impossible to know what there is behind sense-impression, if indeed there can "be" anything;[10] he therefore refuses to project his conceptions, atom and ether into the real world of perception until he has perceived them there. They remain for him valid ideals so long as they continue to economise his thought.

That the conceptions of geometry and physics immensely economise thought is an instance of that wonderful power to which I have previously referred in this work, namely, the power the reasoning faculty possesses of resuming in conceptions and brief formulæ the relationships and sequences it finds in the material presented to it by the perceptive faculty. As our knowledge grows, as our sense becomes keener under the action of evolution and with the guidance of science, so we are compelled to widen our concepts, or to add additional ones. This process does not as a rule signify that the original concepts are invalid, but merely that they form a basis, which is only sufficient for classifying and describing certain phases of sense-impression, certain aspects of phenomena. As we grow cognisant of other phases and aspects, we are forced to adopt new concepts, or to modify and extend the old. We may ultimately reach perceptions of space which cannot be described by the geometry of Euclid, but none the less that geometry will remain perfectly valid as an analysis and classification of the wide range of perceptions to which it at present applies. If the reader will bear in mind the views here expressed with regard to the concepts of science, he will never consider that science reduces the universe to a "dead mechanism" by asserting a reality for atom or ether or force as the basis of sense-impression. Science, as I have so often reiterated, takes the universe of perceptions as it finds it, and endeavours briefly to describe it. It asserts no perceptual reality for its own shorthand.

One word more before we leave this space of conception, separated

---

[10] Our notion of "being" is essentially associated with space and time, and it may well be questioned whether it is intelligible to use the word except in association with these modes of perception.

by continuous boundaries in the eve of the geometrician, peopled with atoms and ether by the mind of the physicist. How, if geometrical surface, if atom and ether have no perceptual reality, has the mind of man historically reached them? I believe by carrying to a limit in conception processes which have no such limit in perception. Preliminary stages in comparison show apparent sameness and continuity, where more exact and final stages show no such limit; hence arises the conception of continuous boundaries. The atom again is a conceptual limit to the "moving bodies" of perception; while the ether possesses properties, which we have never met with in the physical media of our perceptual experience, but which are purely conceptual limits to the types of media with which we are directly acquainted. These concepts themselves are a product of the imagination, but they are suggested, almost insensibly by what we perceive in the world of phenomena.

Chapter 10
ÉMILE BOUTROUX
(1845–1921)

INTRODUCTION

*E*MILE BOUTROUX was born in Montrouge, France (near Paris), on July 28, 1845. He studied at the École Normale Supérieure in Paris, then later became professor of philosophy in Montpellier (1874–1876), Nancy (1876–1877), at the École Normale Supérieure (1877–1886), and finally at the Sorbonne (1886–1914). Boutroux is especially well known for his criticism of positivism and is generally considered to be the founder of the French movement called *critique of science*, a movement to which Poincaré and Duhem belonged, also. Within this general school Boutroux was the advocate of the spiritualist trend, which was strongly opposed to mechanism and materialism. He died in Paris on November 22, 1921.

Boutroux's most important contribution was his work in connection with establishing the "critique of natural science." In 1874 he published his epoch-making book *De la contingence des lois de la nature*, in which he asked the question of whether the deterministic world view propagated by the natural sciences might be maintained and justified. In Boutroux's view one of the first phenomena which manifests itself to man is his freedom, to which corresponds a certain analogous form of freedom in nature. Thus he was led to a careful consideration of whether natural science necessarily leads to determinism. It was his view that in none of the natural sciences is a complete necessity found, inasmuch as one finds everywhere a considerable leeway for contingency, spontaneity, and therefore freedom, and that which makes freedom possible. One must realize that natural science does not manifest the very essence of things to us, but merely teaches the means necessary to influence and govern nature. He believed a complete explanation of reality to be impossible without an explicit appeal to freedom—man's as well as God's—which is adumbrated everywhere in nature in an analogous way. Boutroux's ideas had some influence later on the philosophies of Bergson and Brunschvigs. Similar ideas related to the biological sciences can be found in A. Cournot (1801–1877).

Boutroux published an extremely important study on natural laws, *De l'idée de loi naturelle dans la science et la philosophie contemporaine* (1895). The work constitutes a study of the idea of natural law as it is found in modern times and is an attempt to render a philosophical interpretation of this. In trying to develop his ideas Boutroux began with the fact that neither the rationalist philosophy of Descartes, Leibniz, or Kant, nor the empiricist philosophy of Bacon, Locke, or Hume, had adequately accounted for the nature and meaning of physical laws. In Boutroux's opinion it was difficult for the human mind to conceive the laws of nature as being universal and "real" at the same time. For example, when one stresses the universality of natural laws, their aprioric synthetic character, then their validity for a world of real events seems to slip from one's grasp; however, if one stresses their validity for the world of real events, and tries to account for their universality on the basis of induction, then the universal and aprioric character seems to escape one's hold. One might be inclined here to attempt to unite rationalism and empiricism, but upon closer investigation it would soon become evident that a conjunction of these two opposite viewpoints would lead only to juxtaposition, not genuine synthesis.

It was Boutroux's claim that what for philosophers from Bacon to Hegel had been only an ideal and a "problem," is actually materialized by science the moment it makes a harmonious combination of mathematics and experience, thereby supplying laws which are both universal and concrete, founded in experience and intelligible. He noted that science has tried to discover appropriate, positive principles for each realm of realities. Newton, for example, supplied the general type of scientific explanation by basing celestial mechanics on the law of gravitation, which is radically distinct from purely geometrical laws. Then in physics, chemistry, and biology people tried to imitate Newton's discovery in the realm of mechanics with the consequence that, one after the other, different sciences were freed to develop as autonomous disciplines, helped by special and mutually irreducible principles. Thus although a distinction was made between mechanical, physical, chemical, and biological principles, nevertheless in each science there has been an attempt to maintain a certain analogy with mechanics in order to preserve a similarity with the mathematical sciences.

Boutroux held that in order to study the idea of natural law it is necessary to start with the different sciences (mathematics, physics, chemistry, and biology), while at the same time appealing to philosophy for general indications as to the way of interpreting the principles and results of these sciences. Boutroux wished to take the laws just as he found them in the sciences, divided into different groups, and study them independently, repeatedly asking himself questions about the nature, objectivity, and meaning of such laws: How far and in what sense are the laws of nature intelligible? Do different laws of nature refer to philosophically irreducible principles or are there differences only as to generality and complexity? Do the laws constitute the substance of natural things or do they govern only the mode in which phenomena appear to us? Are they true elements or mere symbols of reality? Is determinism a typical characteristic of nature, or does it simply represent the way in which we connect things in order to make them intelligible? (*Natural Laws*, Chap. I, pp. 11–19)

The following is a complete chapter from Boutroux's investigation concerning the laws of mechanics.

SELECTIVE BIBLIOGRAPHY

Boutroux, É., *The Contingency of the Laws of Nature* (1874), trans. F. Rothwell. London, 1916.

——, *Natural Law in Science and Philosophy* (1895), trans. F. Rothwell. New York, 1914.

——, *La nature et l'esprit* (2 vols.). Paris, 1904–1905.

——, *Science et religion dans la philosophie contemporaine*. Paris, 1908.

Galtier, P., *Les maîtres de la pensée française: Paul Hervieu, Émile Boutroux, Henri Bergson et Maurice Barres*. Paris, 1921.

Albèrgamo, F., *Storia della logica delle scienze empiriche*. Bari, 1952. Pp. 225–230.

La Fontaine, A. P., *La philosophie d'É. Boutroux*. Paris, 1920.

Schyns, M., *La philosophie d'Émile Boutroux*. Paris, 1924.

Crawford, L. S., *The Philosophy of Émile Boutroux*. New York, 1924.

THE MECHANICAL LAWS

*O*UR present object is to examine critically the idea we have of the laws of nature, in the hope of extracting information as to the relation these laws bear to reality, and the position of the human individual in nature herself. It is our ultimate end to know whether, in the present state of the sciences, we may yet regard ourselves as possessed of any power to act freely, any reality as persons. Along these lines, we have examined the logical and mathematical laws, which, after all are more than laws and express the most general relations, the conditions of all the rest. We have shown that the laws of real logic cannot actually be reduced exactly to the only principle most certainly known *à priori*, namely A *is* A, and that concept, judgment, syllogism, all imply a new element: the many as contained in the one, or again the relation of the explicit to the implicit. Mathematics also introduces new elements which the mind cannot thoroughly assimilate: it creates relations of adjustment; it diversifies the identical with the aid of intuition; more than that, in its generalizations, it cannot dispense with a mode of reasoning which may be called apodeictic induction. If both the mathematical and the logical laws do not proceed immediately from the nature of the mind, neither are they deduced from experience. Indeed, were this the case, they would have to coincide with parts or aspects of reality: now, this is not so. Neither the universals of logic nor the infinite number of mathematics are given to us. We cannot even conceive how they could be. Thus, logic and mathematics are solely derived neither from knowledge *à priori*, nor from knowledge *à posteriori:* they represent the work of the mind which, incited by things to exert itself, creates a mass of symbols in order to subject these things to necessity and thus make them capable of being assimilated by itself. The logical and the mathematical laws testify to the mind's need of conceiving things as being necessarily determined; but it cannot be known

This article is from **Natural Law in Science and Philosophy** by Émile Boutroux, trans. F. Rothwell (New York, 1914), pp. 46–78. Copyright © 1914 by The Macmillan Company.

à priori how far reality conforms with these mind-imagined symbols: we must appeal to observation and analysis of the real if we would know whether mathematics, in effect, rules throughout the universe. All that can be admitted, previous to this experimental study, is that there is probably a certain analogy between our intellectual nature and the nature of things. Were it not so, man would be isolated in the universe. This, however, is but conjecture. A consideration of the concrete sciences will alone enable us to say what degree of reality we must attribute to logic and to mathematics.

The laws of reality given to us as approaching nearest to mathematical relations, are the mechanical laws The essential and characteristic element of these laws is the notion of *force*. In order to understand the formation and the present state of this notion, we will now study its historical evolution.

In antiquity, and especially in the times of Plato and Aristotle, what seems, above all else, to strike the human mind, is the difference between motion and rest. This opposition is made the point of departure, and it is admitted that matter, in itself, is in a state of rest. What has then to be explained is the transition from rest to motion. To solve the question, the production of motion in man is considered. Now, motion appears, in man, as resulting from the action of the mind on the body. Thus, above matter, there is assumed a separate force, resembling a soul more or less, and as such, suited for acting upon bodies. This view may readily be connected with the teleological conception, in virtue of which God rules and moves the totality of things; thus, it shows itself favourable to ethics and religion. On the other hand, it opposes the progress of science. How, in effect, are we to gauge and foresee the action of an immaterial force, called upon to exert itself from aesthetic and ethical reasons? As a matter of fact, the science of the real made little progress, so long as it regarded things from this point of view.

At the time of the *Renaissance,* a totally different conception grew up. Instead of contrasting motion with rest, Galileo looked upon them as analogous: matter is self-sufficient, both in motion and in rest. Of itself and apart from supernatural intervention, it continues indefinitely in a uniform, rectilinear motion; of itself, it can pass neither from rest to motion nor from motion to rest: it is the principle of inertia. No doubt, if we wish to bring before the mind the first origin of motion, we must presuppose a first impulsion, a fillip,— *chiquenaude,* as Pascal called it;—but as regards its present state, which alone is the object of science, matter contains within itself the explanation both of its motion and of its rest. From this idea of inertia, it was at first thought possible to infer the abolition of force as a separate idea. Thus, Descartes thought he could explain all physical phenomena by the one law of the conservation of the quantity of motion, a corollary of the principle of inertia. Force, as such, is banished from his system. This philosophy might have been developed deductively, like mathematics, of which it formed the continuation; but there came a time when it was confronted with facts and then found to be inadequate. Newton, in order to account for the motions of the heavenly bodies, regarded it as necessary to re-establish the idea of force. He started with the principle of inertia, according to which a body retains its uniform, rectilinear motion for an indefinite period. The heavenly bodies, however, moved in a curvilinear, non-uniform fashion. To explain this modification of motion, we must admit that some external force acts upon the moving body. This re-appearance of the idea of force, moreover, is not the restoration of the ancient conception. The ancients regarded force as dwelling in a higher, metaphysical form; it acts from above, after the fashion of a soul; it is God Himself, who, by His perfection, produces the motions of the heavenly bodies. Newton, on the other hand, attributed force to matter: an atom has not the power to modify its own motion, but it can modify the motion of other atoms. Thus, without leaving matter, we come to explain modifications in the speed and direction of motion. God is eliminated from the world, in so far, at least, as He is considered to be an artist who produces by separate acts every detail of His work.

Are we not, however, restoring the occult powers of the Schoolmen if we admit the existence of such a force? Newton, as we know by his own declarations, does not regard attraction as a metaphysical force, analogous to a soul's activity. To him, this is but an expression, a kind of metaphor, pointing to a phenomenal relationship. Force, as he interprets it, is none the less the cause of motion. Now, cause must be prior to effect. If, then, this is not an occult power, at least it is something metaphysical and invisible, which logically precedes phenomena. Mathematicians have taken it into account, and so we find them, nowadays, endeavouring to transform the relation between force and motion into a simple mutual dependence, a mere relation of solidarity. It is in this way that force is defined as the *product of the mass into the velocity*. Here, force and motion are two data related to each other, without there being any necessity to inquire whether it is force that is the cause of motion, or motion that is the cause of force: just as, in geometry, we have the relation of diameter to circumference.

Is force, as thus conceived, reduced to a purely mathematical notion, or does it contain some new element? Doubtless, abstract mechanics does not differ from mathematics and consists solely of substitutions of formulæ. But abstract mechanics does not suffice for the realization of the science of nature. This was clearly seen by Newton; he tried to find in experience the mathematical principles of his natural philosophy. Now, what is that element which cannot be found in mathematics and which only experience can give us? It is the measure of the action which bodies exercise upon one another. In mathematics, consequences are analytically deduced from definitions; we start with the identical and then diversify it. But, in nature, we start with things foreign to one another, such as the sun and the planets, and set up a definite, constant dependence between these things. We are really dealing, then, with a mathematical connexion, though it can neither be affirmed nor known *à priori*. Thus, what there is new in the notion of force, is, in short, the idea of natural law, strictly so called. Force is a uniform dependence, experimentally known, between things exterior to one another. Consequently it contains an extra-mathematical element.

But then, may it not be said that the affirmation of the natural laws results from a special necessity of the mind? Following on Kant, profound philosophers maintain, even nowadays, that the notion of law is the result of our mental make-up, and consists of a synthetic, *à priori* judgment. These philosophers justify their theory by setting forth how such an idea of causal law is necessary in order to think phenomena, i.e., to reduce them to unity within a consciousness. Phenomena, in themselves, are heterogeneous with regard to one another. The notion of law, by establishing universal and necessary relations between them, gives them the only unity of which a heterogeneous multiplicity admits. This theory, to our mind, is open to objections.

At the outset, it is clear that we have an irresistible need to think phenomena, to reduce them all to unity, to set up between ourselves and them, in an absolute sense, the metaphysical relationship of subject and object? No doubt we have need of unity; but it is difficult to prove that this need takes precedence of all others and governs the whole of our intellectual life. Indeed, the history of philosophy offers us not only minds that aim at an explanation of the uniform by the multiple and the changing, but also logicians enamoured of a reduction to unity. Now, if unity in the conception of being is not necessary, neither are the means of obtaining it.

But we may go farther. Even granting that we feel this absolute, imperious need to think things, is it certain that the categories of the understanding realize the end assigned to them, viz. the assimilation of things by the mind? It would appear as though this point had been too readily granted to the Kantian doctrine. Indeed, to think things is to understand their particular affinities and connexions, to see how they unify and group together of themselves. Kant's categories, however, leave things as they find them, exterior and alien to one another. They bring them together artificially, as stones are brought together in building a house. They reconcile nature—which unites beings according to their consanguinity or kinship—with art—which brings them together in accordance with its own ideas of fitness. Is a bundle of sensations a thought taking possession of things?

This is not all, and we may enquire whether the position adopted by Kant can be maintained as ultimate, or whether it must not of necessity be transcended in one direction or the other. The objection is frequently urged that, if Kant's categories are purely subjective, it is inexplicable that nature should conform to them. Stated thus, the objection is perhaps not well founded; for, in Kantianism, what we call nature is already the work of the mind, not perhaps of the individual thought, but of the universal human thought identical in each individual consciousness, and the individual mind only recognizes empirically and successively that which reason builds up and unifies à priori. But it would seem that a somewhat similar objection might be raised. Either the laws, we may say, that the mind brings forward, will find analogous matter, obedient to their action, in which case how shall we know that the notion of these laws comes from ourselves rather than from the observation of things, that they are known to us à priori rather than à posteriori? Or else things will not conform to these laws, and in that case shall we claim that it is ourselves who are right and nature who is wrong? It is clear that, as soon as it is proved that facts do not fit in with the limits we wish to impose upon them, we shall make it our object to free ourselves of these limits and form conceptions more in accord with facts.

Thus, the mechanical laws are not an analytical succession of mathematical truths; neither do they rest on synthetic à priori judgments. Are they derived from experience? The ancients claimed to obtain from experience only the general and the probable, i.e. what happens in the ordinary way of things, ὡς ἐπὶ τὸ πολύ; what they wanted it to give them was universal and necessary rules, not laws. For the moderns, however, induction is a kind of magic word, in virtue of which, fact is transmuted into law. By so-called scientific induction, which evidently has scarcely anything in common with ancient induction, it is claimed that the universal can be deduced from the contingent, the necessary from the particular. Still, however productive and methodical modern induction may be, it will never succeed, without superseding experience, in bringing us to true laws. For instance, we cannot possibly, by experience, become acquainted with inertia and force; to do this, we should have had to be present at the creation. We never observe the exactly uniform and rectilinear motion of a moving body, removed from all extraneous influence, any more than the continuance in rest of a body that has received no impulsion. The duality of inertia and force, the action of multiple forces, and the composition of these forces, are abstractions that cannot be verified.

We may go even further and say that induction cannot even account for the most general characteristics of the mechanical laws. In fact, we observe only moments separate from one another, i.e. discontinuity, and yet our laws give us continuity. Secondly, these laws imply precision, whereas experience gives us only approximations. Afterwards, we assume, as fundamental, definite relations between such and such phenomena, whereas experience offers us an infinite number of relations between which there is neither priority nor separation. Finally, we attribute to our laws, fixity, as an essential characteristic. Now, in this we cannot say that we are judging the future by the past, for only to an insignificant extent do we know the past. It is most seriously alleged nowadays that species are not eternal, but have a history of their own. Why also should not laws, those types of the relations that exist between phenomena, be subject to change? The fixity we attribute to them is a characteristic that we add on to the data of experience, one that cannot be revealed to us from without.

Nevertheless, if the mechanical laws are known neither *à priori* nor *à posteriori,* in their distinctive form, it does not therefore follow that they are fictitious. The concept of law results from the effort we make to adapt things to the mind. Law represents the characteristic we must attribute to things in order that they may be expressed by the symbols at our disposal, the matter that physics must offer to mathematics, so that mathematics may unite with it. And the result proves that certain phenomena of nature comply with this requirement, the consequence being that the notion of mechanical law dominates the whole of scientific research, as a guiding idea, at all events.

We have inquired into the nature of the mechanical laws; now we must examine their objectivity and signification, i.e. we must see how far we are justified in believing that things realize mechanism, and to what extent we are included in this mechanism. . . .

We have seen that the mechanical laws are not a mere development and complication of mathematics; in reality, they imply a new element which cannot be reduced to mathematical intuition, viz, the solidarity of fact, the regular, constant dependence, empirically given and unknowable *à priori*, between two different magnitudes. We have shown that these laws are not purely experimental truths either. They result from the collaboration of mind and things; they are products of mental activity and apply to extraneous matter; they represent the effort which the mind makes to set up a coincidence between things and itself. Now we must inquire in what way the mechanical laws may be regarded as realized in nature.

The first step taken by the creators of scientific mechanism was to grant objective existence to those laws that enable us to explain things in so rigorous a fashion, and the first doctrine we find on this subject is dogmatism. According to this doctrine, the mechanical laws, as such, are inherent in things considered *per se*, apart from the mind at work upon them. Descartes teaches this metaphysical mechanism; he regards matter and motion, which are themselves capable of expression in terms of space, as representing the entire essence of things other than mind, and so the mechanical laws exist as such in nature; more than this, they are the fundamental laws of the whole of nature.

Still, Cartesianism lends itself to serious objections. On what is it grounded? On the clearness peculiar to the idea of extension. But, given that clearness of the idea of extension, does it follow that extension is the essence of matter, as Descartes states it to be? Descartes himself succeeds only by having recourse to divine truth, as to some *deus ex machina*. But how are we to see in motion a thing that exists *per se?* Motion is not self-sufficient. Common sense tells us that it presupposes something that moves, and common sense is right. To establish a connexion between the different positions of which motion consists, we must have either a permanent subject such as matter, or a mind that contains the representations of these positions in one and the same consciousness. In a word, motion, of itself, does not involve the principle of unity, of which it stands in need in order to be real.

Newton corrected Descartes' mechanism, though he remained dogmatic. When he says *Hypotheses non fingo,* he means that he is not satisfied, as Descartes is, with merely possible explanations, but that he aspires to find out the real, effective causes of things, the laws which God himself had in mind when he created and planned the universe. Newton introduces into nature that material subject which was lacking in Cartesian mechanism; he admits of bodies, endowed with forces, as a condition of motion, and thereby thinks he is securing, far better than Cartesianism did, the objectivity of the mechanical laws. And so he acknowledges the existence of real motion, whereas, according to Descartes, there existed only relative motions. We must carefully distinguish between Newtonianism as science and Newtonianism as metaphysics. Newtonianism as science is satisfied, as far, almost, as the human mind can be, with experimental or mathematical notions. But if we would convert this science into a knowledge of nature as existing *per se,* we must realize space, mechanical causality, force, atoms, and even attraction, or any other mode in which the cause of motion operates. And here arise the difficulties so well demonstrated by Berkeley, whose system, from the very outset, is a refutation of Newtonianism regarded as metaphysics. If space, matter, atoms, mechanical causality, force, attraction and repulsion, says Berkeley, are looked upon as objective realities, we must first acknowledge that they are things which the human mind is incapable of knowing. It is only by a process of artificial abstraction that we detach them from the sensations of which we are conscious. They are never presented to us in themselves; they cannot be. Nor is this all. Not only are such things, for us, if they exist, as though they were not, but we cannot even conceive that they do exist, in themselves. In fact, these concepts, set up as things *per se,* become contradictory. Infinite and homogeneous space devoid of quality, the extended, indivisible atom, mechanical causality, wherein that which is powerless over itself, possesses power over something else, resulting in progression *ad infinitum,* the action of one crude body upon another, in whatever way this action is brought before the mind: all these symbols, taken as absolute realities, become unintelligible; nor need we be surprised if we remember that these concepts, when analysed, present elements with which thought cannot deal.

A third form of dogmatism is that professed by Leibnitz. According to him, there is everywhere at the same time both the mechanical and the metaphysical; the mechanical laws exist, though not separately and in themselves, as the mechanician conceives them. Their reality consists in the fact that they are well founded, i.e. supported by a reality distinct from themselves, but one that really exists and contains the *requisita* of mechanical action. This subject of mechanical phenomena is force, i.e. a metaphysical essence which, at bottom, offers a certain analogy with our souls. But this system also raises difficulties. The mathematical formulæ of the mechanicians, from the time of Descartes right on to our own, have been so purged of all psychological or metaphysical content, that we no longer see any connexion between force as metaphysics understands it and force as assumed by science. The latter is nothing but a measure of motions. It might with equal justice be conceived both as a consequence and as a condition of motion. The transition, then, from scientific to metaphysical force is lacking. Leibnitz's metaphysics is superimposed from without upon science strictly so called. True or false, it is no longer scientific mechanism that it sets up as a reality.

The mechanical laws, therefore, cannot be considered realized, as such, in the nature of things. The concepts of which they are constituted become unintelligible when converted into beings. Must they therefore be denied all genuine reality, and regarded, along with idealism, as nothing else than a symbolical expression, a projection of the laws of mind itself?

Interpreted idealistically, the concepts of which the mechanical laws consist, avoid the contradictions that appear when they are interpreted realistically. Thus, space, a form of sensibility, is no longer contradictory, like space that exists *per se*. Mechanical causality, connecting representations with one another, no longer lends itself to the objections raised by this causality, conceived as connecting things. But idealism is unable to stand its ground; and, the more closely it pursues the problem, the more it is compelled to recognize destructive elements within itself. As a principle, idealism consists in explaining the unconscious by the conscious, things by thought. But the history of philosophy shows us that, in order to explain the given, idealism is forced to appeal to the unconscious and allot this latter a place alongside of—or even above—the conscious. In the case of Kant, deep within the mind appears the synthetic judgment *à priori*, which the intellect is compelled to accept as a sort of metaphysical fact, without really understanding it. Beneath the conscious self, Fichte places the absolute self, the activity of which precedes the intellect, and it is this activity which, when subjected to an inexplicable im-

pact, explains the self as the not-self. In Schelling, the absolute becomes the identity of the self and the not-self; in Hegel, it becomes
the identity of the contradictories, that offence and stumbling-block
to thought. Thus the self is more and more driven to leave itself and
have recourse to some heterogeneous principle; idealism more and
more abjures itself and approaches realism.

If then the mechanical laws do not exist objectively, neither are
they mere projections of the conscious mind. They witness to the
existence of something different from mind, and yet which must not
be altogether separated from it. We are foiled when we try to determine the substantial nature of things; all the same, we cannot abolish
them. All we can say is that, in things, there is a mode of being which
suggests to our minds the invention of the mechanical laws. In reality,
how do things act in nature? We can form a conjecture of this only
by analogy, when we consider what takes place within ourselves. In
short, consciousness is the only sense of being that we have at our
disposal. Now, the phenomena which, in man, affect the mind in its
most intimate union with the body, are the phenomena of habit, and
it would really seem as though its effects bore a certain resemblance
to mechanical causality. At first we have mental activity, in certain
cases, at all events; actions are related to thought, as their generating
cause. By degrees, they fall away from thought, and jostle one another,
as it were. Thus, in certain cases and with certain men, words follow
one another without being determined by thought; and so we find
inertia and mechanical force in the persistence of our states of consciousness and in their mutual influence. This view may not follow
from an induction based on the results of science, it is but a simple
analogy, still, it constitutes the only way in which we can point to the
reality of mechanical action. To our mind, it is the degradation of
true action, it is activity as represented by a link between its products,
and thereby released and set free for new tasks. If such actions exist,
the mechanical laws are the form we attribute to them, for the purpose of subjecting them to mathematical calculation. And so we see
that the scientist can nowhere find the conditions of science accurately
realized in phenomena.

One final question: do the mechanical laws establish an absolute
determinism?

There are few men, even amongst metaphysicians, who accept
mechanical determinism as absolute. It is generally believed that man
can produce movements in conformity with his volitions. In the very
countries where determinism is professed by eminent philosophers,

teachers and all who appeal to conscience and claim to regulate conduct, affirm the existence of free-will and of its power over things. We find this the case in England as well as in Germany. It is more difficult, however, to prove one's opinion, than to convince oneself of its correctness.

How do we reason when we attempt to lay aside mechanical necessity? Common sense acknowledges that the soul is capable of producing movements; but that is simply appearance which will not bear investigation. The soul, it is said, is a force; but this is a much abused word. We pass—without saying what right we have to do so—from the notion of moral or metaphysical force to that of mechanical force. If the soul is a force, in the sense in which it must be for imparting motion to a body by virtue of the principle of inertia, it must modify the quantity of force wherever it intervenes. But this in itself is strange and contrary to experience and induction, which show us that the quantity of force in nature is constant. Are we to say that the soul cancels a quantity of force exactly equal to that it produces? Such a conception would appear an entirely arbitrary one.

We find philosophers offering us a subtler explanation: that the action of the soul upon the body is real, although of a metaphysical, not a mechanical nature. Descartes acknowledges that the quantity of motion remains constant throughout nature, but that the soul may change the direction of the motion. The mechanical laws remain secure, since, according to Descartes, they do not determine direction, which latter must come from some other source. In spite of the objections of Leibnitz, which in all probability are not decisive, this expedient, interpreted in more or less complicated ways, has frequently been reproduced. Of recent years, M. Cournot, ascertaining that the amount of power necessary for the starting of a machine may be indefinitely diminished, recognizes a limit where this power would be nil. Then it would be replaced by a guiding force, belonging, for instance, to organisms or to thought. M. Boussinesq admitted that there were cases in which the initial state of a system does not wholly determine the course which the phenomenon must take. There would then be a greater or less number of bifurcations, making possible the intervention of a guiding force. Here the action attributed by Claude Bernard to life as a guiding idea, would find a place; life does not violate the mechanical laws, but it communicates to movements a direction they would not of themselves have taken.

This extremely seductive theory was upheld, as we see, by scientists of the first rank. It cannot be said, however, that it succeeded in becoming adopted. As regards passing to the limit, that is an expedient which offends the reason, and one which, in spite of appearances, does not seem to be authorized by mathematics. This latter declares A equal to B—in so far as their difference may be made smaller than any given quantity—only when A and B are both given as fixed, determinate quantities. A distinction is made between the true and the false use of the method of limits. Now, however small the force necessary for starting a machine may be conceived as being, this force is always required, it never becomes nil. The strange solutions of M. Boussinesq have been disputed by several mathematicians, and it would seem rash to regard the efficacy of freewill as depending upon speculations the proof of which is not perfectly evident.

An important distinction, however, appears to dominate the whole question. As long as, with Descartes and even Leibnitz, we confine ourselves to laying down laws of invariability or constancy regarding quantity in general, there is necessarily room for indetermination. The constant as a rule may always be secured in several ways. Newton, however, looked upon the mechanical laws as eliminating this element of indetermination. Indeed, Newton is not satisfied with an abstract law, he determines the quantity and direction of the motion which is to be realized in each case. He envelops the law of conservation in a concrete law which indicates the mode of its application. If motion, then, is modified, it can only be by a formal derogation of law, by a miracle.

There is a particularly metaphysical way of escaping mechanical determinism, and that consists—whilst admitting this determinism for external phenomena—in breaking the bond which links to these phenomena the higher forms of existence. We are given a relation between organic movements and intellectual states. Now, if a determinate movement corresponds with each thought, and if movements are necessarily linked to one another, the consequence is that thoughts also are necessarily linked to one another. It is this dependence of thought as regards movement that certain philosophers endeavour to weaken or to destroy altogether. According to this view, Descartes acknowledged that when a passion is brought into being within us as the result of some external action, we are not condemned to become wedded to the thoughts called up by this passion. He maintained that we have the power to summon before the mind different thoughts, and to hold them there by means of attention. For instance, when the physical body impresses on us an impulse of anger, we can summon before the mind ideas of justice, moderation and duty, to replace ideas of vengeance. Thought, then, is not indissolubly connected with the physical organism. In one sense, Leibnitz goes much farther than Descartes; he breaks off all communication between body and soul, and maintains that the life of souls would remain the same even if all bodies were annihilated. On the other hand, he recognizes that there exists pre-established harmony and exact parallelism between bodies and minds. The mind, however, is not therefore made dependent on the body. It is the contrary to this that Leibnitz has in view, for he regards efficient causes as dependent on final causes. Kant simply abolishes all connexion between the moral subject and the world of motion; he regards the noumenon, which is entirely free from the fetters of mechanism, as having power to determine itself in an absolutely autonomous fashion.

These various theories are either ingenious or profound, still, hypothesis has a large share in them. In the first place, how is it known that the bond between the mechanical order and the higher orders is loose, or liable to be broken? Then, too, who is to guarantee that the orders of things, thus superimposed on the mechanical order, will not themselves also be determinisms, different, it may be, but equally inflexible? Still, even though this system were admitted, it would give us but little satisfaction, for it would leave quite out of our control the world of motion in space, i.e. the world in which we are living, after all, and on which it is primarily important that we should be able to act.

Mechanical conjunction, it must be recognized, is the most perfect form of determinism, for it represents the coincidence of experimental reality with mathematics. But what we have to discover is whether this determinism should be transferred from the explanation of the phenomena it governs to the very beings whose manifestations we are endeavouring to systematize. When we ask ourselves if the mode in which bodies act on one another compromises our freedom, we are misstating the question. Bodies do not act on one another. It is by a process of artificial construction and abstraction that we isolate a world of atoms and mechanical forces, and regard it as self-sufficient. In reality, this world is not self-sufficient. Not only cannot atoms and mechanical causality be conceived without a mind to think them, but mechanical movements themselves cannot be isolated from the physical and organic phenomena that exist in nature. Do we know whether the mechanical laws are the cause or the consequence of the other laws? If by chance, they were the consequence, could we still affirm that they are rigorous and immutable? If there really are activities in nature, they are quite different from the so-called action of one body on another, which is nothing but a numerical relation. And as there is nothing to prove that the real support of so-called mechanical phenomena is itself mechanical and subject to determinism, there is no chain to be broken in order to enable a moral influence to permeate what is called the world of matter and motion. Bodies, in their reality, resemble us already, otherwise they are not for us. The distinction between laws or relations and phenomena or elements, copied from that between precepts and will, is a mental artifice for the reduction to ideas of the greatest possible amount of given reality. In being itself, this distinction disappears, and with it the determinism which implies it.

Chapter 11
HEINRICH RUDOLF HERTZ
(1857–1894)

INTRODUCTION

*H*EINRICH RUDOLF HERTZ was born in Hamburg on February 22, 1857. During his elementary and secondary education he was recognized as a very bright student with a broad interest ranging over languages (including Greek, Arabic, and Sanskrit), literature, philosophy, mathematics, and physics. He first studied engineering after finishing high school, but soon gave up this field in favor of physics, going to Berlin to study under Kirchhoff and von Helmholtz. He became *Privatdocent* in physics in Kiel in 1883, also beginning his studies of Maxwell's electromagnetic theory at that time. By 1885 he was professor of physics at the Higher Institute for Technology at Karlsruhe, where for the next four years he continued his investigations in connection with the production, propagation, and reception of radio waves and other electromagnetic phenomena. In 1889 Hertz was appointed to succeed Clausius as ordinarius professor of physics in Bonn; shortly after his arrival there he became seriously ill. An unsuccessful operation left him unable to pursue his experimental investigations, and so he decided to devote himself to the study of the basic concepts and principles of physics, a course which ultimately led to his famous work, *The Principles of Mechanics* (1894). After a long and painful illness he died in Bonn on January 1, 1894.

In addition to *The Principles of Mechanics* Hertz wrote many scientific papers on different topics; these were collected and re-edited by Philipp Lenard between 1892 and 1895 under the title *Gesammelte Schriften* (3 vols). They were translated into English almost immediately, appearing under the titles *Electric Waves* (1893), *Miscellaneous Papers* (1896), and *The Principles of Mechanics* (1899).

The Introduction to Hertz's *The Principles of Mechanics* is generally held to be of great importance for the philosophy of science. Although its main theme consists in a penetrating analysis of the systematic character of physical theories, it also contains vital remarks about the observational basis of physics and the logical status of laws.

Hertz found his starting point in the remark that all physicists agree that the main problem of physics consists in reducing the phenomena of nature to simple laws of mechanics. What these simple laws are, however, is by no means clear. According to the generally accepted view they are Newton's laws of motion. Their physical meaning depends upon the assumption that the forces they speak of are of a simple nature and possess simple properties, but there is no agreement among physicists as to the nature and properties of these forces. Instead of trying to clarify the vague and obscure concept of force, Hertz attempted to reconstruct the theories of mechanics by taking as fundamental only those concepts about which there was agreement, namely, space, time, and mass.

The Introduction begins by stating that the most important problem for science consists in the anticipation of future events. In attempting to anticipate the future we take a starting point in our experience of past events, and then try to draw inferences as to the future from the past by forming images of external objects in such a way that the necessary consequents of the images in our thought are always the images of the necessary consequents in nature of the things so pictured. In this we presuppose a necessary conformity between nature and our thought, but this conformity is also founded in experience, insofar as experience teaches us that such a conformity in fact does exist, at least as long as we limit ourselves to that particular aspect which is necessarily contained in our presupposition. For we do not have any means of knowing whether or not our conceptions of things as expressed in our images are in conformity with things in any other than this one particular and vital aspect (pp. 1–2).

It is quite possible that one can form more than one image of external objects, each of which may fulfill the requirement that the consequents of the images must be the images of the consequents. In Hertz's opinion one can nevertheless dismiss immediately all images or models which do not fulfill the following three requirements: (1) the models must be *logically permissible,* that is consistent with the laws of our thought; (2) they must be *correct,* that is, their relations must not conflict with the relations between things; (3) they must be *simple and distinct,* in the sense of containing the fewest possible superfluous or empty relations (p. 2).

To these postulates, which govern the constitution of images or models, Hertz maintains we must now add three postulates which determine the scientific representation of the images. First of all these representations must be *appropriate*, that is, the pictures must be represented with the help of the most appropriate notations, definitions, and abbreviations; then they must also be *correct*, that is, the pictures must be represented in terms of immediate data of experience; finally, they must be *permissible* insofar as they may not contain anything which contradicts the laws of thought as determined by the very nature of our mind (pp. 2–3). The influence of the Kantian philosophy manifests itself here, in the formulation of these postulates.

Assisted by these conditioning postulates, one can determine the value of physical theories as well as the value of their representations. In his Introduction Hertz limits himself to considering the representations of the principles of mechanics as developed in the past (p. 3). The term "principle" is to be understood, Hertz claims, as meaning any selection from among mechanical statements such that the whole of mechanics can be deductively derived from them without any further appeal to experience (p. 4). He then briefly examines three possible representations of images (= theories), two taken from history, and the third his own, as defended in the balance of his book.

The first is a theory usually found in most mechanics textbooks. It follows closely the history of mechanics and explicitly introduces *four* fundamental concepts: space, time, mass, and force. For Hertz this representation of our images of things was unsatisfactory. Although one could seriously doubt the permissibility of this representation, insofar as the concept of force lacks all clarity, Hertz nonetheless thought that this lack of clarity was due mainly to the unessential characteristics which we ourselves have arbitrarily introduced into the essential content given by nature (p. 8). Actually he does not so much object to the content of the image as to the form in which this content is represented (*Ib.*); the logical value of the separate statements is not defined with sufficient clearness (p. 13). But this model is not only inappropriate, it is also incorrect insofar as the motions considered in mechanics do not exactly coincide with motions in nature, and that many relations which are studied in the former are probably absent in the latter.

The second type of theory Hertz discusses is of more recent origin and was proposed by many progressive physicists in the last quarter of the nineteenth century. It takes as fundamental concepts space, time, mass, and energy, and in so doing is able to avoid innumerable actions at a distance between atoms and bodies, basing them on transformations of energy. In this view force is only introduced by definition in order to facilitate calculations; and by claiming that energy depends merely on positions or velocities, one is able to show that all basic concepts depend only upon immediate experience (p. 18). In Hertz's view this kind of theory was superior to the first in appropriateness and perhaps in correctness, too, but showed weakness in its lack of logical permissibility. For in this view energy is conceived of as a substance, a fact which immediately leads to difficulties when potential energy is considered, either as negative potential energy or as infinite potential energy of a finite quantity of matter. These were some of the difficulties which prompted Hertz to look for a better model (pp. 21–22). His own view is described in the following selection.

SELECTIVE BIBLIOGRAPHY

Hertz, H., *Miscellaneous Papers* (1895), trans. D. E. Jones. London, 1896.

———, *Electric Waves* (1892), trans. D. E. Jones. London, 1893.

———, *The Principles of Mechanics* (1894), trans. D. E. Jones and J. T. Walley. London, 1899.

Lodge, O., *The Work of Hertz and Some of His Successors.* London, 1894.

Cohen, R. S., "Hertz's Philosophy of Science: An Introductory Essay," in *The Principles of Mechanics.*

Helmholtz, H. von, "Preface to *The Principles*," in *The Principles of Mechanics.*

Smart, J. J., "Heinrich Hertz and the Concept of Force," *Australian J. Phil.*, **29**(1951) 36–45 and **29**(1951) 175–180, **30**(1952) 124–132.

Alexander, P., "The Philosophy of Science, 1850–1910," in *A Critical History of Western Philosophy*, ed. D. J. O'Connor. New York, 1965. Pp. 409–413.

Mach, E., *The Science of Mechanics.* La Salle (Ill.), 1960. Pp. 317–324.

Poincaré, H., "Hertz on Classical Mechanics," in Danto, A. and Morgenbesser, S., eds., *Philosophy of Science.* New York, 1964. Pp. 366–373.

ON THE APPROPRIATENESS, CORRECTNESS, AND PERMISSIBILITY OF SCIENTIFIC THEORIES

A THIRD arrangement of the principles of mechanics is that which will be explained at length in this book. Its principal characteristics will be at once stated, so that it may be criticised in the same way as the other two. It differs from them in this important respect, that it only starts with three independent fundamental conceptions, namely, those of time, space, and mass. The problem which it has to solve is to represent the natural relations between these three, and between these three alone. The difficulties have hitherto been met with in connection with a fourth idea, such as the idea of force or of energy; this, as an independent fundamental conception, is here avoided. G. Kirchhoff has already made the remark in his *Text-book of Mechanics* that three independent conceptions are necessary and sufficient for the development of mechanics. Of course the deficiency in the manifold which thus results in the fundamental conceptions necessarily requires some complement. In our representation we endeavour to fill up the gap which occurs by the use of an hypothesis, which is not stated here for the first time; but it is not usual to introduce it in the very elements of mechanics. The nature of the hypothesis may be explained as follows.

If we try to understand the motions of bodies around us, and to refer them to simple and clear rules, paying attention only to what can be directly observed, our attempt will in general fail. We soon become aware that the totality of things visible and tangible do not form an universe conformable to law, in which the same results always follow from the same conditions. We become convinced that the manifold of the actual universe must be greater than the manifold of the universe which is directly revealed to us by our senses. If we wish to obtain an image of the universe which shall be well-rounded, complete, and conformable to law, we have to presuppose, behind the things which we see, other, invisible things—to imagine confederates concealed beyond the limits of our senses. These deep-lying influences we recognised in the first two representations; we imagined them to be entities of a special and peculiar kind, and so, in order to represent them in our image, we created the ideas of force and energy. But another way lies open to us. We may admit that there is a hidden something at work, and yet deny that this something belongs to a special category. We are free to assume that this hidden something is nought else than motion and mass again,—motion and mass which differ from the visible ones not in themselves but in relation to us and to our usual means of perception. Now this mode of conception is just our hypothesis. We assume that it is possible to conjoin with the visible masses of the universe other masses obeying the same laws, and of such a kind that the whole thereby becomes intelligible and conformable to law. We assume this to be possible everywhere and in all cases, and that there are no causes whatever of the phenomena other than those hereby admitted. What we are accustomed to denote as force and as energy now become nothing more than an action of mass and motion, but not necessarily of mass and motion recognisable by our coarse senses. Such explanations of force from processes of motion are usually called dynamical; and we have every reason for saying that physics at the present day regards such explanations with great favour. The forces connected with heat have been traced back with certainty to the concealed motions of tangible masses. Through Maxwell's labours the supposition that electro-magnetic forces are due to the motion of concealed masses has become almost a conviction. Lord Kelvin gives a prominent place to dynamical explanations of force; in his theory of vortex atoms he has endeavoured to present an image of the universe in accordance with this conception. In his investigation of cyclical systems von Helmholtz has treated the most important form of concealed motion fully, and in a manner that admits of general application; through him "concealed mass" and "concealed motion" have become current as technical expressions in German.[1] But if this hypothesis is capable of gradually eliminating the mysterious forces from mechanics, it can

[1] [*Verborgene Masse; verborgene Bewegung.*]

also entirely prevent their entering into mechanics. And if its use for the former purpose is in accordance with present tendencies of physics, the same must hold good of its use for the latter purpose. This is the leading thought from which we start. By following it out we arrive at the third image, the general outlines of which will now be sketched.

We first introduce the three independent fundamental ideas of time, space, and mass as objects of experience; and we specify the concrete sensible experiences by which time, mass, and space are to be determined. With regard to the masses we stipulate that, in addition to the masses recognisable by the senses, concealed masses can by hypothesis be introduced. We next bring together the relations which always obtain between these concrete experiences, and which we have to retain as the essential relations between the fundamental ideas. To begin with, we naturally connect the fundamental ideas in pairs. Relations between space and time alone form the subject of kinematics. There exists no connection between mass and time alone. Experience teaches us that between mass and space there exists a series of important relations. For we find certain purely special connections between the masses of nature: from the very beginning onwards through all time, and therefore independently of time, certain positions and certain changes of position are prescribed and associated as possible for these masses, and all others as impossible. Respecting these connections we can also assert generally that they only apply to the relative position of the masses amongst themselves; and further that they satisfy certain conditions of continuity, which find their mathematical expression in the fact that the connections themselves can always be represented by homogeneous linear equations between the first differentials of the magnitudes by which the positions of the masses are denoted. To investigate in detail the connections of definite material systems is not the business of mechanics, but of experimental physics: the distinguishing characteristics which differentiate the various material systems of nature from each other are, according to our conception, simply and solely the connections of their masses. Up to this point we have only considered the connections of the fundamental ideas in pairs: we now address ourselves to mechanics in the stricter sense, in which all three have to be considered together. We find that their general connection, in accordance with experience, can be epitomised in a single fundamental law, which exhibits a close analogy with the usual law of inertia. In accordance with the mode of expression which we shall use, it can be represented by the statement:—Every natural motion of an independent material system consists herein, that the system follows with uniform velocity one of its straightest paths. Of course this statement only becomes intelligible when we have given the necessary explanation of the mathematical mode of expression used; but the sense of the law can also be expressed in the usual language of mechanics. The law condenses into one single statement the usual law of inertia and Gauss's Principle of Least Constraint. It therefore asserts that if the connections of the system could be momentarily destroyed, its masses would become dispersed, moving in straight lines with uniform velocity; but that as this is impossible, they tend as nearly as possible to such a motion. In our image this fundamental law is the first proposition derived from experience in mechanics proper: it is also the last. From it, together with the admitted hypothesis of concealed masses and the normal connections, we can derive all the rest of mechanics by purely deductive reasoning. Around it we group the remaining general principles, according to their relations to it and to each other, as corollaries or as partial statements. We endeavour to show that the contents of mechanics, when arranged in this way, do not become less rich or manifold than its contents when it starts with four fundamental conceptions; at any rate not less rich or manifold than is required for the representation of nature. We soon find it convenient to introduce into our system the idea of force. However, it is not as something independent of us and apart from us that force now makes its appearance, but as a mathematical aid whose properties are entirely in our power. It cannot, therefore, in itself have anything mysterious to us. Thus according to our fundamental law, whenever two bodies belong to the same system, the motion of the one is determined by that of the other. The idea of force now comes in as follows. For assignable reasons we find it convenient to divide the determination of the one motion by the other into two steps. We thus say that the motion of the first body determines a force, and that this force then determines the motion of the second body. In this way force can with equal justice be regarded as being always a cause of motion, and at the same time a consequence of motion. Strictly speaking, it is a middle term conceived only between two motions. According to this conception the general properties of force must clearly follow as a necessary consequence of thought from the fundamental law; and if in possible experiences we see these properties confirmed, we can in no sense feel surprised, unless we are sceptical as to our fundamental law. Precisely the same is true of the idea of energy and of any other aids that may be introduced.

What has hitherto been stated relates to the physical content of the image, and nothing further need be said with regard to this; but it will be convenient to give here a brief explanation of the special mathematical form in which it will be represented. The physical content is quite independent of the mathematical form, and as the content differs from what is customary, it is perhaps not quite judicious to present it in a form which is itself unusual. But the form as well as the content only differ slightly from such as are familiar; and moreover they are so suited that they mutually assist one another. The essential characteristic of the terminology used consists in this, that instead of always starting from single points, it from the beginning conceives and considers whole systems of points. Every one is familiar with the expressions "position of a system of points," and "motion of a system of points." There is nothing unnatural in continuing this mode of expression, and denoting the aggregate of the positions traversed by a system in motion as its path. Every smallest part of this path is then a path-element. Of two path-elements one can be a part of the other: they then differ in magnitude and only in magnitude. But two path-elements which start from the same position may belong to different paths. In this case neither of the two forms part of the other: they differ in other respects than that of magnitude, and thus we say that they have different directions. It is true that these statements do not suffice to determine without ambiguity the characteristics of "magnitude" and "direction" for the motion of a system. But we can complete our definitions geometrically or analytically so that their consequences shall neither contradict themselves nor the statements we have made; and so that the magnitudes thus defined in the geometry of the system shall exactly correspond to the magnitudes which are denoted by the same names in the geometry of the point,—with which, indeed, they always coincide when the system is reduced to a point. Having determined the characteristics of magnitude and direction, we next call the path of a system straight if all its elements have the same direction, and curved if the direction of the elements changes from position to position. As in the geometry of the point, we measure curvature by the rate of variation of the direction with position. From these definitions we at once get a whole series of relations; and the number of these increases as soon as the freedom of motion of the system under consideration is limited by its connections. Certain classes of paths which are distinguished among the possible ones by peculiar simple properties then claim special attention. Of these the most important are those paths which at each of their positions have the least possible curvature: these we shall denote as the straightest paths of the system. These are the paths which are referred to in the fundamental law, and which have already been mentioned in stating it. Another important type consists of those paths which form the shortest connection between any two of their positions: these we shall denote as the shortest paths of the system. Under certain conditions the ideas of straightest and shortest paths coincide. The relation is perfectly familiar in connection with the theory of curved surfaces; nevertheless it does not hold good in general and under all circumstances. The compilation and arrangement of all the relations which arise here belong to the geometry of systems of points. The development of this geometry has a peculiar mathematical attraction; but we only pursue it as far as is required for the immediate purpose of applying it to physics. A system of n points presents a $3n$-manifold of motion,— although this may be reduced to any arbitrary number by the connections of the system. Hence there arise many analogies with the geometry of space of many dimensions; and these in part extend so far that the same propositions and notations can apply to both. But we must note that these analogies are only formal, and that, although they occasionally have an unusual appearance, our considerations refer without exception to concrete images of space as perceived by our senses. Hence all our statements represent possible experiences; if necessary, they could be confirmed by direct experiments, viz. by measurements made with models. Thus we need not fear the objection that in building up a science dependent upon experience, we have gone outside the world of experience. On the other hand, we are bound to answer the question how a new, unusual, and comprehensive mode of expression justifies itself; and what advantages we expect from using it. In answering this question we specify as the first advantage that it enables us to render the most general and comprehensive statements with great simplicity and brevity. In fact, propositions relating to whole systems do not require more words or more ideas than are usually employed in referring to a single point. Here the mechanics of a material system no longer appears as the expansion and complication of the mechanics of a single point; the latter, indeed, does not need independent investigation, or it only appears occasionally as a simplification and a special case. If it is urged that this simplicity is only artificial, we reply that in no other way can simple relations be secured than by artificial and well-considered adaptation of our ideas to the relations which have to be represented. But in this objection there may be involved the imputation that the mode of expression is not only artificial, but far-fetched and unnatural. To this we reply that there may be some justification for regarding the consideration of whole systems as being more natural and obvious than the consideration of single points. For, in reality, the material particle is simply an abstraction, whereas the material system is presented directly to us. All actual experience is obtained directly from systems; and it is only by processes of reasoning that we deduce conclusions as to possible experiences with single points. As a second merit, although not a very important one, we specify the advantage of the form in which our mathematical mode of expression enables us to state the fundamental law. Without this we should have to split it up into Newton's first law and Gauss's principle of least constraint. Both of these together would represent accurately the same facts; but in addition to these facts they would by implication contain something more, and this something more, would be too much. In the first place they suggest the conception, which is foreign to our system of mechanics, that the connections of the material system might be destroyed; whereas we have denoted them as being permanent and indestructible throughout. In the second place we cannot, in using Gauss's principle, avoid suggesting the idea that we are not only stating a fact, but also the cause of this fact. We cannot assert that nature always keeps a certain quantity, which we call constraint, as small as possible, without suggesting that this quantity signifies something which is for nature itself a constraint,—an uncomfortable feeling. We cannot assert that nature acts like a judicious calculator reducing his observations, without suggesting that deliberate intention underlies the action. There is undoubtedly a special charm in such suggestion; and Gauss felt a natural delight in giving prominence to it in his beautiful discovery, which is of fundamental importance in our mechanics. Still, it must be confessed that the charm is that of mystery; we do not really believe that we can solve the enigma of the world by such half-suppressed allusions. Our own fundamental law entirely avoids any such suggestions. It exactly follows the form of the customary law of inertia, and like this it simply states a bare fact without any pretence of establishing it. And as it thereby becomes plain and unvarnished, in the same degree does it become more honest and truthful. Perhaps I am prejudiced in favour of the slight modification which I have made in Gauss's principle, and see in it advantages which will not be manifest to others. But I feel sure of general assent when I state as the third advantage of our method, that it throws a bright light upon Hamilton's method of treating mechanical problems by the aid of characteristic functions. During the sixty years since its discovery this mode of treatment has been well appreciated and much praised; but it has been regarded and treated more as a new branch of mechanics, and as if its growth and development had to proceed in its own way and independently of the usual mechanics. In our form of the mathematical representation, Hamilton's method, instead of having the character of a side branch, appears as the direct, natural, and, if one may so say, self-evident continuation of the elementary statements in all cases to which it is applicable. Further, our mode of representation gives prominence to this: that Hamilton's mode of treatment is not based, as is usually assumed, on the special physical foundations of mechanics; but that it is fundamentally a purely geometrical method, which can be established and developed quite independently of mechanics, and which has no closer connection with mechanics than any other of the geometrical methods employed in it. It has long since been remarked by mathematicians that Hamilton's method contains purely geometrical truths, and that a peculiar mode of expression, suitable to it, is required in order to express these clearly. But this fact has only come to light in a somewhat perplexing form, namely, in the analogies between ordinary mechanics and the geometry of space of many dimensions, which have been discovered by following out Hamilton's thoughts. Our mode of expression gives a simple and intelligible explanation of these analogies. It allows us to take advantage of them, and at the same time it avoids the unnatural admixture of supra-sensible abstractions with a branch of physics.

We have now sketched the content and form of our third image as far as can be done without trenching upon the contents of the book; far enough to enable us to submit it to criticism in respect of its permissibility, its correctness, and its appropriateness. I think that as far as logical permissibility is concerned it will be found to satisfy the most rigid requirements, and I trust that others will be of the same opinion. This merit of the representation I consider to be of the greatest importance, indeed of unique importance. Whether the image is more appropriate than another; whether it is capable of including all future experience; even whether it only embraces all present experience, all this I regard almost as nothing compared with the question whether it is in itself conclusive, pure and free from contradiction. For I have not attempted this task because mechanics has shown signs of inappropriateness in its applications, nor because it in any way conflicts with experience, but solely in order to rid myself of the oppressive feeling that to me its elements were not free from things obscure and unintelligible. What I have sought is not the only image of mechanics, nor yet the best image; I have only sought to find an intelligible image and to show by an example that this is possible and what it must look like. We cannot attain to perfection in any direction; and I must confess that, in spite of the pains I have taken with it, the image is not so convincingly clear but that in some points it may be exposed to doubt or may require defence. And yet it seems to me that of objections of a general nature there is only a single one which is so pertinent that it is worth while to anticipate and remove it. It relates to the nature of the rigid connections which we assume to exist between the masses, and which are absolutely indispensable in our system. Many physicists will at first be of opinion that by means of these connections forces are introduced into the elements of mechanics, and are introduced in a way which is secret, and therefore not permissible. For, they will assert, rigid connections are not conceivable without forces; they cannot come into existence except by the action of forces. To this we reply—Your assertion is correct for the mode of thought of ordinary mechanics, but it is not correct independently of this mode of thought; it does not carry conviction to a mind which considers the facts without prejudice and as if for the first time. Suppose we find in any way that the distance between two material particles remains constant at all times and under all circumstances. We can express this fact without making use of any other conceptions than those of space; and the value of the fact stated, as a fact, for the purpose of foreseeing future experience and for all other purposes, will be independent of any explanation of it which we may or may not possess. In no case will the value of the fact be increased, or our understanding of it improved, by putting it in the form—"Between these masses there acts a force which keeps them at a constant distance from one another," or "Between them there acts a force which makes it impossible for their distance to alter from its fixed value." But it will be urged that this latter explanation, although apparently only a ludicrous circumlocution, is nevertheless correct. For all the connections of the actual world are only approximately rigid; and the appearance of rigidity is only produced by the action of the elastic forces which continually annul the small deviations from the position of rest. To this we reply as follows:—With regard to rigid connections which are only approximately realised, our mechanics will naturally only state as a fact that they are approximately satisfied; and for the purpose of this statement the idea of force is not required. If we wish to proceed to a second approximation and to take into consideration the deviations, and with them the elastic forces, we shall make use of a dynamical explanation for these as for all forces. In seeking the actual rigid connections we shall perhaps have to descend to the world of atoms. But such considerations are out of place here; they do not affect the question whether it is logically permissible to treat of fixed connections as independent of forces and precedent to them. All that I wished to show was that this question must be answered in the affirmative, and this I believe I have done. This being so, we can deduce the properties and behaviour of the forces from the nature of the fixed connections without being guilty of a *petitio principii*. Other objections of a similar kind are possible, but I believe they can be removed in much the same way.

By way of giving expression to my desire to prove the logical purity of the system in all its details, I have thrown the representation into the older synthetic form. For this purpose the form used has the merit of compelling us to specify beforehand, definitely even if monotonously, the logical value which every important statement is intended to have. This makes it impossible to use the convenient reservations and ambiguities into which we are enticed by the wealth of combinations in ordinary speech. But the most important advantage of the form chosen is that it is always based upon what has already been proved, never upon what is to be proved later on: thus we are always sure of the whole chain if we sufficiently test each link as we proceed. In this respect I have endeavoured to carry out fully the obligations imposed by this mode of representation. At the same time it is obvious that the form by itself is no guarantee against error or oversight; and I hope that any chance defects will not be the more harshly criticised on account of the somewhat presumptuous mode of presentation. I trust that any such defects will be capable of improvement and will not affect any important point. Now and again, in order to avoid excessive prolixity, I have consciously abandoned to some extent the rigid strictness which this mode of representation properly requires. Before proceeding to mechanics proper, as dependent upon physical experience, I have naturally discussed those relations which follow simply and necessarily from the definitions adopted and from mathematics; the connection of these latter with experience, if any, is of a different nature from that of the former. Moreover, there is no reason why the reader should not begin with the second book. The matter with which he is already familiar and the clear analogy with the dynamics of a particle will enable him easily to guess the purport of the propositions in the first book. If he admits the appropriateness of the mode of expression used, he can at any time return to the first book to convince himself of its permissibility.

We next turn to the second essential requirement which our image must satisfy. In the first place there is no doubt that the system correctly represents a very large number of natural motions. But this does not go far enough; the system must include all natural motions without exception. I think that this, too, can be asserted of it; at any rate in the sense that no definite phenomena can at present be mentioned which would be inconsistent with the system. We must of course admit that we cannot extend a rigid examination to all phenomena. Hence the system goes a little beyond the results of assured experience; it therefore has the character of a hypothesis which is accepted tentatively and awaits sudden refutation by a single example or gradual confirmation by a large number of examples. There are in especial two places in which we go beyond assured experience: firstly, in our limitation of the possible connections; secondly, in the dynamical explanation of force. What right have we to assert that all natural connections can be expressed by linear differential equations of the first order? With us this assumption is not a matter of secondary importance which we might do without. Our system stands or falls with it; for it raises the question whether our fundamental law is applicable to connections of the most general kind. And yet connections of a more general kind are not only conceivable, but they are permitted in ordinary mechanics without hesitation. There nothing prevents us from investigating the motion of a point where its path is only limited by the supposition that it makes a given angle with a given plane, or that its radius of curvature is always proportional to another given length. These are conditions which are not permissible in our system. But why are we certain that they are debarred by the nature of things? We might reply that these and similar connections cannot be realised by any practical mechanism; and in this respect we might appeal to the great authority of Helmholtz's name. But in every example possibilities might be overlooked; and ever so many examples would not suffice to substantiate the general assertion. It seems to me that the reason for our conviction should more properly be stated as follows. All connections of a system which are not embraced within the limits of our mechanics, indicate in one sense or another a discontinuous succession of its possible motions. But as a matter of fact it is an experience of the most general kind that nature exhibits continuity in infinitesimals everywhere and in every sense: an experience which has crystallised into firm conviction in the old proposition—*Natura non facit saltus*. In the text I have therefore laid stress upon this: that the permissible connections are defined solely by their continuity; and that their property of being represented by equations of a definite form is only deduced from this. We cannot attain to actual certainty in this way. For this old proposition is indefinite, and we cannot be sure how far it applies—how far it is the result of actual experience, and how far the result of arbitrary assumption. Thus the most conscientious plan is to admit that our assumption as to the permissible connections is of the nature of a tentatively accepted hypothesis. The same may be said with respect to the dynamical explanation of force. We may indeed prove that certain classes of concealed motions produce forces which, like actions-at-a-distance in nature, can be represented to any desired degree of approximation as differential coefficients of force-functions. It can be shown that the form of these force-functions may be of a very general nature; and in fact we do not deduce any restrictions for them. But on the other hand it remains for us to prove that any and every form of the force-functions can be realised; and hence it remains an open question whether such a mode of explanation may not fail to account for some one of the forms occurring in nature. Here again we can only bide our time so as to see whether our assumption is refuted, or whether it acquires greater and greater probability by the absence of any such refutation. We may regard it as a good omen that many distinguished physicists tend more and more to favour the hypothesis. I may mention Lord Kelvin's theory of vortex-atoms: this presents to us an image of the material universe which is in complete accord with the principles of our mechanics. And yet our mechanics in no wise demands such great simplicity and limitation of assumptions as Lord Kelvin has imposed upon himself. We need not abandon our fundamental propositions if we were to assume that the vortices revolved about rigid or flexible, but inextensible, nuclei;

and instead of assuming simply incompressibility we might subject the all-pervading medium to much more complicated conditions, the most general form of which would be a matter for further investigation. Thus there appears to be no reason why the hypothesis admitted in our mechanics should not suffice to explain the phenomena.

We must, however, make one reservation. In the text we take the natural precaution of expressly limiting the range of our mechanics to inanimate nature; how far its laws extend beyond this we leave as quite an open question. As a matter of fact we cannot assert that the internal processes of life follow the same laws as the motions of inanimate bodies; nor can we assert that they follow different laws. According to appearance and general opinion there seems to be a fundamental difference. And the same feeling which impels us to exclude from the mechanics of the inanimate world as foreign every indication of an intention, of a sensation, of pleasure and pain,—this same feeling makes us unwilling to deprive our image of the animate world of these richer and more varied conceptions. Our fundamental law, although it may suffice for representing the motion of inanimate matter, appears (at any rate that is one's first and natural impression) too simple and narrow to account for even the lowest processes of life. It seems to me that this is not a disadvantage, but rather an advantage of our law. For while it allows us to survey the whole domain of mechanics, it shows us what are the limits of this domain. By giving us only bare facts, without attributing to them any appearance of necessity, it enables us to recognise that everything might be quite different. Perhaps such considerations will be regarded as out· of place here. It is not usual to treat of them in the elements of the customary representation of mechanics. But there the complete vagueness of the forces introduced leaves room for free play. There is a tacit stipulation that, if need be, later on a contrast between the forces of animate and inanimate nature may be established. In our representation the outlines of the image are from the first so sharply delineated, that any subsequent perception of such an important division becomes almost impossible. We are therefore bound to refer to this matter at once, or to ignore it altogether.

As to the appropriateness of our third image we need not say much. In respect of distinctness and simplicity, as the contents of the book will show, we may assign to it about the same position as to the second image; and the merits to which we drew attention in the latter are also present here. But the permissible possibilities are somewhat more extensive than in the second image. For we pointed out that in the latter certain rigid connections were wanting; by our fundamental assumptions these are not excluded. And this extension is in accordance with nature, and is therefore a merit; nor does it prevent us from deducing the general properties of natural forces, in which lay the significance of the second image. The simplicity of this image, as of the second, is very apparent when we consider their physical applications. Here, too, we can confine our consideration to any characteristics of the material system which are accessible to observation. From their past changes we can deduce future ones by applying our fundamental law, without any necessity for knowing the positions of all the separate masses of the system, or for concealing our ignorance by arbitrary, ineffectual, and probably false hypotheses. But as compared with the second image, our third one exhibits simplicity also in adapting its conceptions so closely to nature that the essential relations of nature are represented by simple relations between the ideas. This is seen not only in the fundamental law, but also in its numerous general corollaries which correspond to the so-called principles of mechanics. Of course it must be admitted that this simplicity only obtains when we are dealing with systems which are completely known, and that it disappears as soon as concealed masses come in. But even in these cases the reason of the complication is perfectly obvious. The loss of simplicity is not due to nature, but to our imperfect knowledge of nature. The complications which arise are not simply a possible, but a necessary result of our special assumptions. It must also be admitted that the co-operation of concealed masses, which is the remote and special case from the standpoint of our mechanics, is the commonest case in the problems which occur in daily life and in the arts. Hence it will be well to point out again that we have only spoken of appropriateness in a special sense— in the sense of a mind which endeavours to embrace objectively the whole of our physical knowledge without considering the accidental position of man in nature, and to set forth this knowledge in a simple manner. The appropriateness of which we have spoken has no reference to practical applications or the needs of mankind. In respect of these latter it is scarcely possible that the usual representation of mechanics, which has been devised expressly for them, can ever be replaced by a more appropriate system. Our representation of mechanics bears towards the customary one somewhat the same relation that the systematic grammar of a language bears to a grammar devised for the purpose of enabling learners to become acquainted as quickly as possible with what they will require in daily life. The requirements of the two are very different, and they must differ widely in their arrangement if each is to be properly adapted to its purpose.

In conclusion, let us glance once more at the three images of mechanics which we have brought forward, and let us try to make a final and conclusive comparison between them. After what we have already said, we may leave the second image out of consideration. We shall put the first and third images on an equality with respect to permissibility, by assuming that the first image has been thrown into a form completely satisfactory from the logical point of view. This we have already assumed to be possible. We shall also put both images on an equality with respect to appropriateness, by assuming that the first image has been rendered complete by suitable additions, and that the advantages of both in different directions are of equal value. We shall then have as our sole criterion the correctness of the images: this is determined by the things themselves and does not depend on our arbitrary choice. And here it is important to observe that only one or the other of the two images can be correct: they cannot both at the same time be correct. For if we try to express as briefly as possible the essential relations of the two representations, we come to this. The first image assumes as the final constant elements in nature the relative accelerations of the masses with reference to each other: from these it incidentally deduces approximate, but only approximate fixed relations between their positions. The third image assumes as the strictly invariable elements of nature fixed relations between the positions: from these it deduces when the phenomena require it approximately, but only approximately, invariable relative accelerations between the masses. Now, if we could perceive natural motions with sufficient accuracy, we should at once know whether in them the relative acceleration, or the relative relations of position or both, are only approximately invariable. We should then know which of our two assumptions is false; or whether both are false; for they cannot both be simultaneously correct. The greater simplicity is on the side of the third image. What at first induces us to decide in favour of the first is the fact that in actions-at-a-distance we can actually exhibit relative accelerations which, up to the limits of our observation, appear to be invariable; whereas all fixed connections between the positions of tangible bodies are soon and easily perceived by our senses to be only approximately constant. But the situation changes in favour of the third image as soon as a more refined knowledge shows us that the assumption of invariable distance-forces only yields a first approximation to the truth; a case which has already arisen in the sphere of electric and magnetic forces. And the balance of evidence will be entirely in favour of the third image when a second approximation to the truth can be attained by tracing back the supposed actions-at-a-distance to motions in an all-pervading medium whose smallest parts are subjected to rigid connections; a case which also seems to be nearly realised in the same sphere. This is the field in which the decisive battle between these different fundamental assumptions of mechanics must be fought out. But in order to arrive at such a decision it is first necessary to consider thoroughly the existing possibilities in all directions. To develop them in one special direction is the object of this treatise,—an object which must necessarily be attained even if we are still far from a possible decision, and even if the decision should finally prove unfavourable to the image here developed.

Chapter 12
LUDWIG BOLTZMANN
(1844–1906)

INTRODUCTION

*L*UDWIG BOLTZMANN was born in Vienna on February 20, 1844. He studied at the University of Vienna, receiving his doctorate in physics in 1866. Between 1869 and 1906 he taught in the fields of mathematics and experimental and theoretical physics at Graz, Munich, Leipzig, and Vienna universities. During the seventies Boltzmann devoted himself to the task of explaining the second law of thermodynamics on the basis of the atomic theory of matter. He was able to show that the second law can be understood in principle by combining the laws of classical mechanics with the theory of probability. From this it became clear that the second law of thermodynamics is essentially a statistical law. That is, each thermodynamic system tends toward a state of thermodynamic equilibrium which is by far the most probable state; the entropy function which describes this propensity toward equilibrium and the maximum of which characterizes the state of equilibrium is itself the measure of probability of the macroscopic state. Boltzmann not only laid the foundation of statistical mechanics during these years, but was also able to develop many of its essential elements. His views were further developed by J. W. Gibbs.

In addition to providing this groundwork on statistical mechanics, Boltzmann made significant contributions to the kinetic theory of gases, Maxwell's theory of electromagnetism, and the theory concerning black body radiation. He also wrote many short essays and delivered numerous speeches on different issues; here his studies on topics associated with the philosophy of science occupy quite an important place. In this realm his main interest revolved around the question concerning the very nature of physical theories.

Boltzmann's work in statistical mechanics was strongly attacked by W. Ostwald and other energeticists who rejected real atoms and tried to base all of physical science upon the concept of energy. In addition Boltzmann's work suffered from misunderstanding on the part of people who substantially agreed with his view, but did not fully understand the statistical character of his reasoning. Although his views were fully justified by discoveries in atomic physics, quite some time was required before the intrinsic value of his investigations was generally accepted. Boltzmann became seriously ill in 1906 and during a period of depression took his own life (September 5, 1906).

The selection chosen here is taken from a lecture series which the scientist gave at Clark University in 1899. The series was entitled *Über die Grundprincipien und Grundgleichungen der Mechanik* ("On the Fundamental Principles and Basic Equations of Mechanics"); it first appeared in *Clark University: 1889–1899, Decennial Celebration,* then was later reprinted in his *Populäre Schriften.* The piece presented here is the first lecture. (Although an English translation of a part of this lecture is available, I have preferred, because of its historical value, to make a new translation of the lecture in its entirety.)

In the lecture series Boltzmann begins by stating that physical theories cannot be developed with a geometrical method such as was used, for example, by Euclid in his *Elements.* However, it is Boltzmann's view that this does not entail the exclusion of all deductive representations of physical theories, but rather that in physics deductive theories must take their starting point from somewhat arbitrary principles which, in principle, are independent of experience. He briefly discusses Hertz's ideas in this regard, showing their strong and weak points. Then he explains that in addition to Hertz's approach other attempts, which remain much closer to classical mechanics, could be made. He next gives an outline of his own view as presented in his book *Vorlesungen über die Principe der Mechanik* ("Lectures on the Principles of Mechanics"), pointing out its strengths and weaknesses.

In the second, third, and fourth lectures, perhaps from a certain viewpoint the most important, Boltzmann showed that in addition to a deductive representation of physical theories an inductive approach is possible and extremely valuable. It was his view, however, that most of the philosophers and scientists who have tried to materialize such an approach did not really succeed in their various enterprises. Boltzmann defended the thesis that a merely inductive approach in the sense of Mill is impossible because in all descriptions of experience, this experience is necessarily transcended. In other words the thesis that science can never transcend experience is wrong; for this rather simplistic view another thesis is to be substituted along the following lines: science must try not to transcend experience by too much, and never introduce any abstractions which cannot immediately afterward be subjected to verification in experience. Using concrete analyses Boltzmann shows that an inductive approach in physics is valuable and necessary, but that two extremes must be avoided: the view of earlier phenomenologists that a physical theory could be developed without abstractions and idealizations, and the arbitrary and uncontrolled introduction of abstractions which are most often found in textbooks. Boltzmann concludes by observing that many authors combine a deductive and inductive method, a fundamentally unacceptable procedure which necessarily leads to much confusion and many wrong conclusions.

SELECTIVE BIBLIOGRAPHY

Boltzmann, L., *Vorlesungen über die Principe der Mechanik*, Leipzig, 1897.

———, "Über die Grundprincipien und Grundgleichungen der Mechanik," in *Clark University: 1889–1899*. Worcester, Mass., 1899.

———, *Populäre Schriften*. Leipzig, 1905.

———, *Wissenschaftliche Abhandlungen* (3 vols.). Leipzig, 1909.

Broda, E., *Ludwig Boltzmann, Mensch, Physiker, Philosoph*. Vienna, 1955.

Dugas, R., *La théorie physique au sens de Boltzmann et ses prolongements modernes*. Neuchâtel, 1959.

ON THE FUNDAMENTAL PRINCIPLES
AND BASIC EQUATIONS OF MECHANICS

ANALYTIC mechanics is a science which its founder Newton already developed with such sagacity and perfection that it is almost without equal in the whole domain of man's scientific knowledge. The great masters who succeeded Newton founded on a still firmer basis the building which he had erected, and it seemed that a more perfect and more homogeneous creation of man's mind than the basic doctrines of mechanics as we encounter them in the works of Lagrange, Laplace, Poisson, Hamilton, and others, was absolutely inconceivable. It is especially the foundation of the first principles that seemed to have been effected by their explorers with a sagacity and logical consistency which definitively set the example one has attempted to imitate in the foundation of the other branches of science, albeit not always with the same success. For a long time it seemed as if it were wholly impossible at all to add something to this foundation or to change it in any way.

It is therefore all the more striking and unexpected that particularly in Germany rather animated controversies have arisen precisely concerning the fundamental principles of analytic mechanics. This, however, does not mean that the reverence and admiration which we hold for the genius of a Newton, a Lagrange, or a Laplace, is somehow to be minimalized by that fact. From the sketchy rudiments which they came upon, these scientists created a queen which will be a model for all times. There was so much of the genuinely new which had to be worked out that they would only have wasted time and damaged the overall impression if they had lingered too long over certain difficulties and obscurities. But since that time our knowledge of facts has increased substantially, our mind is trained, so that many conceptions which still raised difficulties for the scientists

This article is from **Populäre Schriften** by Ludwig Boltzmann (Leipzig, 1905), pp. 253–270. Copyright © 1905 by Johann Ambrosius Barth. The present essay was translated into English expressly for this edition by Joseph J. Kockelmans.

in Newton's time, have become common property to all of us. Because of this one has found time to consider Newton's construction, as it were, with a magnifying glass, and behold, many difficulties arose of the kind that always oppose the human mind precisely and mostly when it attempts to analyze the simplest foundations of knowledge.

To be sure, these difficulties are more philosophical, or as one says today, epistemological in character. We Germans have already been much ridiculed because of our inclination toward philosophical speculation, and in earlier times often rightly so. A philosophy which turns its back upon the facts has never produced anything useful, and is not able to do so. It is an immediately tangible adventure first of all to enlarge our knowledge of facts by means of experience; and our scientific knowledge, too, is above all the most successfully furthered in this way. But notwithstanding all of this the inclination to analyze the simplest concepts and to account for the basic operations of our thought seems to be invincible in man's mind.

Furthermore, the method of this analysis has been considerably improved upon in the course of time, so that today, although in no way yet immediately practically fruitful, it nonetheless is not by a long way as meaningless as the old philosophy. For in the course of history the whole cultural view of mankind is subject to continual and important oscillations, and the Germans are no longer the unpractical dreamers of former times. They have shown this in all the domains of experimental science, engineering, industry, and politics. In the beginning, the Americans were naturally oriented toward the practical activities of industry and technique for the purpose of the settling of their land. But for a long time now they have been far from exclusively dedicated to that cause, and America is able already to show in all the domains of abstract science scientists who are of equal rank with the most prominent researchers of Europe. So, gentlemen, in view of the fact that you have invited a German to lecture in your country, it is for that reason that I venture to enter the realm of epistemology with you.

First of all, I want to return to the objections which have been raised against the fundamental principles of Newton's mechanics, or (better formulated) to those areas where it seems to be in need of further elucidation, to an analysis of its method of argumentation, and an examination of its basic concepts.

In formulating the laws of motion Newton considered the motion of bodies to be absolute in space. However, absolute space is nowhere accessible to our experience. Only the relative changes in the position of bodies are given empirically. Thus it is already at the very beginning in this view that one completely transcends man's experience; and this is surely questionable and hazardous in a science which claims exclusively to describe facts. This difficulty has evidently in no way escaped Newton's genius. He believed, however, that without the concept of absolute space he would not be able to arrive at a simple formulation of the law of inertia, in which he primarily was interested. And I think that in this regard he has proved to be right; for, however much this difficulty has been examined and clarified, scarcely any essential progress has been made.

Instead of Newton's absolute space Neumann introduced a mysterious, ideal reference body; but in so doing he obviously transcends our experience as much as Newton did. Streintz sets himself the task of avoiding such concepts and bodies, and shows how, by means of a gyroscope which is not subject to any forces or which is influenced by well-known forces only, one can decide in a relative way in regard to a chosen system of coordinates whether or not it is a useful frame of reference. However, these considerations of Streintz seem to be of little help for the foundation of mechanics because they presuppose the laws of motion of a rotating gyroscope and a definite opinion about whether or not this top is subject to any forces; but for this the knowledge of Newton's laws of motion is precisely a prerequisite. Lange tried, indeed, to formulate the laws of inertia without introducing any frame of reference whatsoever by merely considering relative motions. But although he has succeeded in so doing, his solution nonetheless is so complicated and involved that one finds it difficult to decide to substitute a law of such little clarity, for the simple formula of Newton. Obviously Mach's proposal to substitute straight lines which are determined by the totality of all masses in the universe, or his suggestion to substitute the light ether for absolute space, both transcend our experience, too, albeit in both cases in a different way. For the first proposal refers again to purely ideal and transcendent concepts, whereas the latter suggestion implies an assertion which possibly could be proved empirically, but as of today this has not been shown. Furthermore, for the light ether a completely different mechanics ought to be in effect and the ether itself ought to be subject to the cause of the law of inertia but not to that law itself.

One encounters a similar difficulty when one introduces the concept of time. This, too, is introduced by Newton as something absolute, whereas such time is never given to us, but rather only the simultaneity of the courses of different events. Yet here it is not too difficult to take remedial measures insofar as one can take his starting point in an event which always repeats itself periodically under completely identical circumstances. True, it is not possible to produce the absolute identity of circumstances, but it is nonetheless possible to secure with the highest possible degree of probability that all circumstances are the same, insofar at least as they ever have an essential influence. One can even further corroborate this by comparing different forms of this property (such as the rotation of the earth, the oscillation of a pendulum, or the spring of a chronometer) with each other. The agreement of all these events in indicating equal times then excludes all doubt as to the usefulness of this method.

A third difficulty applies to the concepts of mass and force. It has long been recognized that Newton's definition of mass as the quantity of matter is meaningless. But there are serious doubts as well in regard to the relationship between force and mass. Is mass the only thing which exists, and is force merely a property of mass or, the other way around, is force the truly existing thing; or must one admit a dualism of two separated real things (mass and force) in such a way that force, existing as separated from mass, is the cause of the former's motion? In recent times in addition to this, the question has come up concerning whether reality should be attributed to energy, also, or whether the latter is even the only thing that exists.

On this point Kirchhoff was the first who objected even to the way the question was formulated. It is very often the case that a problem is half solved once the right method for formulating it is found. Kirchhoff rejected the idea that it is the task of natural science to unravel the true essence of phenomena and to indicate their first metaphysical primary causes. Rather he reduced the task of natural science to describing the phenomena. Kirchhoff still called this a limitation of the task of natural science. However, if one really penetrates into the nature and mode, I would say into the very mechanism of our thought, one might perhaps want to deny even this.

For all our ideas and concepts are but inner thought-pictures or, when uttered, combinations of sounds. The task of our thinking, then, is to use and connect these concepts in such a way that with their help we always and most easily perform the right actions and guide others to right actions, also. Here metaphysics has attached itself to the most sober, practical position: extremes meet. The conceptual signs we form thus have only existence within us: we cannot gauge external phenomena with the measure of our representations.

Formally, therefore, we could raise questions of this kind: does only matter exist and is force one of its properties, or does the latter exist independently of matter; or, conversely, is matter a product of force? None of these questions, however, has any meaning, since all of these concepts are merely thought-pictures that have the purpose of correctly representing phenomena.

Hertz has expressed this with special clarity in his famous book on the principles of mechanics; however, Hertz lays down as his first requirement that the pictures we construct must correspond to the laws of thought. Concerning this requirement I would like to raise certain objections, or at least explain it somewhat more carefully. True, we must certainly carry with us a rich treasure of laws of thought. Without them our experience would be completely useless; we would have no way whatsoever of fixing it by means of inner pictures. These laws are almost without exception innate; however, they nonetheless undergo modifications through education, instruction, and personal experience. They are not completely the same in a child, a simple uneducated man, and a man of learning. We can understand this also when we compare the way of thought of a naïve people such as the Greeks with that of the Scholastics of the Middle Ages and that, in turn, with the way of today. To be sure, there are laws of thought which have held so unexceptionally that we trust them without reservation, that we consider them as a priori and unchangeable principles of thought. But I nonetheless still believe that even they have been the result of a slow development. Their first source was the primitive experiences of mankind in its original state; gradually they were strengthened and became clearer by means of complicated experiences, until finally they took on their present sharp formulation; still, I do not wish to recognize them unconditionally as the highest standards. For we cannot know whether they will not undergo this or that modification. We will remember the certainty with which children or uneducated people are convinced that one must be able to distinguish by means of mere feeling, between the directions of up and down in all places of the universe and how from this they believe themselves able to deduce the impossibility of antipodes. Were such people to write on logic, they would surely take this to be a law of thought that is evident a priori. In the same way, at first various aprioric objections were also raised against Copernicus' theory, and the history of science exhibits numerous cases where assertions were now supported, now refuted, by means of arguments which were at the time held to be self-evident laws of thought, whereas today we are convinced of their futility.

I would therefore like to modify Hertz's demand in such a way that insofar as we possess laws of thought which we have recognized as indubitably correct on the basis of constant confirmation in experience, we can initially test the correctness of our pictures on them; however, the final and sole decision concerning the usefulness of the pictures lies in the fact that they represent our experiences as simply and accurately as possible. And it is precisely in this way that we have the final test for the correctness of our laws of thought.

When we have thus understood the task of thought in general and of science in particular, consequences will follow which at first sight are quite startling. We shall call a representation of nature false when it represents certain facts incorrectly, or if there are obviously simpler conceptions which represent these facts more clearly, especially when the representation contradicts generally confirmed laws of thought. However, it is possible to have theories which represent a great number of facts correctly, but which are incorrect on other points, and which therefore possess a certain relative truth only. Indeed, it may even be that we can construct a system of pictures of the phenomena in different ways. Each of these systems is not equally simple and does not represent the phenomena equally well. Still, it may be doubtful—to a certain extent even a matter of taste— which one we consider to be the simpler one, with which representation of the phenomena we feel more satisfied. In this way science loses its uniform character. Formerly one adhered to the view that there can be only *one* truth; that errors were manifold, but truth unique. From our present standpoint this view must be rejected; but, to be sure, the difference between the new view as compared to the old is merely a formal one. Never has it been in doubt that man would ever be able to know the totality of all truths. Such knowledge is an ideal only. In line with our contemporary conception we have a similar ideal: the most perfect model is that which represents all phenomena in the simplest and most expedient way. Thus, according to the first point of view we turn our glance more to the unattainable ideal which is merely a unitary one; from the other viewpoint we look to the multiplicity of what is attainable.

Now, if we are convinced that science is only an inner picture, a thought-construction, which is never identical with the multiplicity of phenomena, but is capable of representing clearly and distinctly certain of its parts only, how then do we attain such a model? How can we represent it as systematically and clearly as possible? Formerly, a method which imitates the method employed in Euclid's geometry was popular, and might therefore be called the Euclidean method. It proceeds from as few and as evident propositions as possible. In ancient times these were considered to be evident a priori, given directly to our mind; hence they were called axioms. Later, however, they were attributed the character of being merely sufficiently established, experiential propositions. Using the laws of thought only, certain pictures were then deduced from these axioms and it was believed that thus a proof had been found for the thesis that these pictures were the only possible ones; that they therefore could not be replaced by others. As example I may mention the deductions which served for the derivation of the parallelogram of forces and Ampère's law, or those which served as proof that the force acting between two material points acts in the direction of the distance between them and must be a function of that distance.

But the argumentative force of this method of inference gradually became depreciated. The first step in this direction (as previously mentioned) was the transition from a foundation which was a priori evident, to one that was merely verified experientially. It was further recognized that the deductions from this foundation could not be carried out without numerous new hypotheses, also. Finally, Hertz pointed out that particularly in the field of physics our conviction concerning the correctness of a general theory substantially does not depend upon its derivation according to Euclid's method, but rather upon the fact that this theory leads us to correct conclusions in regard to phenomena in all the cases hitherto known. He first made use of this view in his description of Maxwell's fundamental equations of the theory of electricity and magnetism, by proposing not to concern oneself at all with their derivation from certain fundamental principles, but to simply start with them and to look for their justification in the fact that one could subsequently show that they were always in agreement with experience. For experience is and remains the sole judge of the usefulness of a theory whose judgment is inappealable and irrevocable. Indeed, when we go further into themes which are most closely connected with this subject, the law of inertia, the parallelogram of forces, and the other fundamental laws of mechanics, we shall find that the different proofs offered in all the textbooks of mechanics for each and every one of these theorems are by far not as convincing as the fact that all the consequences drawn from the totality of all these theorems have been so splendidly confirmed by experience. The ways by which we reach such models quite often differ very much from each other and depend upon the most diverse coincidences.

Some models, such as those of the mechanical theory of heat, were only gradually developed in the course of centuries by the combined efforts of many scientists. Others were discovered by a single ingenious scientist, though again, often via many complex detours, and only afterward elucidated by others in the most different ways, such as Maxwell's theory of electricity and magnetism, just mentioned. Now there is one mode of representation which possesses quite special advantages, though it also has its weak sides. This mode of representation is characterized by the fact that—mindful of our task of constructing merely inner, conceptual, or imaginary pictures—at first we operate only with mental abstractions. In so doing we do not yet take any possible experiential facts into consideration. We merely make the effort to develop our mental pictures with as much clarity as possible and to draw all possible consequences from them. Only subsequently, after the entire exposition of the model has been completed, do we check its agreement with experiential facts. Thus we justify only afterward why the model had to be chosen in exactly this and in no other way; at the beginning we do not give the slightest indication concerning this. We shall call this procedure the method of deductive representation.

The advantages of this method are obvious. In the first place, no doubt can arise that it intends to present not the things themselves, but only an inner mental picture, and that it attempts only to make this mental picture a skillful representation of the phenomena. Since the deductive method does not constantly blend external experiences which are forced upon us with inner pictures that are arbitrarily chosen by us, it is by far easier for it to develop the latter in a clear and distinct way, free from contradictions. For it is one of the most important requirements of these models that they be perfectly clear, that they never perplex us as to how they should be formed in any particular case, and that each time we are able to deduce the results from them in an unambiguous and indubitable way. Just such clarity suffers when we too hastily blend the model with experience; but it is preserved most securely by the deductive method of representation. On the other hand, what stands out sharply, especially in this way of representing, is the arbitrariness of the pictures, since one begins here with quite arbitrary thought-constructions whose necessity is not motivated at the very outset but is justified only afterward. Not one ounce of proof is offered against the possibility of inventing other thought-pictures which would equally agree with experience. This seems to be a shortcoming, but might nevertheless very well be an advantage, at least for one who defends the view on the nature of theory which was explained previously. However, the fact that the way by which the model in question was arrived at does not become manifest is a genuine defect of the deductive method. But, of course, it is usually the case in the realm of epistemology that the coherence of the conclusions stands out most clearly when they are explained as much as possible in their natural order and regardless of the often winding path by which they were discovered.

Hertz has given us a paradigm example of such a purely deductive representation in the field of mechanics, specifically in the book already cited. I believe that I may assume here that you are familiar with the content of Hertz's book and will therefore limit myself to a brief characterization of his view. Hertz starts with material points which he regards as pure thought-pictures. Mass, too, is defined entirely independently of all experience, by means of a number which we must conceive of as associated with each material point, namely, the number of the simple mass points it contains. From these abstract concepts he constructs a motion which, like the points themselves, is at first of course present only in the mind. From all of this the concept of force is completely absent. Taking its place are certain conditions which are formulated as equations between the differentials of the coordinates of the material points. Now, these material points are fitted with certain initial velocities and at all subsequent times they move according to a very simple law which, as soon as the conditioning equations are given, unambiguously determines their motion for all times. Hertz expresses this as follows: for every moment of time the sum of the masses multiplied by the squares of the deviations of the material points from a straight, uniform motion must be a minimum; or more briefly still: motion must occur along the straightest possible path. This law has a very great resemblance to Gauss's principle of least constraint. Indeed, it is, so to speak, that special case which occurs when Gauss's principle is applied to a system of points which, while subjected to a constraint, are not subjected to any other external forces.

In my book, entitled *Vorlesungen über die Principe der Mechanik* [Lectures on the Principles of Mechanics], I, too, have attempted to develop a merely deductive representation of the fundamental principles of mechanics, but in quite a different way and much more closely connected with the usual treatment of mechanics. Like Hertz, I begin with pure thought-objects—exact material points. I relate their position to a rectangular coordinate system that is also a mere thought-object, and I imagine then a mental picture of the motion which, at first, is constructed in the following way. Every time that two material points are at a distance r from each other, each of them is to experience an acceleration in the direction of r, which is a function $f(r)$ of this distance; later, this function can be disposed of at will. Furthermore, the accelerations of both points must have a numerical relationship that remains unchanged at all times and which defines the relation between the masses of the two material points. How we are to imagine the motion of all material points is then unambiguously determined by the specification that the actual acceleration of each point is the vector sum of all accelerations found for it by means of the previous rule, which can be added to the velocity of the point which is already given, in the same way as vector quantities are to be added. Where these accelerations come from, and just why I give the instruction that the model should be constructed in this way, is not further discussed. It suffices that the model is a perfectly clear one which, by means of calculations, can be worked out in detail for a sufficient number of cases. It finds its justification only in the fact that the function $f(r)$ can in all cases be determined such that the thought motion of the imagined material point becomes a faithful copy of actual phenomena.

By means of this method, which we have called the purely deductive one, we have of course not solved the questions concerning the nature of matter, mass, and force; however, we have evaded these questions by making it superfluous that one begins with them. In our model these concepts are fully determined numbers and instructions for geometric constructions about which we know how we are to think and perform them in order that we may obtain a useful model of the world of phenomena. What the real cause for the fact that the world of phenomena runs its course in just this way may be, and what might be hidden behind the world of phenomena, propelling it as it were—to examine such questions we do not consider to be the task of natural science. We may leave it here completely out of account whether it is or could be the task of another science or whether, in analogy with other word combinations, we have here merely concatenated words which in these combinations do not express a clear thought.

By means of this deductive method we have just as little solved the problem of absolute space and of absolute motion; yet this question, too, no longer creates any pedagogical difficulties. We need no longer raise it at the beginning of our development of the laws of mechanics, but can discuss it only after we have deduced all its laws. For, since initially we introduce only thought-constructions, a coordinate system that is merely thought, too, does not at all cut a strange figure among them. It is just one of the different understandable and familiar means of construction with which we compose our thought-pictures. It is no more and no less abstract than the material points whose motion we represent relative to our coordinate system and for which alone we first formulate laws and give them a mathematical form. By checking with experience, we then find that a coordinate system which is invariably associated with the fixed stars is in practice perfectly adequate to insure agreement with experience. What kind of coordinate system we will one day have to take as our basis, once we are able to express the motion of the fixed stars by means of mechanical formulas, that question stands at the bottom of the list in our repertory. And now that we have all the laws of mechanics at our disposal we can also easily discuss all the hypotheses of Streintz, Mach, Lange, and so on, which were mentioned at the outset. We are no longer embarrassed as before, when we had to presuppose those complicated considerations concerning the development of the law of inertia. To be sure, in the deductive method we must again deliver a proof for this law, a proof which in earlier methods was superfluous. Since in the latter we took our starting point immediately from the phenomena, it was obvious that in that theory the laws governing the phenomena did not depend upon the choice of the system of coordinates which was merely added by thought; and the fact that these laws receive a different and much more complicated form when one introduces a rotating frame of reference even ought to have been striking in that view. In the deductive method, however, we have in our model, from the very outset, attributed the same function to the system of coordinates as to the material points. It is an integral part of a model, and one should not be surprised that this will be different in the event that one chooses a different framework. On the contrary, here we must derive from the model itself the argument that this model will not change when we introduce an arbitrary, different system of coordinates, as long at least as these do not rotate relative to each other and do not move with acceleration in regard to each other.

We will now compare the manner of representation as found in my book which was mentioned previously, with that proposed by Hertz. Mr. Classen has interpreted my explanation as a kind of polemic against Hertz and pictured the situation in such a way that it looks as if I flatter myself with the belief of having proposed something which is positively better than Hertz has done. Nothing is more untrue than this. Without any reservation I recognize the advantages of Hertz's model; nevertheless, according to the principle that it is possible and desirable to develop more models for one and the same class of phenomena, I do believe that my model still has its specific meaning in addition to Hertz's insofar as my model possesses certain advantages which are lacking in Hertz's.

The principles of mechanics which Hertz has formulated are of an extraordinary simplicity and beauty. Obviously they are not completely free of abitrariness, but I may say that the arbitrariness is reduced to a minimum. The model which Hertz has constructed independently from experience has a certain inner perfection and evidence. In itself it contains only very few arbitrary elements. In this regard my model obviously falls short by far. The latter contains many more traits which bear the stamp of not being determined by an inner necessity, but of being added merely in order, later on, to make the agreement with experience possible. It also contains a completely arbitrary function, and of the many pictures which result when this function has been given all its possible forms, there are only very few which correspond to real events. Whereas in Hertz's model one sees immediately that, if any at all, then there are only very few other pictures possible (which in that case participate in a similar simplicity and inner perfection), my model immediately gives the impression that there still might be many others which will represent the phenomena with equal perfection. This not withstanding there are however other points in which my model is superior to the one proposed by Hertz. True, from his model Hertz is able to explain some phenomena in a direct way; or as we prefer to formulate it, he is able to represent them by means of that model—for example, the motion of a material point in a specified plane or on a similar curve, or the rotation of a rigid body around a fixed point, both, that is, only as long as there are no unknown external forces involved. However, one runs into difficulties as soon as one wants to represent the most common events occurring in our everyday experience in which there are forces at work. Let us first consider one of the most universal and most important forces of nature, gravitation.

From Hertz's point of view we are evidently not allowed to conceive of this force as one working at a distance. True, many attempts have been made to explain gravitation mechanically by means of the action of a medium. However, it is known also that none of these endeavors has led to a well determined and decisive result. One of the best known attempts is the theory of molecular collisions, first proposed by Lesage and later taken up again by Lord Kelvin, Isenkrahe, and others. Notwithstanding the fact that its exact implementation is still doubtful, for Hertz's theory, it is of no use because already the explanation of one single elastic collision implies the same difficulties, as we will see shortly. One were thus to first develop a wholly new theory in order to explain the action of the gravitational force, for instance, through vortices, pulsations, or something of that sort, in which the particles of the respective medium similarly cannot be connected by forces of the form proposed by Hertz. Even if one were to succeed in so doing, even then this would mean only resorting to an absolutely arbitrary model for which in the course of time in all probability another would have to be substituted. The objection which Hertz raises against the old mechanics, namely, that it delivers a model which is much too broad insofar as from all possible functions $f(r)$, which represent the relevant forces, only very few can have a practical application, can be raised even in a stronger form against his own model as well, as soon as one tries to apply it to concrete cases. Already in the application to gravitation one has to choose a determinate medium from all the possible ones which allow for an action at a distance. But in so doing there is found more vagueness and arbitrariness than in the choice of certain functions $f(r)$.

In his earlier works Maxwell has, as you know, explained the electrical and magnetic forces by means of the action of a medium. Yet, in spite of the fact that this medium had a highly complicated structure and abounded in properties which bore the stamp of arbitrariness and of being merely provisional in character, the medium would again in no way be useful to Hertz, because its parts likewise are kept together by forces in the sense of the older mechanics. And what is more, the properties of the elastic, liquid, and gaseous bodies were to be replaced by new pictures, too, since the former without exception are founded upon the acceptance of forces which are supposed to be at work between the particles. That is why one has the following choice only: either one leaves the nature of the mechanism which has to bring about gravitation, the electrical and magnetic phenomena, and the like, undetermined and arbitrary (but this leads to an unacceptable lack of clarity inasmuch as one is obliged to operate with equations of which only some very general properties are known, but whose specific form is completely unknown); or one endeavors to choose a determinate mechanism through which one again becomes involved in as many other arbitrary assumptions and difficulties.

However, I still want to show, with the help of a much simpler example, the difficulties which the application of Hertz's fundamental law runs into even in the most trivial case.

Let it be given that there are three masses, m_1, u, m_2, for which the condition holds that the distance m_1 u as well as the distance u m_2 will always be equal to the same quantity a. If we we now let the mass u get smaller and smaller, then we get a case which is in complete accordance with the spirit of Hertz's mechanics and which gives us an accurate description of the following natural phenomena. Suppose that in an elastic hollow sphere of mass m_2 a small elastic solid ball is moving and that the difference between their radiuses be equal to $2a$. Then we have here an example of one and the same natural phenomenon which can be explained in two completely different ways: on the one hand through the molecular theory, on the other according to the method indicated by Hertz. However, not all phenomena are constituted that way. Already the very trivial case of the collision of two elastic solid balls can be derived from Hertz's schema only by means of rather arbitrarily chosen mechanisms or complicated suppositions in regard to the medium between them, since indeed Hertz's model excludes inequalities. Thus, even in the simplest case does Hertz's method lead to the greatest complications.

I emphasize again that these explanations in no way are meant to deny the great value of Hertz's model, which, as you know, consists in the logical simplicity of its fundamental principles. For it is quite possible that in the distant future one will be able to explain all actions by means of media whose properties will not be chosen in a fantastic way, but will be offered by the nature of the thing itself in an obvious and unambiguous way. It is not impossible that the particles of these media would exert forces upon each other which are different from those of classical mechanics, and that one would be able to manage things with the help of conditioning equations between the coordinates of the elementary particles in the sense that Hertz proposes. From that moment onward the mechanics of Hertz would undoubtedly have gained the victory, and all other explanations would have a merely historical meaning. Whether or not one considers the future appearance of such a moment of time probable and likely is evidently a matter of taste. The very possibility of such a development of our knowledge has not even been shown. That is why from our present standpoint we will look up to that ideal with admiration, but also do our part for the advancement of our approximation of that ideal. Meanwhile, however, we will not be able to dispense with such simple and immediately useful models which at this moment can be worked out in further detail, in addition to Hertz's model.

Part III
THE FIRST DECADE OF THE TWENTIETH CENTURY

Chapter 13
HENRI JULES POINCARÉ
(1854–1912)

INTRODUCTION

*H*ENRI JULES POINCARÉ, recognized as one of the greatest mathematicians of his era, was born in Nancy, France, April 29, 1854. He studied first at the École Polytechnique before entering the Écoles des Mines to prepare for a career in mining engineering. In the years which followed, however, he became more interested in pure mathematics, and in 1879 accepted an appointment in mathematics at the University of Caen. Two years later he was called to the University of Paris where he lectured for many years on pure and applied mathematics, physical mechanics, astronomical mechanics, and mathematical physics. Poincaré wrote thirty books and some five hundred essays in these areas.

Poincaré's first great achievement came in pure mathematics, where he generalized the idea of functional periodicity in his famous theory of automorphic functions, which he called Fuchsian functions in honor of the German mathematician I. L. Fuchs (1833–1902). In the field of celestial mechanics he made substantial contributions to the theory of orbits, in particular as related to the three-body problem. He developed important new mathematical tools during these investigations. The results of his research on new mathematical methods in astronomy were systematized in the work *Les Méthodes nouvelles de la mécanique céleste,* which appeared in three volumes between 1892 and 1899. Poincaré also made valuable contributions to the theory of the figures of equilibrium of rotating fluid masses, but his most important contribution to mathematical physics was his paper on the dynamics of the electron (1906) in which, independently of Einstein, he derived many of the results of the special theory of relativity.

In addition to these contributions Poincaré developed many ideas of importance for philosophy of the mathematical and natural sciences. He was a forerunner of the Brouwer intuitionist school in philosophy of mathematics, having made a substantial contribution to this discipline in his analysis of the psychology of mathematical discovery and invention. In philosophy of the natural sciences Poincaré placed great emphasis on the role of convention in the development of physical theories. The results of his investigations in this area were published in his famous books *La science et l'hypothèse* (1902), *La valeur de la science* (1905), and *Science et méthode* (1909). Poincaré was elected to the Académie Française in 1908. Four years after receiving this distinction he died, in Paris (July 17, 1912).

Poincaré's position in the realm of philosophy of natural science can be perhaps most accurately indicated by the term "conventionalism." This term, which indeed adequately characterizes his position as influenced by Kant and Mach, is however subject to misunderstanding in that it may suggest insights incompatible with Poincaré's view.

For Poincaré natural science is essentially inductive in its basic dependence upon generalizations from observed concrete facts. As such then, it must presuppose a general order in the universe independent from man. In Poincaré's view this presupposition is no more than a kind of belief, in the Humean sense of the term, which in turn is why natural science in principle lacks the certainty so characteristic of the mathematical science (*Science and Hypothesis* p. 13).

According to Poincaré, the method of natural science consists in observation and experiment. Because one cannot observe all facts one must make a selection, and this calls for some criterion of selectivity. Scientifically the most valuable facts are those which have an undoubted chance of recurring, and such facts are clearly the simplest ones. What is supposed to be a simple fact depends in its determination to a large extent upon our familiarity with such facts: what appears regularly is simple for us because we are accustomed to it. For the different disciplines of natural science the simple is found in various directions; for physics and chemistry an atom is a simple thing; for an astronomer a celestial body is a simple body (*Science and Method*, pp. 15–19).

Once man has found regularities in nature, Poincaré maintains, he becomes immediately fascinated by the seeming exceptions, because it is the exceptions which are puzzling, demand further investigations and explanation, and are most instructive. When a scientist comes upon an irregularity he attempts to find a rule for it. Science, however, not only enumerates regularities and differences, but also shows how similarities underlie the differences (*Ib.*, pp. 21–23).

Poincaré came to the conclusion that the aim of science was generality, the discovery of laws governing as many diverse facts as possible. To accomplish this man uses ordering principles which to a very high degree are conventional and not determined by either the nature of things or the laws of thought. To illustrate this point he compared geometry with physics. Concerning the former it is clear from non-Euclidean geometries that there is a certain convention at the origin of our choice of the fundamental axioms and postulates. By saying that the axioms of geometry are conventional, however, we do not necessarily say that they are arbitrary, since there are many good reasons for choosing one set of axioms rather than another. But there is no possibility of giving an absolute privilege to one of these geometries on the basis of either experience or laws of thought (*Science and Hypothesis,* pp. 79 and 89).

Poincaré defended an analogous view in regard to natural science. Here, too, he found an important conventional element. It is this view which constitutes the main topic of the selection which follows. The material presented is the last part of a critical appraisal of LeRoy's point of view on the subject. Poincaré agrees with LeRoy that science is conventional, but definitely rejects the nominalism and anti-intellectualism which LeRoy connects with the view. LeRoy held that science is the artificial work of the scientist and can teach us nothing of the truth; he believed further that our intellect, particularly in its discourse, deforms all it touches. Poincaré attempts to maintain his conventionalism while rejecting these skeptical consequences.

SELECTIVE BIBLIOGRAPHY

Poincaré, H. *Science and Hypothesis* (1902), trans. W. J. Greenstreet. New York, 1952.

———, *The Value of Science* (1905), trans. Bruce Halsted. New York, 1958.

———, *Science and Method* (1909), trans. F. Maitland. New York, 1958.

———, *Dernières Pensées* (1912). Paris, 1912.

Frank, P., *Modern Science and Its Philosophy.* New York, 1961. Pp. 13–61, and *passim.*

Alexander, P., "The Philosophy of Science, 1850–1910," in *A Critical History of Western Philosophy*, ed. D. J. O'Connor. New York, 1965. Pp. 413–417.

Albèrgamo, F., *Storia della Logica delle Scienze Empiriche*. Bari, 1952. Pp. 245–255.

Boutroux, É., *Nouvelles études d'histoire de la philosophie*. Paris, 1927.

Slosson, E. E., *Major Prophets of Today*. Boston, 1914.

Dantzig, T., *Henri Poincaré, Critic of Crisis. Reflections on his Universe of Discourse*. New York, 1954.

Poirier, R., "Henri Poincaré et le problème de la valeur de la science," *Rev. Phil. France Etrang.*, **79**(1954) 485–513.

Cecchine, A., *Il Concetto di Convenzione Matematica in Henri Poincaré*. Turin, 1951.

Volterra, V., "Henri Poincaré," *Book of the Opening of the Rice Institute*. Houston, 3(1912) 899–928.

Hadamard, J. S., *The Early Scientific Work of Henri Poincaré*. Houston, 1922 (*Rice Institute Pamphlet 9, no. 3*).

————, *The Later Scientific Work of Henri Poincaré*. Houston, 1933 (*Rice Institute Pamphlet 20, no. 1*).

Meyerson, E., *Identity and Reality*. New York, 1962 (*passim*).

SCIENCE AND REALITY

CONTINGENCE AND DETERMINISM

I DO not intend to treat here the question of the contingence of the laws of nature, which is evidently insoluble, and on which so much has already been written. I only wish to call attention to what different meanings have been given to this word, contingence, and how advantageous it would be to distinguish them.

If we look at any particular law, we may be certain in advance that it can only be approximative. It is, in fact, deduced from experimental verifications, and these verifications were and could be only approximate. We should always expect that more precise measurements will oblige us to add new terms to our formulas; this is what has happened, for instance, in the case of Mariotte's law.

Moreover the statement of any law is necessarily incomplete. This enunciation should compromise the enumeration of *all* the antecedents in virtue of which a given consequent can happen. I should first describe *all* the conditions of the experiment to be made and the law would then be stated: If all the conditions are fulfilled, the phenomenon will happen.

But we shall be sure of not having forgotten *any* of these conditions only when we shall have described the state of the entire universe at the instant t; all the parts of this universe may, in fact, exercise an influence more or less great on the phenomenon which must happen at the instant $t + dt$.

Now it is clear that such a description could not be found in the enunciation of the law; besides, if it were made, the law would become incapable of application; if one required so many conditions, there would be very little chance of their ever being all realized at any moment.

Then as one can never be certain of not having forgotten some

This article is from **The Value of Science** by Henri Poincaré, trans. Bruce Halsted (New York, 1958), pp. 129–142. Reprinted by permission of Dover Publications, Inc.

essential condition, it can not be said: If such and such conditions are realized, such a phenomenon will occur; it can only be said: If such and such conditions are realized, it is probable that such a phenomenon will occur, very nearly.

Take the law of gravitation, which is the least imperfect of all known laws. It enables us to foresee the motions of the planets. When I use it, for instance, to calculate the orbit of Saturn, I neglect the action of the stars, and in doing so, I am certain of not deceiving myself, because I know that these stars are too far away for their action to be sensible.

I announce, then, with a quasi-certitude that the coordinates of Saturn at such an hour will be comprised between such and such limits. Yet is that certitude absolute? Could there not exist in the universe some gigantic mass, much greater than that of all the known stars and whose action could make itself felt at great distances? That mass might be animated by a colossal velocity, and after having circulated from all time at such distances that its influence had remained hitherto insensible to us, it might come all at once to pass near us. Surely it would produce in our solar system enormous perturbations that we could not have foreseen. All that can be said is that such an event is wholly improbable, and then, instead of saying: Saturn will be near such a point of the heavens, we must limit ourselves to saying: Saturn will probably be near such a point of the heavens. Although this probability may be practically equivalent to certainty, it is only a probability.

For all these reasons, no particular law will ever be more than approximate and probable. Scientists have never failed to recognize this truth; only they believe, right or wrong, that every law may be replaced by another closer and more probable, that this new law will itself be only provisional, but that the same movement can continue indefinitely, so that science in progressing will possess laws more and more probable, that the approximation will end by differing as little as you choose from exactitude and the probability from certitude.

If the scientists who think thus were right, must it still be said that *the* laws of nature are contingent, even though *each* law, taken in particular, may be qualified as contingent? Or must one require, before concluding the contingence *of the* natural laws, that this progress have an end, that the scientist finish some day by being arrested in his search for a closer and closer approximation and that, beyond a certain limit, he thereafter meet in nature only caprice?

In the conception of which I have just spoken (and which I shall call the scientific conception), every law is only a statement, imperfect and provisional, but it must one day be replaced by another, a superior law, of which it is only a crude image. No place therefore remains for the intervention of a free will.

It seems to me that the kinetic theory of gases will frunish us a striking example.

You know that in this theory all the properties of gases are explained by a simple hypothesis; it is supposed that all the gaseous molecules move in every direction with great velocities and that they follow rectilineal paths which are disturbed only when one molecule passes very near the sides of the vessel or another molecule. The effects our crude senses enable us to observe are the mean effects, and in these means, the great deviations compensate, or at least it is very improbable that they do not compensate; so that the observable phenomena follow simple laws such as that of Mariotte or of Gay-Lussac. But this compensation of deviations is only probable. The molecules incessantly change place and in these continual displacements the figures they form pass successively through all possible combinations. Singly these combinations are very numerous; almost all are in conformity with Mariotte's law, only a few deviate from it. These also will happen, only it would be necessary to wait a long time for them. If a gas were observed during a sufficiently long time, it would certainly be finally seen to deviate, for a very short time, from Mariotte's law. How long would it be necessary to wait? If it were desired to calculate the probable number of years, it would be found that this number is so great that to write only the number of places of figures employed would still require half a score places of figures. No matter; enough that it may be done.

I do not care to discuss here the value of this theory. It is evident that if it be adopted, Mariotte's law will thereafter appear only as contingent, since a day will come when it will not be true. And yet, think you the partisans of the kinetic theory are adversaries of determinism? Far from it; they are the most ultra of mechanists. Their molecules follow rigid paths, from which they depart only under the influence of forces which vary with the distance, following a perfectly determinate law. There remains in their system not the smallest place either for freedom, or for an evolutionary factor, properly so-called, or for anything whatever that could be called contingence. I add, to avoid mistake, that neither is there any evolution of Mariotte's law itself; it ceases to be true after I know not how many centuries;

but at the end of a fraction of a second it again becomes true and that for an incalculable number of centuries.

And since I have pronounced the word evolution, let us clear away another mistake. It is often said: Who knows whether the laws do not evolve and whether we shall not one day discover that they were not at the Carboniferous epoch what they are to-day? What are we to understand by that? What we think about the past state of our globe, we deduce from its present state. And how is this deduction made? It is by means of laws supposed known. The law being a relation between the antecedent and the consequent, enables us equally well to deduce the consequent from the antecedent, that is, to foresee the future, and to deduce the antecedent from the consequent, that is, to conclude from the present to the past. The astronomer who knows the present situation of the stars can from it deduce their future situation by Newton's law, and this is what he does when he constructs ephemerides; and he can equally deduce from it their past situation. The calculations he thus can make can not teach him that Newton's law will cease to be true in the future, since this law is precisely his point of departure; not more can they tell him it was not true in the past. Still in what concerns the future, his ephemerides can one day be tested and our descendants will perhaps recognize that they were false. But in what concerns the past, the geologic past which had not witnesses, the results of his calculation, like those of all speculations where we seek to deduce the past from the present, escape by their very nature every species of test. So that if the laws of nature were not the same in the Carboniferous age as at the present epoch, we shall never be able to know it, since we can know nothing of this age only what we deduce from the hypothesis of the permanence of these laws.

Perhaps it will be said that this hypothesis might lead to contradictory results and that we shall be obliged to abandon it. Thus, in what concerns the origin of life, we may conclude that there have always been living beings, since the present world shows us always life springing from life; and we may also conclude that there have not always been, since the application of the existent laws of physics to the present state of our globe teaches us that there was a time when this globe was so warm that life on it was impossible. But contradictions of this sort can always be removed in two ways; it may be supposed that the actual laws of nature are not exactly what we have assumed; or else it may be supposed that the laws of nature actually are what we have assumed, but that it has not always been so.

It is evident that the actual laws will never be sufficiently well known for us not to be able to adopt the first of these two solutions and for us to be constrained to infer the evolution of natural laws.

On the other hand, suppose such an evolution; assume, if you wish, that humanity lasts sufficiently long for this evolution to have witnesses. The *same* antecedent shall produce, for instance, different consequents at the Carboniferous epoch and at the Quaternary. That evidently means that the antecedents are closely alike; if all the circumstances were identical, the Carboniferous epoch would be indistinguishable from the Quaternary. Evidently this is not what is supposed. What remains is that such antecedent, accompanied by such accessory circumstance, produces such consequent; and that the same antecedent, accompanied by such other accessory circumstance, produces such other consequent. Time does not enter into the affair.

The law, such as ill-informed science would have stated it, and which would have affirmed that this antecedent always produces this consequent, without taking account of the accessory circumstances, this law, which was only approximate and probable, must be replaced by another law more approximate and more probable, which brings in these accessory circumstances. We always come back, therefore, to that same process which we have analyzed above, and if humanity should discover something of this sort, it would not say that it is the laws which have evoluted, but the circumstances which have changed.

Here, therefore, are several different senses of the word contingence. M. LeRoy retains them all and he does not sufficiently distinguish them, but he introduces a new one. Experimental laws are only approximate, and if some appear to us as exact, it is because we have artificially transformed them into what I have above called a principle. We have made this transformation freely, and as the caprice which has determined us to make it is something eminently contingent, we have communicated this contingence to the law itself. It is in this sense that we have the right to say that determinism supposes freedom, since it is freely that we become determinists. Perhaps it will be found that this is to give large scope to nominalism and that the introduction of this new sense of the word contingence will not help much to solve all those questions which naturally arise and of which we have just been speaking.

I do not at all wish to investigate here the foundations of the principle of induction; I know very well that I shall not succed; it is as difficult to justify this principle as to get on without it. I only wish to show how scientists apply it and are forced to apply it.

When the same antecedent recurs, the same consequent must likewise recur; such is the ordinary statement. But reduced to these terms this principle could be of no use. For one to be able to say that the same antecedent recurred, it would be necessary for the circumstances *all* to be reproduced, since no one is absolutely indifferent, and for them to be *exactly* reproduced. And, as that will never happen, the principle can have no application.

We should therefore modify the enunciation and say: If an antecedent A has once produced a consequent B, an antecedent A', slightly different from A, will produce a consequent B', slightly different from B. But how shall we recognize that the antecedents A and A' are "slightly different"? If some one of the circumstances can be expressed by a number, and this number has in the two cases values very near together, the sense of the phrase "slightly different" is relatively clear; the principle then signifies that the consequent is a continuous function of the antecedent. And as a practical rule, we reach this conclusion that we have the right to interpolate. This is in fact what scientists do every day, and without interpolation all science would be impossible.

Yet observe one thing. The law sought may be represented by a curve. Experiment has taught us certain points of this curve. In virtue of the principle we have just stated, we believe these points may be connected by a continuous graph. We trace this graph with the eye. New experiments will furnish us new points of the curve. If these points are outside of the graph traced in advance, we shall have to modify our curve, but not to abandon our principle. Through any points, however numerous they may be, a continuous curve may always be passed. Doubtless, if this curve is too capricious, we shall be shocked (and we shall even suspect errors of experiment), but the principle will not be directly put at fault.

Furthermore, among the circumstances of a phenomenon, there are some that we regard as negligible, and we shall consider A and A' as slightly different if they differ only by these accessory circumstances. For instance, I have ascertained that hydrogen unites with oxygen under the influence of the electric spark, and I am certain that these two gases will unite anew, although the longitude of Jupiter may have changed considerably in the interval. We assume, for instance, that the state of distant bodies can have no sensible influence on terrestrial phenomena, and that seems in fact requisite, but there are cases where the choice of these practically indifferent circumstances admits of more arbitrariness or, if you choose, requires more tact.

One more remark: The principle of induction would be inapplicable if there did not exist in nature a great quantity of bodies like one another, or almost alike, and if we could not infer, for instance, from one bit of phosphorus to another bit of phosphorus.

If we reflect on these considerations, the problem of determinism and of contingence will appear to us in a new light.

Suppose we were able to embrace the series of all phenomena of the universe in the whole sequence of time. We could envisage what might be called the *sequences*, I mean relations between antecedent and consequent. I do not wish to speak of constant relations or laws, I envisage separately (individually, so to speak) the different sequences realized.

We should then recognize that among these sequences there are no two altogether alike. But, if the principle of induction, as we have just stated it, is true, there will be those almost alike and that can be classed alongside one another. In other words, it is possible to make a classification of sequences.

It is to the possibility and the legitimacy of such a classification that determinism, in the end, reduces. This is all that the preceding analysis leaves of it. Perhaps under this modest form it will seem less appealing to the moralist.

It will doubtless be said that this is to come back by a detour to M. LeRoy's conclusion which a moment ago we seemed to reject: we are determinists voluntarily. And in fact all classification supposes the active intervention of the classifier. I agree that this may be maintained, but it seems to me that this detour will not have been useless and will have contributed to enlighten us a little.

OBJECTIVITY OF SCIENCE

I arrive at the question set by the title of this article: What is the objective value of science? And first what should we understand by objectivity?

What guarantees the objectivity of the world in which we live is that this world is common to us with other thinking beings. Through the communications that we have with other men, we receive from them ready-made reasonings; we know that these reasonings do not come from us and at the same time we recognize in them the work of reasonable beings like ourselves. And as these reasonings appear to fit the world of our sensations, we think we may infer that these reasonable beings have seen the same thing as we; thus it is we know we have not been dreaming.

Such, therefore, is the first condition of objectivity; what is objective must be common to many minds and consequently transmissible from one to the other, and as this transmission can only come about by that "discourse" which inspires so much distrust in M. LeRoy, we are even forced to conclude: no discourse, no objectivity.

The sensations of others will be for us a world eternally closed. We have no means of verifying that the sensation I call red is the same as that which my neighbor calls red.

Suppose that a cherry and a red poppy produce on me the sensation A and on him the sensation B and that, on the contrary, a leaf produces on me the sensation B and on him the sensation A. It is clear we shall never know anything about it; since I shall call red the sensation A and green the sensation B, while he will call the first green and the second red. In compensation, what we shall be able to ascertain is that, for him as for me, the cherry and the red poppy produce the *same* sensation, since he gives the same name to the sensations he feels and I do the same.

Sensations are therefore intransmissible, or rather all that is pure quality in them is intransmissible and forever impenetrable. But it is not the same with relations between these sensations.

From this point of view, all that is objective is devoid of all quality and is only pure relation. Certes, I shall not go so far as to say that objectivity is only pure quantity (this would be to particularize too far the nature of the relations in question), but we understand how some one could have been carried away into saying that the world is only a differential equation.

With due reserve regarding this paradoxical proposition, we must nevertheless admit that nothing is objective which is not transmissible, and consequently that the relations between the sensations can alone have an objective value.

Perhaps it will be said that the esthetic emotion, which is common to all mankind, is proof that the qualities of our sensations are also the same for all men and hence are objective. But if we think about this, we shall see that the proof is not complete; what is proved is that this emotion is aroused in John as in James by the sensations to which James and John give the same name or by the corresponding combinations of these sensations; either because this emotion is associated in

John with the sensation A, which John calls red, while parallelly it is associated in James with the sensation B, which James calls red; or better because this emotion is aroused, not by the qualities themselves of the sensations, but by the harmonious combination of their relations of which we undergo the unconscious impression.

Such a sensation is beautiful, not because it possesses such a quality, but because it occupies such a place in the woof of our associations of ideas, so that it can not be excited without putting in motion the 'receiver' which is at the other end of the thread and which corresponds to the artistic emotion.

Whether we take the moral, the esthetic or the scientific point of view, it is always the same thing. Nothing is objective except what is identical for all; now we can only speak of such an identity if a comparison is possible, and can be translated into a 'money of exchange' capable of transmission from one mind to another. Nothing, therefore, will have objective value except what is transmissible by 'discourse,' that is, intelligible.

But this is only one side of the question. An absolutely disordered aggregate could not have objective value since it would be unintelligible, but no more can a well-ordered assemblage have it, if it does not correspond to sensations really experienced. It seems to me superfluous to recall this condition, and I should not have dreamed of it, if it had not lately been maintained that physics is not an experimental science. Although this opinion has no chance of being adopted either by physicists or by philosophers, it is well to be warned so as not to let oneself slip over the declivity which would lead thither. Two conditions are therefore to be fulfilled, and if the first separates reality[1] from the dream, the second distinguishes it from the romance.

Now what is science? I have explained in the preceding article, it is before all a classification, a manner of bringing together facts which appearances separate, though they were bound together by some natural and hidden kinship. Science, in other words, is a system of relations. Now we have just said, it is in the relations alone that objectivity must be sought; it would be vain to seek it in beings considered as isolated from one another.

To say that science can not have objective value since it teaches us only relations, this is to reason backwards, since, precisely, it is relations alone which can be regarded as objective.

---

[1] I here use the word real as a synonym of objective; I thus conform to common usage; perhaps I am wrong, our dreams are real, but they are not objective.

External objects, for instance, for which the word *object* was invented, are really *objects* and not fleeting and fugitive appearances, because they are not only groups of sensations, but groups cemented by a constant bond. It is this bond, and this bond alone, which is the object in itself, and this bond is a relation.

Therefore, when we ask what is the objective value of science, that does not mean: Does science teach us the true nature of things? but it means: Does it teach us the true relations of things?

To the first question, no one would hesitate to reply, no; but I think we may go farther; not only science can not teach us the nature of things; but nothing is capable of teaching it to us and if any god knew it, he could not find words to express it. Not only can we not divine the response, but if it were given to us, we could understand nothing of it; I ask myself even whether we really understand the question.

When, therefore, a scientific theory pretends to teach us what heat is, or what is electricity, or life, it is condemned beforehand; all it can give us is only a crude image. It is, therefore, provisional and crumbling.

The first question being out of reason, the second remains. Can science teach us the true relations of things? What it joins together should that be put asunder, what it puts asunder should that be joined together?

To understand the meaning of this new question, it is needful to refer to what was said above on the conditions of objectivity. Have these relations an objective value? That means: Are these relations the same for all? Will they still be the same for those who shall come after us?

It is clear that they are not the same for the scientist and the ignorant person. But that is unimportant, because if the ignorant person does not see them all at once, the scientist may succeed in making him see them by a series of experiments and reasonings. The thing essential is that there are points on which all those acquainted with the experiments made can reach accord.

The question is to know whether this accord will be durable and whether it will persist for our successors. It may be asked whether the unions that the science of to-day makes will be confirmed by the science of to-morrow. To affirm that it will be so we can not invoke any *à priori* reason; but this is a question of fact, and science has already lived long enough for us to be able to find out by asking its history whether the edifices it builds stand the test of time, or whether they are only ephemeral constructions.

Now what do we see? At the first blush it seems to us that the theories last only a day and that ruins upon ruins accumulate. To-day the theories are born, to-morrow they are the fashion, the day after to-morrow they are classic, the fourth day they are superannuated, and the fifth they are forgotten. But if we look more closely, we see that what thus succumb are the theories, properly so called, those which pretend to teach us what things are. But there is in them something which usually survives. If one of them has taught us a true relation, this relation is definitively acquired, and it will be found again under a new disguise in the other theories which will successively come to reign in place of the old.

Take only a single example: The theory of the undulations of the ether taught us that light is a motion; to-day fashion favors the electromagnetic theory which teaches us that light is a current. We do not consider whether we could reconcile them and say that light is a current, and that this current is a motion. As it is probable in any case that this motion would not be identical with that which the partisans of the old theory presume, we might think ourselves justified in saying that this old theory is dethroned. And yet something of it remains, since between the hypothetical currents which Maxwell supposes there are the same relations as between the hypothetical motions that Fresnel supposed. There is, therefore, something which remains over and this something is the essential. This it is which explains how we see the present physicists pass without any embarrassment from the language of Fresnel to that of Maxwell. Doubtless many connections that were believed well established have been abandoned, but the greatest number remain and it would seem must remain.

And for these, then, what is the measure of their objectivity? Well, it is precisely the same as for our belief in external objects. These latter are real in this, that the sensations they make us feel appear to us as united to each other by I know not what indestructible cement and not by the hazard of a day. In the same way science reveals to us between phenomena other bonds finer but not less solid; these are threads so slender that they long remained unperceived, but once noticed there remains no way of not seeing them; they are therefore not less real than those which give their reality to external objects; small matter that they are more recently known since neither can perish before the other.

It may be said, for instance, that the ether is no less real than any external body; to say this body exists is to say there is between the color of this body, its taste, its smell, an intimate bond, solid and persistent; to say the ether exists is to say there is a natural kinship between all the optical phenomena, and neither of the two propositions has less value than the other.

And the scientific syntheses have in a sense even more reality than those of the ordinary senses, since they embrace more terms and tend to absorb in them the partial syntheses.

It will be said that science is only a classification and that a classification can not be true, but convenient. But it is true that it is convenient, it is so not only for me, but for all men; it is true that it will remain convenient for our descendants; it is true finally that this can not be by chance.

In sum, the sole objective reality consists in the relations of things whence results the universal harmony. Doubtless these relations, this harmony, could not be conceived outside of a mind which conceives them. But they are nevertheless objective because they are, will become, or will remain, common to all thinking beings.

This will permit us to revert to the question of the rotation of the earth which will give us at the same time a chance to make clear what precedes by an example.

THE ROTATION OF THE EARTH

". . . . Therefore," have I said in *Science and Hypothesis*, "this affirmation, the earth turns round, has no meaning . . . or rather these two propositions, the earth turns round, and, it is more convenient to suppose that the earth turns round, have one and the same meaning."

These words have given rise to the strangest interpretations. Some have thought they saw in them the rehabilitation of Ptolemy's system, and perhaps the justification of Galileo's condemnation.

Those who had read attentively the whole volume could not, however, delude themselves. This truth, the earth turns round, was put on the same footing as Euclid's postulate, for example. Was that to reject it? But better; in the same language it may very well be said: These two propositions, the external world exists, or, it is more convenient to suppose that it exists, have one and the same meaning.

So the hypothesis of the rotation of the earth would have the same degree of certitude as the very existence of external objects.

But after what we have just explained in the fourth part, we may go farther. A physical theory, we have said, is by so much the more true, as it puts in evidence more true relations. In the light of this new principle, let us examine the question which occupies us.

No, there is no absolute space; these two contradictory propositions: 'The earth turns round' and 'The earth does not turn round' are, therefore, neither of them more true than the other. To affirm one while denying the other, *in the kinematic sense,* would be to admit the existence of absolute space.

But if the one reveals true relations that the other hides from us, we can nevertheless regard it as physically more true than the other, since it has a richer content. Now in this regard no doubt is possible.

Behold the apparent diurnal motion of the stars, and the diurnal motion of the other heavenly bodies, and besides, the flattening of the earth, the rotation of Foucault's pendulum, the gyration of cyclones, the trade-winds, what not else? For the Ptolemaist all these phenomena have no bond between them; for the Copernican they are produced by the one same cause. In saying, the earth turns round, I affirm that all these phenomena have an intimate relation, and *that is true,* and that remains true, although there is not and can not be absolute space.

So much for the rotation of the earth upon itself; what shall we say of its revolution around the sun? Here again, we have three phenomena which for the Ptolemaist are absolutely independent and which for the Copernican are referred back to the same origin; they are the apparent displacements of the planets on the celestial sphere, the aberration of the fixed stars, the parallax of these same stars. Is it by chance that all the planets admit an inequality whose period is a year, and that this period is precisely equal to that of aberration, precisely equal besides to that of parallax? To adopt Ptolemy's system is to answer, yes; to adopt that of Copernicus is to answer, no; this is to affirm that there is a bond between the three phenomena and that also is true although there is no absolute space.

In Ptolemy's system, the motions of the heavenly bodies can not be explained by the action of central forces, celestial mechanics is impossible. The intimate relations that celestial mechanics reveals to us between all the celestial phenomena are true relations; to affirm the immobility of the earth would be to deny these relations, which would be to fool ourselves.

The truth for which Galileo suffered remains, therefore, the truth, although it has not altogether the same meaning as for the vulgar, and its true meaning is much more subtile, more profound and more rich.

SCIENCE FOR ITS OWN SAKE

Not against M. LeRoy do I wish to defend science for its own sake; maybe this is what he condemns, but this is what he cultivates, since he loves and seeks truth and could not live without it. But I have some thoughts to express.

We can not know all facts and it is necessary to choose those which are worthy of being known. According to Tolstoi, scientists make this choice at random, instead of making it, which would be reasonable, with a view to practical applications. On the contrary, scientists think that certain facts are more interesting than others, because they complete an unfinished harmony, or because they make one foresee a great number of other facts. If they are wrong, if this hierarchy of facts that they implicitly postulate is only an idle illusion, there could be no science for its own sake, and consequently there could be no science. As for me, I believe they are right, and, for example, I have shown above what is the high value of astronomical facts, not because they are capable of practical applications, but because they are the most instructive of all.

It is only through science and art that civilization is of value. Some have wondered at the formula: science for its own sake; and yet it is as good as life for its own sake, if life is only misery; and even as happiness for its own sake, if we do not believe that all pleasures are of the same quality, if we do not wish to admit that the goal of civilization is to furnish alcohol to people who love to drink.

Every act should have an aim. We must suffer, we must work, we must pay for our place at the game, but this is for seeing's sake; or at the very least that others may one day see.

All that is not thought is pure nothingness; since we can think only thought and all the words we use to speak of things can express only thoughts, to say there is something other than thought, is therefore an affirmation which can have no meaning.

And yet—strange contradiction for those who believe in time— geologic history shows us that life is only a short episode between two eternities of death, and that, even in this episode, conscious thought has lasted and will last only a moment. Thought is only a gleam in the midst of a long night.

But it is this gleam which is everything.

Chapter 14
CHARLES SANDERS PEIRCE
(1839–1914)

INTRODUCTION

*C*HARLES SANDERS PEIRCE was born in Cambridge, Massachusetts, on September 10, 1839. His father, Benjamin Peirce, a mathematician, was largely responsible for his son's education prior to the latter's entering Harvard University in 1855 to undertake studies in physics and chemistry for which he received his degree in 1859. Peirce joined the United States Coast and Geodetic Survey in 1861 and with the exception of a few interventions remained in that position for some thirty years. During that time he lectured on logic at Johns Hopkins University and also gave some lectures on philosophy at Harvard and at Lowell Institute in Boston, never, however, obtaining a regular university appointment. Peirce did not enjoy much success as a teacher, partly because of his rigorous insistence upon clarity and precision with the difficult and highly personal terminology involved, and also because of his somewhat eccentric behavior.

Peirce was a voluminous writer, but during his lifetime most of his philosophical works appeared only in the form of articles and critical reviews. Only after the posthumous publication of his *Collected Papers* has it been possible to appreciate genuinely his many contributions to the various fields of philosophy. His scientific reputation was secured in 1878 by the appearance of his book *Photometric Researches;* subsequent studies on sign theory and symbolic logic earned him a permanent place in the history of logic.

Toward the end of his life Peirce was ill and in poverty but continued, nevertheless, to carry on his scientific work, even writing some of his best studies under these difficult circumstances. He died in Milford, Pennsylvania, on April 19, 1914.

Peirce is generally known as the founder of American pragmatism for which he gave a brief outline in two papers, "The Fixation of Belief" (1877), and "How to Make Our Ideas Clear" (1878). He formulated the basic principles of pragmatism as follows: "Consider what effects, that might conceivably have practical bearings, we conceive the object of our conception to have. Then our conception of these effects is the whole of our conception of the object" (*Collected Papers*, 5, p. 402).

Although Peirce's formulation of this principle is not very clear, it nonetheless indicates that in his view philosophy must attempt to secure "clarity of apprehension" or *meaning* by replacing unclear concepts with clear ones, that is, by replacing our original conception of an object with a conception of the conceivable practical effects of that object. Pragmatism, often called a new theory of meaning, is thus designated by Peirce first and foremost as a method of determining the meaning of our intellectual concepts, "that is of those upon which reasonings may turn" (*Ib.*, p. 8)

In making these statements Peirce assumed a starting point in some central issues of British empiricism which he then changed substantially. In his view the weakest point of classical empiricism is to be found in its interpretation of sense data and ideas, and in its attempt to reduce ideas to sense impressions. Peirce does not deny that our ideas and beliefs finally have their origin in experience, but he holds that "the experiential causes of ideas and beliefs," though necessary conditions for their occurrence, are certainly not sufficient conditions insofar as their meaning is concerned. As he saw it, the genuine meaning of ideas and beliefs does not depend merely upon their experiential origin, but rather for the most part upon their conceivable consequences. By conceivable consequences Peirce does not mean concrete and particular experiences, actions, or events. The justification of our ideas and beliefs is to be found in the general form of operations and their results; not in particular actions or events, but in rules of action. Here the influence of Kantian philosophy manifests itself quite clearly, although Peirce explicitly rejected the most fundamental ideas of that philosophy.

Peirce, James, and others agreed upon the fundamental principle of pragmatism despite wide differences on other important points. James wished to use the pragmatist method to elucidate *and* vindicate our basic moral and religious beliefs; Peirce raised objections at this point because he did not believe that moral and religious concepts are of a kind that can be made clear. In James's opinion a belief, view, or theory is true insofar as it "works." Even in its optimal interpretation this view was unacceptable to Peirce, however, because he thought that a higher form of truth could be reached by the pragmatic method. In addition, James had a strong prejudice in favor of whatever is concrete and particular, whereas Peirce claimed that a depreciation of the abstract and the general is ruinous for science.

Finally, for James the pragmatic method was necessarily connected with a certain metaphysical view which he called "radical empiricism," but in Peirce's view a doctrine of logic cannot be preferred for its metaphysical affinities, and thus pragmatism is not at all metaphysics.

Perhaps Peirce's most important contributions have been made in the realm of logic. He always had a keen interest in logical questions and carefully followed developments in this field. His most significant contributions to the development of modern logic consist in that he was able to formulate the first definition of simple order (1881); he wrote the first treatment of the propositional calculus as a calculus of two truth-values (1885); he gave the first definition of the symbol =: things are identical where one can be substituted for the other with preservation of truth; finally, he gave an adequate definition of the finiteness of a class. In addition to these, in 1881 he began to apply logical insights to the foundations of arithmetic in a manner further developed by Dedekind and Peano some years later. He turned his attention also to the classical logical paradoxes, and did important work on the theory of the categorical syllogism, proposing in this regard to give existential import to particular, not general, propositions, whereas in the classical theory affirmative propositions are to be taken as implying that their subjects exist while the negatives do not.

As far as philosophy of science is concerned, it is to be noted that Peirce never developed his conception of science in a systematic way. He never even gave a "description" of the methods of scientific investigation, but limited himself instead to a logical and philosophical reflection of concepts which he believed to be fundamental to a philosophical understanding of scientific methods and their results. Such concepts include, for example, inquiry, reality, meaning, probability, induction, deduction, abduction, chance, law, theory,' and so on. Peirce himself considered most of his written works in this realm as contributions to logic.

The selection which follows contains Peirce's criticism of Mill's conception of induction, a theme which occupied him in several papers.

SELECTIVE BIBLIOGRAPHY

Peirce, C. S., *Collected Papers*. Vols. I–VI, ed. C. Hartshorne and P. Weiss; Vols. VII–VIII, ed. A. Burks, Cambridge, Mass., 1931–1958.

————, *Essays in the Philosophy of Science,* ed. Vincent Thomas. Indianapolis, 1957.

Goudge, T. A., *The Thought of C. S. Peirce,* Toronto, 1950.

————, "Peirce's Treatment of Induction," *Phil. Sci.,* 7 (1940) 56–68.

Buchler, J., *Charles Peirce's Empiricism.* London, 1939.

Madden, Edw. H., "Charles Sanders Peirce's Search for a Method," in *Theories of Scientific Method: The Renaissance Through the Nineteenth Century,* ed. Edward H. Madden. Seattle, 1960. Pp. 248–262.

Freeman, E., *The Categories of Charles Peirce.* Chicago, 1934.

Feibleman, J., *An Introduction to Peirce's Philosophy.* New York, 1946.

Thompson, M., *The Pragmatic Philosophy of C. S. Peirce.* Chicago, 1953.

Wiener, P. P. and Young, F. H., *Studies in the Philosophy of C. S. Peirce.* Cambridge, Mass., 1952.

MILL ON INDUCTION

IN regard to the theory of the validity of Induction the great majority still follow the *System of Logic* set forth in 1843 by John Stuart Mill, who was certainly a clear thinker, and apparently a remarkably candid thinker, in spite of his long training in writing for one of the old "quarterlies," and his consequent unfortunate taste for and skill in controversy, which, combined with his having imbibed his father's sterilizing nominalism with his mother's milk, rendered him, for example, incapable of appreciating Whewell, whose acquaintance with the processes of thought of science was incomparably greater than his own. J. S. Mill's beautiful style, of truly French perfection, together with the bulk of the two volumes, prevent all but the keenest readers from perceiving that he unconsciously wavers between three (not to say four) incompatible theories of the validity of induction. The first (stated in Bk. III, Chap. 3, Sec. 1) is [that] the whole force of Induction is the same as that of a syllogism of which the major premiss is the same for all inductions, being a certain "Axiom of the uniformity of the course of nature" (so described in the table of "Contents"). This was substantially Whately's theory of 1826. The second theory (which seems to be usually uppermost in Mill's mind; especially in Bk. II, Chap. 3, Sec. 7 and in Bk. III, Chap. 4, Sec. 2), is that induction proceeds *as if* upon the principle that a predicate which throughout a more or less extensive experience has been uniformly found to be true of all the members of a given class that have been examined in this respect may, with little risk, be presumed to be true of every member of that class, without exception; and that while it is not necessary that the inductive reasoner should have this principle clearly in mind, the *logician*, whose business it partly is to explain why inductions turn out to be true, must recognize the fact that nature is sufficiently uniform to render that *quasi* principle true,

This article is reprinted by permission of the publishers from **The Collected Papers of Charles Sanders Peirce**, ed. Charles Hartshorne & Paul Weiss, Cambridge, Mass.: The Belknap Press of Harvard University Press, Vol. II (1932, 1960), pp. 483–494. Copyright by the President and Fellows of Harvard College.

and must recognize that [nothing] else renders induction a safe and justifiable procedure. This theory is little more than the old maxim that "we must judge of the future by the past," which Mill—into such unfairness can an inclination toward controversy betray even an eminently fair mind!—attacks as if it merely meant that future history will repeat past history, instead of what it has meant, that future experience must be presumed to resemble past experience under sufficiently similar conditions. The third theory (see Bk. III, Chap. 3, Sec. 3), is that nature as a whole is not absolutely uniform, variety being a far more prominent characteristic of it; and that such uniformity as there is, is "a mere tissue of partial regularities," each consisting in the fact that some classes of objects show a greater, and some a less, tendency to a resemblance of all their members in respect to certain lines of characters; and that whoever knows this "has solved the problem of Induction." This theory was original with Mill; and though it is not the sole, nor the main, support of induction, it certainly does bring a powerful additional support to many inductions. But it is curious that Mill should have chanced to say, whoever might be acquainted with this theory "knows more of the philosophy of logic than the wisest of the ancients." For a quarter of a century later Gomperz[1] published so much as remained of the contents of a papyrus from Herculaneum, which was a defence of induction and a theory of its validity by the Epicurean Philodemus, under whose instruction Cicero studied; and the theory of Philodemus, like that of Mill, is that this kind of reasoning (the only valid reasoning in his opinion) derives its validity from the existence in nature of special uniformities. Only, the uniformities that attracted the attention of Philodemus, instead of characterizing certain classes, characterize certain characters, and consist in their having a special tendency to be present (or to be absent) throughout all the members of certain kinds of classes. In fact, still other types of uniformities may affect the strength of inductions.

A yet fourth theory of induction, that of Laplace, received, by implication, the assent of Mill; and since this theory is taught as correct in all the textbooks of the Doctrine of Chances, it behooves me, in adopting another, to state, with the utmost brevity wherein Laplace's theory is false and harmful. I shall also give my explanation of Mill's assenting to it.

[1] Theodor Gomperz, *Philodemi de ira liber* (1864); *Herkulanische Studien* (1865–6).

If, upon any occasion, we were to devise a method of forming a numerous sample of any class, say the S's, which should be suitable for use in determining to a given degree of approximation what proportion of future experiences of S's would, in the long run, be found to have the character of being P, in case existing general conditions should undergo no alteration, then in case there were any definite reason to expect that, among S's coming to our attentive experience from any particular sub-class, say among the S's that belong to the sub-class of T's, a markedly different proportion would turn out to be P from the proportion among the S's that should not be T's, then that method of forming the sample, since we have supposed it to be "suitable" for showing the proportion of P's among all future experiences of S's, must needs insure that the proportion of S's that are T's should be nearly the same in the sample as it was destined to be among all the S's of our future experience; though, as I need not repeat again, this would be so only under the supposition of unchanged general circumstances, and need not be more precisely true than would suffice to keep the errors of the concluded proportion of subsequently experienced S's that should be P within the intended limit of approximation. Moreover, should there be any serious reason to suspect that any identifiable S presenting itself for admission to the sample was so connected with any S already admitted as to have a special liability whether to being like or to being unlike that already admitted instance in respect to being or not being P (under the same limitation that is not to be repeated), then our "suitable" method would have to exclude that instance from the sample. And once again, should there happen to be any reason to suspect that an instance had attracted our attention owing to causes connected, whether directly or indirectly, with its being P, or to such causes as should be connected with its *not* being P, then our suitable method must exclude that instance. Furthermore, our suitable method must so operate that the sample shall contain a sufficient number of instances to give the intended degree of approximation. For instance, if it will suffice that the figure next following the decimal point in the decimal expression of the proportion among all the S's of such as are P should be exact, 9 instances may first be taken, and if these make the ratio less than 0.05 or greater than 0.95, they will suffice. If not, 14 more instances may be collected; and if the whole 23 make the ratio less than 0.15 or greater than 0.85, they will suffice. If not, then if 11 more instances being taken, the whole 34 make the ratio less than 0.25 or greater than 0.75, they will suffice. If not, add 7 more, and if the ratio appears as less than 0.35 or more than 0.65, the 41 will suffice. If not, take 4 more and if the ratio then appears as less than 0.45 or greater than 0.55, the 45 will suffice. If not, one more instance will in any case be enough. If the first two figures of the decimal fraction must be correct, a hundred times as many instances will be requisite.

Every person of common-sense must, upon reflexion, acknowledge, what is familiar to everybody habituated to inductive reasoning, that all the above precautions are requisite, except that the concluding rule need not have been so detailed, and that, if the instances were sufficiently multiplied, it would suffice that the other rules should not be too frequently and grossly violated and that they should not prevailingly be violated in the same direction. Let all such diminutions from them be made, and it still remains true for sound reason, that *such an induction does not follow merely from the fact that* P *is true of such and such of the* S's *of a collection of* S's, *but that it is necessary to take account of the manner in which these* S's *were brought to the inquirer's attention.* This fixes a great gulf between Induction and Deduction. It is quite true that we may describe the general conditions of a valid quantitative induction, and may convince ourselves that if the sample be drawn strictly at random from among the S's and be made sufficiently numerous, then, the general conditions remaining unchanged, it necessarily follows that future experience, *under the same general conditions,* will on the average of an indefinite multitude of such inductions, bear out the inductive conclusion. Still, this in no wise suffices to reduce the quantitative induction to any kind of induction [deduction?]. For even if I were to grant that the truth of the inductive conclusion would necessarily follow if the conditions of a fair sample were to be ideally fulfilled, which, for a reason that I will presently state, I find myself unable to do, still the person who really draws the inductive inference cannot possibly have any demonstrative evidence that those conditions are fulfilled even to the imperfect degree that is needful for an approximation to the true ratio. He knows, if you will, that [he] has made strenuous efforts to make his sample a fair one; but he cannot be quite sure that deep down in the caverns of his heart there may not lurk, unsuspected by him, a determination to force himself to believe in a certain value for the ratio, nor that this has not frustrated all his efforts to make the sample a fair one; and if he cannot be absolutely certain even of his own honesty, how can he so much as approach certainty as to the correctness of his concluded approximation to the ratio not having been destroyed by external circumstances? A theoretician—or rather, a papyrobite, a man whose vitality is that of sentences written down or imagined—may reply that the contingency is covered by proviso that general conditions remain sufficiently unchanged. But that is to overlook the principal end of inquiry, as regards human life. What is the chief end of man? *Answer:* To actualize ideas of the immortal, ceaselessly prolific kind. To that end it is needful to get beliefs that the believer will take satisfaction in acting upon, not mere rules set down on paper, with lethal provisos attached to them. The inductive reasoner cannot possibly find any strictly demonstrative reasoning that could take the place of his induction, since every demonstrative is strictly limited to the field of that part of its copulate premiss that corresponds to the minor premiss of a syllogism; while to serve his purpose, that of forming a basis for conduct, it must transcend that limit in concluding future from past experience. Now every valid mathematical reasoning is demonstrative and is limited to an ideal state of things. The reasoning of the calculus of probabilities consists simply of demonstrations concerning "probabilities," which, in all useful applications of the calculus, are *real* probabilities, or ratios of frequency in the "long run" of experiences of designated species among experiences designated, or obviously designable, genera over those species; which real probabilities are ascertained by quantitative inductions from statistics laboriously collected and critically tabulated. But the phrase "the probability of an event," which is perpetually recurring in the treatises, and which is not free from objection, even when the real probability is meant (because it seems to refer to a singular experience considered by itself, and because it does not mention that *two* classes of experiences are essentially concerned), is used in various different senses, owing to the ambiguity of the word "probability"; and the writers of the mathematical treatises on the subject have not had sufficient power of logical analysis to found any useful theory upon it. . . .

Laplace maintains that it is possible to draw a necessary conclusion regarding the probability of a particular determination of an event based on not knowing anything at all about [it]; that is, based on nothing. When a man thinks himself to know nothing at all as to which of a number of alternatives is the truth, his mind can no more incline toward or against any one of them or any combination of them than a mathematical point can have an inclination toward any point of the compass. Suppose the question concerns the color of an object which we know has a high color, but are otherwise in a state of blank ignorance [about it]. Then, according to Laplace, if one were to draw two lines across a map of the spectrum, it would be probable that the color did not match any part of the spectrum included between those lines; no matter how nearly they might include the whole spectrum. Laplace holds that for every man there is one law (and necessarily but one) of dissection of each continuum of alternatives so that all the parts shall seem to that man to be "*également possibles*" in a quantitative sense, antecedently to all information. But he presents not the slightest reason for thinking this to be so, and seems to admit that to different men different modes of dissection will seem to give alternatives that are *également possibles*. It is only by basing the theory of probability upon this doctrine, and thus rendering probability without interest except to a student of human eccentricities, that it is possible to assign any mathematical probability to an inductive conclusion. Much might be added in refutation of Laplace's position.

In the first edition of his *Logic*,[2] Mill presents arguments against Laplace's view; but in his third, without answering his former arguments, as far as I can see, he abandons them, and thus assents to all that is necessary for calculating a necessary probability for the inductive conclusion, without any regard to the manner in which the instances have been collected.

I will now sketch one or more ways of refuting each of Mill's three professed theories of induction.

To the first theory, that an induction is equivalent to a syllogism whose major premiss is the axiom of the uniformity of nature, while its minor premiss states the observed facts about the instances, the conclusion being identical with that of the induction, each of the following objections is conclusive: *first* that an induction, unlike a demonstration, does not rest solely upon the facts observed, but upon the manner in which those facts have been collected; *secondly*, that a syllogism infers its conclusion apodictically, while an induction does not; *thirdly*, that a syllogism enriches our knowledge of ideas, but not our information, which is what Kant meant in saying that it only explicates but does not amplify knowledge, while an induction does amplify our knowledge; *fourthly*, that the proposed syllogism would be fallacious, because its major premiss is vague, so that it could be fairly thrown into the form of a fallacy of undistributed middle, since all we really know of the general uniformity of nature is that *some*

[2] Bk. III, ch. 18.

pairs of phenomena (an apparently infinitesimal proportion of all pairs) are connected as logical antecedent and consequent; *fifthly,* because a sound syllogism must not conclude beyond the breadth, or logical extension of its minor premiss (when this is suitably stated), while to represent a true induction it must do so. There are other objections, fully as strong as these five; but it seems needless to mention them.

The second theory correctly describes the procedure of the mind in crude inductions, but in no others; and Mill's celebrated four methods (chiefly based on the *Novum Organum*), though they may be of some help to minds that need such aids, yet furnish nothing but crude inductions, after all. The principle of this theory also sufficiently explains how it is that we meet such frequent opportunities to draw crude inductions as we do. But the moment the attempt is made to apply this theory to *justifying,* or explaining the validity even of crude inductions (and it is still worse with other kinds of induction), it lays itself open to all the objections to the first method, including the five that were specified above. For this second theory, which is the point where Mill's vain attempt to make reasoning able to get along without generalization becomes the most futile, and verges closely upon overt absurdity, differs from the first merely in not allowing, as essential to induction, that it should have any of such force as it might derive from employing the uniformity of experience as a premiss. Now this point of difference cannot confer upon induction as explained by the second theory any validity that it would not have if it were explicable by the first theory.

The third theory presents two decided advantages. For it may remove entirely the vagueness of the general principle of uniformity; and in some cases makes the special uniformity predicate a probability, so as to render the refutation of the theory, on the ground that induction does not conclude apodictically, considerably more difficult in those cases. Moreover, this theory does correctly state a part of the argument for very many inductive conclusions. But this part of the argument is not inductive but deductive. For these special uniformities (such, for example, as that every chemical element has the same combining weight, no matter from what mineral or from what part of the globe it has come), have only become known by induction, often only by elaborate investigations, and are not logical principles, so that they need to be stated as premisses when the argument is to be set forth in full. The special uniformities, when they become known, enable us to dispense with certain inductive inquiries that would otherwise be requisite. But they leave other inductions (such as that which led Mendeléeff to enunciate his periodic law), quite untouched, not explaining them in any sense.

The true guarantee of the validity of induction is that it is a method of reaching conclusions which, if it be persisted in long enough, will assuredly correct any error concerning future experience into which it may temporarily lead us. This it will do not by virtue of any deductive necessity (since it never uses all the facts of experience, even of the past), but because it is manifestly adequate, with the aid of retroduction and of deductions from retroductive suggestions, to discovering any *regularity* there may be among experiences, while *utter irregularity is not surpassed in regularity by any other relation of parts to whole,* and is thus readily discovered by induction to exist where it does exist, and the amount of departure therefrom to be mathematically determinable from observation where it is imperfect. The doctrine of chances, in all that part of it that is sound, is nothing but the science of the laws of irregularities. I do not deny that God's beneficence is in nothing more apparent than in how in the early days of science Man's attention was particularly drawn to phenomena easy to investigate and how Man has ever since been led on, as through a series of graduated exercises, to more and more difficult problems; but what I do say is that there is no possibility of a series of experiences so wanting in uniformity as to be beyond the reach of induction, provided there be sufficiently numerous instances of them, and provided the march of scientific intelligence be unchecked.

Quantitative induction approximates gradually, though in an irregular manner to the experiential truth for the long run. The antecedent probable error of it at any stage is calculable as well as the probable error of that probable error. Besides that, the probable error can be calculated from the results, by a mixture of induction and theory. Any striking and important discrepancy between the antecedent and *a posteriori* probable errors may require investigation, since it suggests some error in the theoretical assumptions. But the fact which is here important is that Quantitative Induction always makes a gradual approach to the truth, though not a uniform approach.

Qualitative Induction is not so elastic. Usually either this kind of induction confirms the hypothesis or else the facts show that some alteration must be made in the hypothesis. But this modification may be a small detail.

Experiments which I have conducted in great numbers and great elaboration have convinced me of the extremely important advantages of making use, in Qualitative Induction, of numbers in place of such adverbs of comparison of the intensity of feelings as, "slightly," "a little," "somewhat," "tolerably," "moderately," "considerably," "much," "greatly," "excessively," etc. It is not necessary to use the adverbs; but in some cases I have found it convenient to employ a few of them. What is necessary is to get certain feelings so fixed in one's mind that they can be exactly and severally reproduced in the imagination at any time, these feelings forming such a series of ten or so, beginning with the zero of intensity and running up to high intensities; and further being such that any one of them being contemplated by the investigator and compared with the next intenser in the series, the interval of intensity between them shall appear, to the contemplator's feeling, to be equal to the interval between any other one of the series and the member next intenser than it. It is certain that this can be done, since all sidereal astronomers since Ptolemy have practised this; and many psychologists beside me have done something similar for other feelings than that of luminosity. It has been demonstrated that a series of positive numbers, integer and fractional, expresses *in itself* nothing more than an order of succession. But this scale is made to express, besides, a feeling of a difference of feeling in one respect; and the experiments of many persons prove conclusively that people generally can form such a scale; and further that the scales of different persons are concordant to a pretty high degree. The next step has been executed by but few persons so far as I know; but the experiments of these few render it all but certain that all normal persons can do so with good accord. This consists in comparing a difference of feeling in one respect with a difference in a single other respect; such as luminosity and pressure-feelings, or the relative bitterness of two solutions of quassia and that of self-blame for two former actions. Such comparisons as these last are, to be sure, of no direct applicability so far as I am aware; but they are good exercises in that prescissive abstraction of intensity from its subject which is required for estimating the equality of two differences of intensity. Such estimations enable us to add and take the arithmetical mean of intensities referred to the same standard; and not only the practice of all photometricians, both astronomers and gas-examiners, but also very many thousands of experiments by me upon a wide variety of qualities of sensation, establishes, to my full satisfaction, the great utility of such applications of number in giving a control over qualitative inductions. I have not found multiplications of such numbers useful, for example, in establishing the laws of such comparisons as the relative photometric value of two lights of different colors, where I need not say that it is one thing to ask what intensity of a light A, of fixed hue and chroma but variable luminosity best matches a light B, that is altogether fixed, and quite another and independent question what photometric intensity of B, if this be made to vary, best matches an A of given fixed intensity. The meaning of the product of two differences of intensity which refer in general to different qualities is obvious enough: it is the number to be attached as a measure to a phenomenon which involves two feelings of the intensities indicated by the multiplicand and multiplier, these two feelings being [in] a certain fixed relation to one another in which they are as independent as possible. But there is no advantage in attaching any single measure to such a complex phenomenon, unless there are different ways of analyzing it, more or less similar to the different systems of coördinates in geometry. I mean that, for example, different horizontal areas are not only measured by the sum of the parallelograms into which [they] may be cut up, each parallelogram having its sides in the directions of ENE and N by W. Were that the sole method of measurement, nothing would be gained by combining the linear measures in the two directions, but rather the reverse. But in fact we may measure the area by parallelograms in any other two dimensions; and the ratio between any two areas will be the same by any two such methods of measurement. Moreover, we may employ polar, in place of Cartesian, coördinates, and cut the area up into a circle and broken, concentric, and very thin rings. The area will be the sum of the areas, each of which will be $X \times Y$, where X is the difference of the two radii, while Y is the proportion of their sum, the proportion being that of the entire ring which forms a part of the area.

Chapter 15
PIERRE MAURICE MARIE DUHEM
(1861–1916)

INTRODUCTION

*P*IERRE MAURICE MARIE DUHEM was born in Paris on June 10, 1861. He studied at the École Normale Supérieure where he concentrated particularly on mathematics and physics. Almost immediately his interest centered on thermodynamics and its applications. Studying the works of such pioneers of thermodynamics as Lord Kelvin, Clausius, and Gibbs, Duhem was especially impressed by the analogy between Lagrange's methods of analytical mechanics and those of thermodynamics. After a careful study of the analogy he introduced in a general sense the notion of thermodynamic potential. His views on this subject were published in 1886 in a book entitled *Le Potentiel thermodynamique et ses applications à la mécanique chimique et à la théorie des phénomènes électriques.* When his formal education was completed in 1887 Duhem became a lecturer in physics at the University of Lille. Six years later he went to the University of Bordeaux as professor of physics and remained in this capacity until his death on September 14, 1916.

Duhem devoted a large segment of his life to the study and teaching of theoretical physics; much of that time was spent in an attempt to construct a general energetics designed to include both classical mechanics and abstract thermodynamics. Here he used purely axiomatic methods, rigorously objecting to the use of concrete and mechanical models in general energetics. This effort at axiomatization and consistent deduction is characteristic not only of his first work in the area, *Commentaires sur la thermodynamique*, but also of his main work *Traité d'énergétique générale*. Duhem made some original contributions to other branches of physics and chemistry, also; his work in the realm of the propagation of waves of impact in fluids, for example, is of lasting value. His opposition to mechanical models prevented his understanding of the importance of Lorentz' theory of electrons and atomic physics in general.

Duhem has many publications on the history of science to his credit, especially in the fields of mechanics, astronomy, and physics. One of the principal theses he attempted to defend in these historical works was his view that the great revival of mechanics, astronomy, and physics of the sixteenth and seventeenth centuries was rooted in the work of late medieval philosophers and "scientists." Two Duhem publications of especial significance in this area are *Études sur Léonard de Vinci* in three volumes, and *Le système du monde* which remained incomplete, "only" five volumes having been published during his lifetime.

Duhem's work in theoretical physics and investigations in the history of science were extremely good preparations for his studies in philosophy of science. One of the most important of his many publications was *The Aim and Structure of Physical Theory* which appeared in 1906.

Although Duhem did not have a negative attitude toward classical metaphysics in general, he nonetheless made a clear separation between physics and philosophy, a fact which placed his philosophy of science in the realm of positivism. In his view physical theories were not able to teach anything about the very nature of reality; they did not give a genuine explanation of natural phenomena. A physical theory is, rather, a system of mathematical propositions deduced from a small number of principles with the aim of representing as simply, as completely, and as exactly as possible a whole domain of experimental laws. Duhem's radical separation of physics and metaphysics made it possible for him to accept the substance of Poincaré's conventionalism and Mach's phenomenistic positivism without in any sense being obliged to subscribe to all the radical conclusions usually drawn by positivists from this point of view as regards physical theories. In elaborating this general view Duhem paid very little attention to the problems of observation and perception closely connected with the verification of theories, an aspect studied particularly by Mach; like Poincaré and Hertz he was more interested in the logical and systematic characteristics of theories, which happened to be in complete harmony with his own investigations in theoretical physics.

According to Duhem, a distinction between four successive operations must be made in the construction of any physical theory. A brief description of the process follows:

1. We begin with selecting from among the physical properties we want to represent, those which can be considered to be simple. We then substitute for these simple properties certain mathematical symbols, namely, numbers and magnitudes which are to be obtained by means of measurement. These symbols do not have any intrinsic connection with the properties they represent, but bear to the latter only the relation of sign to thing signified.

2. We then connect these numbers and magnitudes into a small number of propositions which serve as the principles of deduction. These principles, called hypo-theses in the etymological sense of the word, do not claim to state any real relations between the real properties of things, but are rather arbitrary. They are controlled only by the requirement of logical consistency among the terms of one principle, or among the different principles of one theory.

3. Then we try to combine the different hypotheses of one theory by logical and mathematical means, and no real relations between properties are implied by these combinations, nor is it necessary that the operations one performs agree or correspond to real physical transformations.

4. Finally we draw consequences from these hypotheses and translate them into statements about the physical properties of bodies, the methods of defining and measuring these properties serving as a "dictionary," permitting one to make the translation. By comparing the resulting statements with the results of experiment or experience, one can decide whether or not the theory is good. Agreement with experience is the only criterion of "truth" for a physical theory.

The selection which follows shows where this conception of theory leads insofar as physical laws are concerned.

SELECTIVE BIBLIOGRAPHY

Duhem, P., Études sur Léonard de Vinci (3 vols.). Paris, 1906.

———, Le système du monde (10 vols.). Paris, 1913–1958.

———, The Aim and Structure of Physical Theory (1906), trans. P. P. Wiener. New Jersey, 1954.

Lowinger, A., *The Methodology of Pierre Duhem*. New York, 1941.

H. P. Duhem, *Un savant français*. Paris, 1936.

Alexander, P., "The Philosophy of Science, 1850–1910," in *A Critical History of Western Philosophy*, ed. D. J. O'Connor. New York, 1965. Pp. 417–420

Humbert, P., *Pierre Duhem*. Paris, 1923.

Jammer, M., *Concepts of Force*. Cambridge (Mass.), 1957 (*passim*).

Ginzburg, B., "Duhem and Jordanus Nemorarius," *Isis*, 25 (1936) 341–362.

PHYSICAL LAW

THE LAWS OF PHYSICS ARE SYMBOLIC RELATIONS

Just as the laws of common sense are based on the observation of facts by means natural to man, so the laws of physics are based on the results of physical experiments. Of course, the profound differences which separate the non-scientific ascertainment of a fact from the result of a physical experiment will also separate the laws of common sense from the laws of physics; thus, nearly everything we have said about the experiments of physics will extend to the laws that science states.

Let us consider one of the simplest and most certain of common-sense laws: All men are mortal. This law surely relates two abstract concepts, the abstract idea of man in general, rather than the concrete idea of this or that man in particular, and the abstract idea of death, rather than the concrete idea of this or that form of death; indeed, it is only on this condition, viz., that the concepts related are abstract, that the law can be general. But these abstractions are in no way theoretical symbols, for they merely extract what is universal in each of the particular cases to which the law applies. Thus, in each of the particular cases where we apply the law, we shall find concrete objects in which these abstract ideas are realized; each time we might wish to ascertain that all men are mortal we shall find ourselves aware of a certain individual man embodying the general idea of man, and of a certain particular death implying the general idea of death.

Let us take another law, quoted as an example by Milhaud when he expounded these ideas[1] expressed by us a little earlier. It is a law

---

[1] G. Milhaud, "La Science rationelle," *Revue de Métaphysique et de Morale,* IV (1896), 280. Reprinted in *Le Rationnel* (Paris, 1898), p. 44.

---

about an object belonging to the domain of physics, but it retains the form that the laws of physics had when this branch of knowledge existed only as a dependency of common sense without yet having acquired the dignity of a rational science.

Here is the law: We see the flash of lightning before we hear thunder. The ideas of lightning and thunder which this statement ties together are abstract and general ideas, but these abstractions are drawn so instinctively and naturally from particular data that with each bolt of lightning we perceive a glare and a rumbling in which we recognize immediately the concrete form of our ideas of lightning and thunder.

This is not, however, true of the laws of physics. Let us take one of these laws, Mariotte's law,[2] and examine its formulation without caring for the moment about the accuracy of this law. At a constant temperature, the volumes occupied by a constant mass of gas are in inverse ratio to the pressures they support; such is the statement of the law of Mariotte. The terms it introduces, the ideas of mass, temperature, pressure, are still abstract ideas. But these ideas are not only abstract; they are, in addition, symbolic, and the symbols assume meaning only by grace of physical theories. Let us put ourselves in front of a real, concrete gas to which we wish to apply Mariotte's law; we shall not be dealing with a certain concrete temperature embodying the general idea of temperature, but with some more or less warm gas; we shall not be facing a certain particular pressure embodying the general idea of pressure, but a certain pump on which a weight is brought to bear in a certain manner. No doubt, a certain temperature corresponds to this more or less warm gas, and a certain pressure corresponds to this effort exerted on the pump, but this correspondence is that of a sign to the thing signified and replaced by it, or of a reality to the symbol representing it. This correspondence is by no means immediately given; it is established with the aid of instruments and measurements, and this is often a very long and very complicated process. In order to assign a definite temperature to this more or less warm gas, we must have recourse to a thermometer; in order to evaluate in the form of a pressure the effort exerted by the pump, we must use a manometer, and the use of the thermometer and manometer imply, as we have seen in the preceding chapter, the use of physical theories.

The abstract terms referred to in a common-sense law being no

[2] Translator's note: Boyle's law.

more than whatever is general in the concretely observed objects, the transition from the concrete to the abstract is made in such a necessary and spontaneous operation that it remains unconscious; placed in the presence of a certain man or of a certain case of death, I associate them immediately with the general idea of man and with the general idea of death. This instinctive and unreflective operation yields unanalyzed general ideas, abstractions taken grossly, so to speak. No doubt, the thinker may analyze these general and abstract ideas, he may wonder what man is, what death is, and seek to penetrate the deep and full sense of these words. This inquiry will lead him to a better understanding of the reasons for the law, but it is not necessary to do that in order to understand the law; it is sufficient to take the terms related in their obvious sense in order to understand this law, which is clear to us whether we are philosophers or not.

The symbolic terms connected by a law of physics are, on the other hand, not the sort of abstractions that emerge spontaneously from concrete reality; they are abstractions produced by slow, complicated, and conscious work, i.e., the secular labor which has elaborated physical theories. If we have not done this work or if we do not know physical theories, we cannot understand the law or apply it.

According to whether we adopt one theory or another, the very words which figure in a physical law change their meaning, so that the law may be accepted by one physicist who admits a certain theory and rejected by another physicist who admits some other theory.

Take a peasant who has never analyzed the notions of man or of death and a metaphysician who has spent his life analyzing them; take two philosophers who have analyzed and adopted different, irreconcilable notions of man and of death; for all, the law "All men are mortal" will be equally clear and true. In the same way, the law "We see the flash of lightning before we hear thunder" has for the physicist who knows thoroughly the laws of disruptive electrical discharge the same clarity and certainty as it had for the Roman plebeian who saw in a stroke of lightning the anger of Capitoline Jupiter.

On the other hand, let us consider the following physical law: "All gases contract and expand in the same manner"; and let us ask different physicists whether this law is or is not violated by iodine vapor. The first physicist professes theories according to which iodine vapor is a single gas, and draws from the foregoing law the consequence that the density of iodine vapor relative to air is a constant. Now, experiment shows that this density depends on the temperature and pressure; therefore, our physicist concludes that iodine vapor is not subject to the law stated. A second physicist will have it that iodine vapor is not a single gas but a mixture of two gases which are polymers of each other and capable of being transformed into each other; consequently, the law mentioned does not require the iodine-vapor density relative to air to be constant, but claims this density varies with the temperature and pressure according to a certain formula established by J. Willard Gibbs. This formula represents, indeed, the results of experimental determinations; our second physicist concludes that iodine vapor is not an exception to the rule which states that all gases contract and expand in the same manner. Thus our two physicists have entirely different opinions concerning a law which both enunciate in the same form: one finds fault with it because of a certain fact, the other finds that it is confirmed by that very fact. That is because the different theories they hold do not determine uniquely the meaning suited to the words "a single gas," so that though they both pronounce the same sentence, they mean two different propositions; in order to compare his proposition with reality each makes different calculations, so that it is possible for one to verify this law which the other finds contradicted by the same facts. This is plain proof of the following truth: A physical law is a symbolic relation whose application to concrete reality requires that a whole group of laws be known and accepted.

A LAW OF PHYSICS IS, PROPERLY SPEAKING, NEITHER TRUE NOR FALSE BUT APPROXIMATE

A common-sense law is merely a general judgment; this judgment is either true or false. Take, for instance, the law that everyday observation reveals: In Paris, the sun rises every day in the east, goes up in the heavens, then comes down and sets in the west. There you have a true law without conditions or restrictions. On the other hand, take this statement: The moon is always full. That is a false law. If the truth of a common-sense law is questioned, we can answer this question by yes or no.

Such is not the case with the laws that a physical science, come to full maturity, states in the form of mathematical propositions; such laws are always symbolic. Now, a symbol is not, properly speaking, either true or false; it is, rather, something more or less well selected to stand for the reality it represents, and pictures that reality in a more or less precise, a more or less detailed manner. But applied to a symbol the words "truth" and "error" no longer have any meaning; so, the logician who is concerned about the strict meaning of words will have to answer anyone who asks whether physics is true or false, "I do not understand your question." Let us comment on this answer which may seem paradoxical but the understanding of which is necessary for anyone who claims to know what physics is.

The experimental method, as practiced in physics, does not make a given fact correspond to only one symbolic judgment, but to an infinity of different symbolic judgments; the degree of symbolic indetermination is the degree of approximation of the experiment in question. Let us take a sequence of analogous facts; finding the law for these facts means to the physicist finding a formula which contains the symbolic representation of each of these facts. The symbolic indetermination corresponding to each fact consequently entails the indetermination of the formula which is to unite these symbols; we can make an infinity of different formulas or distinct physical laws correspond to the same group of facts. In order for each of these laws to be accepted, there should correspond to each fact not *the* symbol of the fact but some one of the symbols infinite in number, which can represent the fact; that is what is meant when the laws of physics are said to be only approximate.

Let us imagine, for example, that we refuse to be satisfied with the information supplied by the common-sense law about the sun's rising in the east, climbing the sky, descending, and setting in the west every day in Paris; we address ourselves to the physical sciences in order to have a precise law of the motion of the sun seen from Paris, a law indicating to the observer in Paris what place the sun occupies in the sky at each moment. In order to solve the problem, the physical sciences are not going to use sensed realities, say of the sun just as we see it shining in the sky, but will use symbols through which irregularities of its surface, despite the enormous protuberances it has, will be replaced in their theories by a geometrically perfect sphere, and it is the position of the center of this ideal sphere that these theories will try to determine; or rather, they will seek to determine the position that this point would occupy if astronomical refraction did not deviate the rays, and if the annual aberration did not modify the apparent position of the heavenly bodies. It is, therefore, a symbol that is substituted for the sole sensible reality offered to our observation, for the shiny disk that our lens may sight. In order to make the symbol correspond to the reality, we must effect complicated measurements, we must make the edges of the sun coincide with the hairlines of a lens equipped with a micrometer, we must make many readings on divided circles, and subject these readings to diverse corrections; we must also develop long and complex calculations whose legitimacy depends on admitted theories, on the theory of aberration, and on the theory of atmospheric refraction.

The point symbolically called the center of the sun is not yet obtained by our formulas; they tell us only the coordinates of this point, for instance, its longitude and latitude, coordinates whose meaning cannot be understood without knowing the laws of cosmography, and whose values do not designate a point in the sky that you can indicate with your finger or that a telescope can sight except by virtue of a group of preliminary determinations: the determination of the meridian of the place, its geographical coordinates, etc.

Now, can we not make a single value for the longitude and a single value for the latitude of the sun's center correspond to a definite position of the solar disk, assuming the corrections for aberration and refraction to have been made? Indeed not. The optical power of the instrument used to sight the sun is limited; the diverse operations and readings required of our experiment are of a limited sensitivity. Let the solar disk be in such a position that its distance from the next position is small enough, and we shall not be able to perceive the deviation. Admitting that we cannot know the coordinates of a fixed point on the celestial sphere with a precision greater than 1', it will suffice, in order to determine the position of the sun at a given instant, to know the longitude and latitude of the sun's center to approximately 1'. Hence, to represent the path of the sun, despite the fact that it occupies only one position at each instant, we shall be able to give for each instant not one value alone for the longitude and only one value for the latitude, but an infinity of values for each, except that for a given instant two acceptable values of the longitude or two acceptable values of the latitude will not differ by more than 1'.

We now proceed to seek the law of the sun's motion, that is to say, two formulas permitting us to calculate at each instant of a period the value of the longitude and latitude, respectively, of the center of the sun. Is it not evident that, in order to represent the path of the longitude as a function of the time, we shall be able to adopt not a single formula, but an infinity of different formulas, provided that for a given instant all these formulas give us values for the longitude differing by less than 1'? And is not the same evident for the latitude? We shall then be able to represent equally well our observations on the path of the sun by an infinity of different laws; these diverse laws will be expressed by equations which algebra regards as incompatible, by equations such that if one of them is verified, no other is. They will each trace a different curve on the celestial sphere, and it would be absurd to say that the same point describes two of these curves at the same time; yet, to the physicist all these laws are equally acceptable, for all determine the position of the sun with a closer approximation than can be observed with our instruments. The physicist does not have the right to say that any of these laws is true to the exclusion of the others.

No doubt the physicist has the right to choose between these laws, and generally he will choose; but the motives which will guide his choice will not be of the same kind or be imposed with the same imperious necessity as those which compel him to prefer truth to error.

He will choose a certain formula because it is simpler than the others; the weakness of our minds constrains us to attach great importance to considerations of this sort. There was a time when physicists supposed the intelligence of the Creator to be tainted with the same debility, when the simplicity of these laws of nature was imposed as an indisputable dogma in the name of which any experimental law expressing too complicated an algebraic equation was rejected, when simplicity seemed to confer on a law a certainty and scope transcending those of the experimental method which supplied it. It was then that Laplace, speaking of the law of double refraction discovered by Huygens, said: "Until now this law has been only the result of observation, approximating the truth within the limits of error to which the most exact experiments are subject. Now the simplicity of the law of action on which it depends should make us consider it a rigorous law."[3] That time no longer exists. We are no longer dupes of the charm which simple formulas exert on us; we no longer take that charm as the evidence of a greater certainty.

The physicist will especially prefer one law to another when the first follows from the theories he admits; he will, for example, ask the theory of universal attraction to decide which formulas he should prefer among all those which could represent the motion of the sun. But physical theories are only a means of classifying and bringing

[3] P. S. Laplace, *Exposition du système du monde*, I, IV, ch. XVIII: "De l'attraction moléculaire."

together the approximate laws to which experiments are subject; theories, therefore, cannot modify the nature of these experimental laws and cannot confer absolute truth on them.

Thus, every physical law is an approximate law. Consequently, it cannot be, for the strict logician, either true or false; any other law representing the same experiments with the same approximation may lay as just a claim as the first to the title of a true law or, to speak more precisely, of an acceptable law.

EVERY LAW OF PHYSICS IS PROVISIONAL AND RELATIVE BECAUSE IT IS APPROXIMATE

What is characteristic of a law is that it is fixed and absolute. A proposition is a law only because once true, always true, and if true for this person, then also for that one. Would it not be contradictory to say that a law is provisional, that it may be accepted by one person and rejected by another? Yes and no. Yes, certainly, if we mean by "laws" those that common sense reveals, those we can call true in the proper sense of the word; such laws cannot be true today and false tomorrow, and cannot be true for you and false for me. No, if we mean by "laws" the laws that physics states in mathematical form. Such laws are always provisional; not that we must understand this to mean that a physical law is true for a certain time and then false, but at no time is it either true or false. It is provisional because it represents the facts to which it applies with an approximation that physicists today judge to be sufficient but will some day cease to judge satisfactory. Such a law is always relative; not because it is true for one physicist and false for another, but because the approximation it involves suffices for the use the first physicist wishes to make of it and does not suffice for the use the second wishes to make of it.

We have already noticed that the degree of approximation is not something fixed; it increases gradually as instruments are perfected, and as the causes of error are more rigorously avoided or more precise corrections permit us to evaluate them better. As experimental methods gradually improve, we lessen the indetermination of the abstract symbol brought into correspondence with the concrete fact by physical experiment; many symbolic judgments which might have been regarded at one time as adequately representing a definite, concrete fact will no longer be accepted at another time as signifying this fact with sufficient precision. For example, the astronomers of one century will, in order to represent the position of the sun's center at a given instant, accept all the values of the longitude which do not differ from each other by more than 1' and all the values of the latitude confined within the same interval. The astronomers of the next century will have telescopes with greater optical power, more perfectly divided circles, more minute and precise methods of observation; they will require then that the diverse determinations of the longitude and latitude, respectively, of the sun's center at a given instance agree within about 10''; an infinity of determinations which their predecessors were willing to permit would be rejected by them.

As the indetermination of experimental results becomes narrower, the indetermination of the formulas used to condense these results becomes more restricted. One century would accept as the law of the sun's motion any group of formulas which gave for each instant the coordinates of the center of this star within approximately 1'; the next century will impose on any law of the sun's motion the condition that the coordinates of the sun's center be known within approximately 10''; an infinity of laws accepted by the first century will thus be rejected by the second.

This provisional character of the laws of physics is made plain every time we read the history of this science. For Dulong and Arago, Mariotte's [Boyle's] law was an acceptable form of the law of the compressibility of gases because it represented the experimental facts with deviations that remained less than the possible errors of the methods of observation used by them. When Regnault had improved the apparatus and experimental method, this law had to be rejected; the deviations of Mariotte's law from the results of observation were much greater than the uncertainties affecting the new apparatus.

Now, given two contemporary physicists, the first may be in the circumstances Regnault was in, whereas the second may still be working under conditions under which Dulong and Arago worked. The first possesses very precise apparatus and plans to make very exact observations; the second possesses only crude instruments and, in addition, the investigations he is making do not demand close approximation. Mariotte's law will be accepted by the latter and rejected by the former.

More than that, we can see the same physical law simultaneously adopted and rejected by the same physicist in the course of the same work. If a law of physics could be said to be true or false, that would be a strange paradox; the same proposition would be affirmed and denied at the same time, and this would constitute a formal contradiction.

Regnault, for example, is making inquiries about the compressibility of gases for the purpose of finding a more approximate formula to substitute for Mariotte's law. In the course of his experiments he needs to know the atmospheric pressure at the level reached by the mercury in his manometer; he uses Laplace's formula to obtain this pressure, and Laplace's formula rests on the use of Mariotte's law. There is no paradox or contradiction here. Regnault knows that the error introduced by this particular employment of Mariotte's law is much smaller than the uncertainties of the experimental method he is using.

Any physical law, being approximate, is at the mercy of the progress which, by increasing the precision of experiments, will make the degree of approximation of this law insufficient: the law is essentially provisional. The estimation of its value varies from one physicist to the next, depending on the means of observation at their disposal and the accuracy demanded by their investigations: the law is essentially relative.

EVERY PHYSICAL LAW IS PROVISIONAL BECAUSE IT IS SYMBOLIC

Physical law is provisional not only because it is approximate, but also because it is symbolic: there are always cases in which the symbols related by a law are no longer capable of representing reality in a satisfactory manner.

In order to study a certain gas, for example, oxygen, the physicist has created a schematic representation of it which can be grasped in mathematical reasoning and algebraic calculation. He has pictured this gas as one of the perfect fluids that mechanics studies: it has a certain density, is brought to a certain temperature, and is subject to a certain pressure. Among these three elements, density, temperature, and pressure, he has established a certain relation that a certain equation expresses: that is the law of the compressibility and expansion of oxygen. Is this law definitive?

Let the physicist place some oxygen between the plates of a strongly charged electrical condenser; let him determine the density, temperature, and pressure of the gas; the values of these three elements will no longer verify the law of the compressibility and expansion of oxygen. Is the physicist astonished to find his law at fault? Not at all. He realizes that the faulty relation is merely a symbolic one, that it did not bear on the real, concrete gas he manipulates but on a certain logical creature, on a certain schematic gas characterized by its density, temperature, and pressure, and that this schematism is undoubtedly too simple and too incomplete to represent the properties of the real gas placed in the conditions given now. He then seeks to complete this schematism and to make it more representative of reality: he is no longer content to represent oxygen by means of its density, its temperature, and the pressure it supports; he introduces into the construction of the new schematism the intensity of the electrical field in which the gas is placed; he subjects this more complete symbol to new studies and obtains the law of the compressibility of oxygen endowed with dielectric polarization. This is a more complicated law; it includes the former as a special case, but it is more comprehensive and will be verified in cases where the original law would fail.

Is this new law definitive?

Take the gas to which it applies and place it between the poles of an electromagnet; you will see the new law falsified in its turn by the experiment. Do not think that this new falsity upsets the physicist; he knows that he has to deal with a symbolic relation and that the symbol he has created, though a faithful picture of reality in certain cases, cannot resemble it in all circumstances. Hence, without being discouraged, he again takes up the schematism by which he pictures the gas on which he is experimenting. In order to have this sketch represent the facts he burdens it with new features: it is not enough for the gas to have a certain density, a certain temperature, and a certain dielectric power, to support a certain pressure, and to be placed in an electrical field of a given intensity; in addition, he assigns to it a certain coefficient of magnetization; he takes into account the magnetic field in which the gas is and, connecting all these elements by a group of formulas, he obtains the law of the compressibility and expansion of the polarized and magnetized gas, a more complicated and more comprehensive law than those he had at first obtained, a law which will be verified in an infinity of cases where the former would be falsified; and yet it is a provisional law. Some day the physicist expects to find conditions in which this law will in its turn be faulty; on that day, he will have to take up again the symbolic representation of the gas studied, add new elements to it and enounce a more comprehensive law. The mathematical symbol forged by theory applies to reality as armor to the body of a knight clad in iron: the more complicated the armor, the more supple will the rigid metal seem to be; the multiplication of the pieces that are overlaid like shells assures more perfect contact between the steel and the limbs it protects; but no matter how numerous the fragments composing it, the armor will never be exactly wedded to the human body being modelled.

I know what is going to be said in objection to this. I shall be told that the law of compressibility and expansion formulated at the very first has not in any way been upset by the later experiments; that it remains the law according to which oxygen is compressed and dilated when all electrical and magnetic actions are eliminated; that the physicist's later inquiries have taught us only that it was suitable to join to this law, whose validity was unaffected, the law of the compressibility of an ionized gas and the law of the compressibility of a magnetized gas.

These same persons who take things so obliquely ought to recognize that the original law could lead to serious mistakes if taken without caution, for the domain it governs has to be delimited by the following double restriction: the gas studied is removed from all electrical action as well as magnetic action. Now the necessity for this restriction did not appear at first but was imposed by the experiments we have mentioned. Are such restrictions the only ones which should be imposed on the law's statement? Will not experiments done in the future indicate other restrictions as essential as the former? What physicist would dare to pronounce judgment on this and assert that the present statement is not provisional but final?

The laws of physics are therefore provisional in that the symbols they relate are too simple to represent reality completely. There are always circumstances in which the symbol ceases to picture concrete things and to announce phenomena exactly; the statement of the law must then be accompanied by restrictions which permit one to eliminate these circumstances. It is the progress of physics which brings knowledge of these restrictions; never is it permissible to affirm that we possess a complete enumeration of them or that the list drawn up will not undergo some addition or modification.

This task of continual modification by which the laws of physics avoid more and more adequately the refutations provided by experiment plays such an essential role in the development of the science that we may be permitted to insist somewhat further on its importance and to study its course in a second example.

Here is some water in a vessel. The law of universal attraction teaches us what force acts on each of the particles of this water: this force is the weight of the particle. Mechanics indicates to us what shape the water should assume: whatever the nature and shape of the vessel are, the water should be bounded by a horizontal plane. Look closely at the surface bounding the water: horizontal at a distance from the edge of the vessel, it stops being so in the vicinity of the walls of glass, and rises along these walls; in a narrow tube the water rises very high and becomes altogether concave. There you have the law of universal attraction failing. In order to prevent capillary phenomena from refuting the law of gravitation, it will be necessary to modify it: we shall no longer have to regard the formula of the inverse ratio of the square of the distance as an exact formula but as an approximate one; we shall have to suppose that this formula shows with sufficient precision the attraction of two distant material particles but that it becomes very incorrect when the problem is to express the mutual action of two elements very close to each other; we shall have to introduce into the equations a complementary term which, while complicating them, will make them capable of representing a wider class of phenomena and will permit them to include the motions of heavenly bodies and capillary effects under the same law.

This law will be more comprehensive than Newton's law, but will not be, for all that, safe from all contradiction. At two different points of a liquid mass, let us insert the metallic wires coming from two poles of a battery: there you see the laws of capillarity in disagreement with observation. In order to remove this disagreement, we must again take up the formula for capillary action, and modify and complete it by taking into account the electrical charges carried by the fluid's particles and the forces acting among these ionized particles. Thus, this struggle between reality and the laws of physics will go on indefinitely: to any law that physics formulates, reality will oppose sooner or later the harsh refutation of a fact, but indefatigable physics will touch up, modify, and complicate the refuted law in order to replace it with a more comprehensive law in which the exception raised by the experiment will have found its rule in turn.

Physics makes progress through this unceasing struggle and the work of continually supplementing laws in order to include the exceptions. It was because the laws of weight were contradicted by a piece of amber rubbed by wool that physics created the laws of electrostatics, and because a magnet lifted iron despite these same laws of weight that physics formulated the laws of magnetism; it was because Oersted had found an exception to the laws of electrostatics and of magnetism that Ampère invented the laws of electrodynamics and electromagnetism. Physics does not progress as does geometry, which adds new final and indisputable propositions to the final and indisputable propositions it already possessed; physics makes progress because experiment constantly causes new disagreements to break out between laws and facts, and because physicists constantly touch up and modify laws in order that they may more faithfully represent facts.

THE LAWS OF PHYSICS ARE MORE DETAILED THAN THE LAWS OF COMMON SENSE

The laws that ordinary non-scientific experience allows us to formulate are general judgments whose meaning is immediate. In the presence of one of these judgments we may ask, "Is it true?" Often the answer is easy; in any case the answer is a definite yes or no. The law recognized as true is so for all time and for all men; it is fixed and absolute.

Scientific laws based on the experiments of physics are symbolic relations whose meaning would remain unintelligible to anyone who did not know physical theories. Since they are symbolic, they are never true or false; like the experiments on which they rest, they are approximate. The degree of approximation of a law, though sufficient today, will become insufficient in the future through the progress of experimental methods; sufficient for the needs of the physicist, it would not satisfy somebody else, so that a law of physics is always provisional and relative. It is provisional also in that it does not connect realities but symbols, and that is because there are always cases where the symbol no longer corresponds to reality; the laws of physics cannot be maintained except by continual retouching and modification.

The problem of the validity of the laws of physics hence poses itself in an entirely different manner, infinitely more complicated and delicate than the problem of the certainty of the laws of common sense. One might be tempted to draw the strange conclusion that the knowledge of the laws of physics constitutes a degree of knowledge inferior to the simple knowledge of the laws of common sense. We are content to reply to those who would deduce this paradoxical conclusion from the foregoing considerations by repeating for the laws of physics what we have said about scientific experiments: A law of physics possesses a certainty much less immediate and much more difficult to estimate than a law of common sense, but it surpasses the latter by the minute and detailed precision of its predictions.

Take the common-sense law "In Paris the sun rises every day in the east, climbs the sky, then comes down and sets in the west" and compare it with the formulas telling us the coordinates of the sun's center at each instant within about a second, and you will be convinced of the accuracy of this proposition.

The laws of physics can acquire this minuteness of detail only by sacrificing something of the fixed and absolute certainty of common-sense laws. There is a sort of balance between precision and certainty: one cannot be increased except to the detriment of the other. The miner who presents me with a stone can tell me without hesitation or qualification that it contains gold; but the chemist who shows me a shiny ingot, telling me, "It is pure gold," has to add the qualification "or nearly pure"; he cannot affirm that the ingot does not retain minute traces of impurities.

A man may swear to tell the truth, but it is not in his power to tell the whole truth and nothing but the truth. "Truth is so subtle a point that our instruments are too blunt to touch it exactly. When they do reach it, they crush the point and bear down around it, more on the false than on the true."[4]

4 B. Pascal, *Pensées*, ed. Havet, Art. III, No. 3.

Chapter 16
WILHELM OSTWALD
(1853–1932)

INTRODUCTION

Wilhelm Ostwald was born in Riga, Latvia, on September 2, 1853. Chemistry was his major area of concentration at the University of Dorpat in Estonia, where he received his doctorate in 1878. He first taught in a high school, then in 1882 was appointed professor of chemistry at the Riga Polytechnical School. He lectured in chemistry at the University of Leipzig from 1887 to 1909 after which he retired to study and write, mainly in the area of philosophy of science.

Several important studies on physical and analytical chemistry were produced during his academic career; one of the most important was his *Textbook of General Chemistry* (1885–1888). In company with S. Arrhenius and J. van't Hoff he firmly established physical chemistry, making it an almost independent branch of chemistry. He was also chiefly responsible for the foundation in 1887 of the *Zeitschrift für physikalische Chemie*. Ostwald's work in chemistry began with studies on the dynamics of reactions and in electrochemistry. By 1894 he was able to give the first valid definition of a catalyst, after which he proceeded to study a great number of catalytic reactions. He was awarded the Nobel Prize in chemistry in 1909 for his work in catalysis, chemical equilibrium, and reaction velocities.

Throughout his life Ostwald maintained a strong and continuing interest in historical problems connected with the natural sciences. In 1889 he undertook publication of the series *Klassiker der exakten Wissenschaften* in which virtually all his important papers on physics and chemistry were reprinted. Another of his projects was the careful examination of the psychological motives which supposedly led great scientists to their scientific productivity; he published the results of these investigations in 1909 in a book called *Grosse Männer*.

Ostwald's approach to chemistry as well as science in general was always strongly philosophical. In fact his refusal to believe in the reality of atoms and his attempt to explain all natural phenomena in terms of energy were based on philosophical grounds. His views on the subject appeared in various publications between 1888 and 1914: *Die Energie und Ihre Wandlungen* (1888) ("Energy and Its Transformations"); *Die Überwindung des wissenschaftlichen Materialismus* (1895) ("The Conquest of Scientific Materialism"); *Vorlesungen über Naturphilosophie* (1901) ("Lectures on Philosophy of Nature"); *Grundriss der Naturphilosophie* (1908) ("Outline of Philosophy of Nature"); *Die Energien* (1908) ("Energies"); *Energetische Grundlagen der Kulturwissenschaft* (1909) ("Energetism's Foundations of the Science of Culture"); *Moderne Naturphilosophie* (Vol. I): *Ordnungswissenschaften* (1914) ("Modern Philosophy of Nature (Vol. I): Sciences of Order"); and several essays published in *Vorträge und Abhandlungen* (1904) ("Lectures and Essays"). In 1901 Ostwald founded the review *Annalen der Naturphilosophie* (1901–1921); and in 1912 he published *Das monistische Jahrhundert*. Ostwald died at Grossboten, near Leipzig, on April 4, 1932.

Ostwald's philosophy is based on a classification of the sciences quite similar to Comte's: (1) first there are the sciences of order (*Ordnungslehre*): logic, mathematics, and the "pure space and time doctrine"; (2) then there are in the order of complexity of their subject matters: mechanics, physics, chemistry; physiology, psychology, and sociology. Ostwald is similar to Comte in that he believes the concept of order in science as such to be basic, but he steadfastly rejects the latter's materialism and mechanism, substituting for these views a form of energetism which he and Mach founded.

By the nineteenth century most scientists had developed some form of mechanism which ultimately came down to materialism. It involved in general an attempt to explain all natural phenomena by means of local movements of extremely small, extended particles to the exclusion of all qualitative aspects of matter. Ostwald and Mach objected to this view and proposed in its place their energetism. Although energetism was originally intended only as a physical theory, gradually the view expanded into a complete philosophy.

In Ostwald's view the essence of things cannot be known by man; he only knows how a certain "primary matter" manifests itself under the form of energy. There are, to be sure, many forms of energy which differ not only in quantitative, but also qualitative, aspects. As manifestations of the "primary matter" these many forms of energy are positive "realities" which have the ability to perform work. When they do so they are subject to the laws of conservation of energy, degradation of energy, and the principle of least action. Although energy is found in various forms such as potential, actual, kinetic, magnetic, electrical, and such, in all cases there is only one primary reality which manifests itself in these different forms. Energy is that only reality, the genuine substance; matter as distinguished from energy is inconceivable. As Ostwald formulated it: the most typical characteristic of energetism was that it rejected dualism, which had held sway over matter and energy for centuries; now energy was to occupy the central and most general position. Ostwald did not mean that matter must yield its place to energy, but rather that the concept of matter was to be abandoned altogether; what is called matter is no more than a frequently occurring combination of different forms of energy, particularly gravity and inert mass. In addition, the so-called spiritual realities are no more than different combinations of different types of energy. Briefly stated, Ostwald held that "The substance, that which is, is energy; the accidents, or the things as differentiated, that again is energy . . . Energy comprises the complete reality" (*Vorlesungen über Naturphilosophie*, pp. 146, 147, 377).

It is clear that in Ostwald's view there is no real distinction between philosophy and science. They are two roads leading to the same goal, domination of nature by man. Following Comte's positivistic lead, Ostwald claimed that philosophy's task consists in the classification and systematization of all sciences and their results. He did not, however, develop this general positivistic principle into a materialistic mechanism, but went instead in the direction of a monistic energetism.

In the selection presented here Ostwald attempts to show how the major mechanical theories can be "translated" in terms of his physical energetism.

SELECTIVE BIBLIOGRAPHY

Ostwald, W., *Vorlesungen über Naturphilosophie.* Leipzig, 1901.

————,*Vorträge und Abhandlungen.* Leipzig, 1904.

————,*Grundriss der Naturphilosophie.* Leipzig, 1908. (Trans. T. Seltzer, as *Natural Philosophy.* New York, 1910.)

Adler, W., *Die Metaphysik in der Ostwaldschen Energetik.* Berlin, 1905.

Slosson, E. E., *Major Prophets of Today.* Boston, 1914.

Günther, P., *Wilhelm Ostwald.* Leipzig, 1932.

Dochmann, A., *Wilhelm Ostwalds Energetik.* Bern, 1908.

Delbos, V., *Une théorie allemande de la culture: Wilhelm Ostwald et sa philosophie.* Paris, 1916.

Ostwold, G., *Wilhelm Ostwald, Mein Vater.* Stuttgart, 1953.

Meyerson, E., *Identity and Reality.* New York, 1962 (Chapter X and *passim*).

ENERGETISM AND MECHANICS

GENERAL

ALL properties which physical bodies in contradistinction to geometric bodies possess can be traced back to a fundamental concept, which . . . serves to characterize and distinguish the physical structure. For example, the fact that we can distinguish cubes of equal size but of different material, different temperature, and different luminosity, can be traced back always and entirely to the different kinds of energy acting in the geometric space in question. The concept of energy, therefore, plays approximately the same role in the physical sciences as the concept of thing in the formal sciences, and the essentials of this new field of science are the comprehensive knowledge and development of this concept. Because of its great importance it has long been known and applied in individual forms. But the systematization of the physical sciences relative to energy is a matter of only recent date.

MECHANICS

Recently many scientists have taken exception to the traditional division of mechanics into *statics,* or the science of equilibrium and *dynamics,* or the science of motion, because it does not correspond to the essence of the thing, equilibrium being only the limit-case of motion. However, the classic presentations of this science are based on that division, so that it must express an essential difference. This difference we can clearly recognize through the application of the concept of energy to mechanics. We then learn that statics is the science of work, or the energy of position, and that dynamics is the science of living force, or of the energy of motion.

By *work* in the mechanical sense we mean the expenditure of force required for the locomotion of physical bodies. While a cube of lead is geometrically equal to a cube of glass, we experience a great difference between them when we lift them from the floor to a table. We call the cube of lead heavier than the glass cube, and we find it requires more work to raise the former than the latter. For psychologic reasons this judgment becomes especially clear when the work required to lift the lead cube marks the limit of our physical capacity.

Work depends not only upon the difference described above, but also upon the distance through which it is exerted. It increases in proportion as the distance increases. In mechanics work is proportional both to the distance and to that peculiar property which in the given example we call *weight*. But a more general concept has been formed for that property in the mechanical sense, called *force*, of which weight constitutes but a special instance. Whenever there is a resistance combined with a change of place we speak of a force, *and the product of the force and the distance we call work.*

The cause of this kind of concept formation is the following: There are a great number of different machines, all of them possessing the peculiarity that work can be put into them at a definite place and taken out at another place. Now, centuries of experience have shown that it is impossible to obtain more work from such mechanical machines than has been put into them. As a matter of fact, the work obtained is always less than the work put in, and the two approach equality as the machine approaches perfection. It is to such ideal machines, therefore, that *the law of the conservation of work* applies. This law states that, though a given quantity of work may be changed in the most manifold ways as to direction, force, etc., it is impossible to change its *quantity.*

The reason we can judge of this fact with such certainty is because for many centuries a number of the ablest mechanicians have sought for a solution of the problem of perpetual motion, that is, for the construction of a machine from which more work can be gotten than is put into it. All such attempts have failed. But the positive result secured from these apparently futile efforts is the law of the conservation of work. The greatness and importance of this result will become apparent in the further course of our study.

Here for the first time we meet with a law expressing the *quantitative* conservation of a thing, which may none the less undergo the most varied qualitative changes. With the knowledge of this fact we involuntarily combine the notion that it is the "same" thing that passes through all these transformations, and that it only changes its outward form without being changed in its essence. Such ideas, it is true, are widespread, but they have a very doubtful side to them, since they correspond to no distinct concept. If we want to call the quantitative magnitude of the product of the force and distance the "essence" of work, and the determination of the force and the distance according to magnitude and direction, which come under consideration for each special value, as its "form," then, of course, there is no objection to be made to mere nomenclature. But we must bear in mind that the difference obtaining here lies exclusively in the fact that the amount of work measured quantitatively remains unchanged, while its factors undergo simultaneous and opposite changes.

This discovery, that there is a magnitude which can be quantitatively determined, and which, as experience shows, remains unchanged, however much its factors may change, invariably results not only in a very simple and clear formulation of the corresponding natural law, but also corresponds to the general tendency of the human mind to work out conceptually "the permanent in change." If, in accordance with the word-sense, we denote everything which persists under changing conditions by the name of *substance, we encounter in work the first substance* of which we have attained knowledge in our scientific journeys. In the history of the evolution of human thought this substance has been preceded by others, especially by the weight and mass of ponderable bodies (which are also subject to a law of conservation), so that at present we are inclined to connect with the word substance a tacit secondary sense of ponderability. But this is a remnant of the still very widely spread mechanistic theory of the universe, which, though it has almost finished its role in physics, will presumably continue to persist for a long time to come in the popularly scientific consciousness in accordance with the laws of collective thought.

KINETIC ENERGY

The law of the conservation of work is by no means true of all cases in which work is expended or converted, but, as has been said, only of *ideal* machines, that is, of such cases which do not exist in reality. But while in imperfect machines there is at least an approximation to this law, it must also be recognized that there are besides countless normal cases in which we cannot even speak of an approximation. When, for example, a stone falls to the ground from a certain height, a certain quantity of work is expended, which is equal to that by means of which the stone can be raised again to its original height. This quantity of work apparently disappears entirely when the stone remains lying on the ground. We shall discuss this case later. Or the falling of the stone can be so guided that it can lift itself again. This happens, for instance, when, by fastening the stone to a thread, it is forced to move in a curved path, or to perform pendular oscillations. In that case, it is true, the stone will fall to the lowest point which the thread permits, and so will there have lost its work without having done any other work in the meantime. But it has entered a condition by virtue of which it raises itself again, so that (as before, only in the ideal limit-case) it reaches its former height, and so has lost no work. For this moment, too, then, the law of the conservation of work obtains. But in the meantime new relations have arisen.

What distinguishes the stone moving like a pendulum from the stone which simply falls is, that at its lowest point it has not remained lying still, but possesses a certain velocity. By means of this it lifts itself again, and after it has reached its former height, it has lost its velocity. *Therefore, there is a reciprocal relation between the work which it loses and the velocity which it gains,* and the question may therefore be put. How can this relation be represented mathematically? Experience teaches that in every such case a function of the velocity and of another property of the body, called *mass,* can be established in such a way that this function, called the *kinetic energy* of the body, increases precisely as much as the amount of work the body has expended, and *vice versa.* The sum of the kinetic energy of the body and of the *work* is therefore *constant,* and the clearest mode of conceiving of this relation is by assuming *that work can be transformed into kinetic energy and vice versa* in such a way that given amounts of the two magnitudes are equal or equivalent to one another. Naturally, this is only an abbreviated way of expressing the actual relations, for it might just as well be assumed that the work really disappears and the kinetic energy really originates anew, and that the disappearance of the one substance only happens regularly to coincide with the origin of the other. But it is this regular conjunction of phenomena that constitutes the sole ground of every *causal* relation, and in such a sense we are justified *in regarding the disappearing work as the cause of the kinetic energy that arises,* and to designate this relation summarily as a transformation.

By the inclusion of cases in which work is converted into kinetic energy the law of the conservation of work therefore becomes *the law of the conservation of the sum of work and kinetic energy*. We are thereby compelled to extend the concept of substance, which at first contains only work, to the sum of both magnitudes, and to introduce a new name for this enlarged concept.

It will soon appear that all cases of imperfect machines, in which work disappears without giving rise to an equivalent amount of kinetic energy, can, with a corresponding enlargement of the concept, be likewise included in the law of conservation. For experience has shown that in such cases something else arises, heat, light, or electric force, etc. This generalized concept, which embraces all natural processes and permits the sum of all corresponding values to be expressed by a law of conservation, we call *energy*. The law in question, therefore, is:

In all processes the sum of the existing energies remains unchanged.

The principle of the conservation of work in perfect machines proves to be an ideal special instance of this general law. A perfect machine is one in which work changes into nothing but *work* of another kind, and not into a different kind of energy. Then each side of the equation which expresses the general law of energy, namely,

Energy that has disappeared = energy that has arisen,

contains only the magnitude of the work, and expresses the law of the conservation of work. If, on the other hand, as in the case of the pendulum, the work increasingly changes part by part into kinetic energy, and *vice versa,* the equation during the first period is:

Work that has disappeared = kinetic energy that has arisen,

and during the second period in which the pendulum rises again,

Kinetic energy that has disappeared = work that has arisen.

Thus, while work can be called a substance only in a limited sense, since its conservation is limited only to perfect machines, we may call energy a substance unqualifiedly, since in every instance of which we know the principle has been maintained *that a quantity of any energy never disappears unless an equivalent quantity of another energy arises*. Accordingly, this law of the conservation of energy must be taken as a fundamental law of the physical sciences. But not only do all the phenomena of physics, including chemistry, occur within the limits of the law of conservation, but until the contrary is proved the law of conservation must also be regarded as operative in all the later sciences, that is, in all the activities of organisms, so that all the phenomena of life must also take place within the limits of the law of conservation. This corresponds to the general fact, which I have emphasized a number of times, that all the laws of a former science find application in all the following sciences, since the latter can only contain concepts which by specialization, that is, by the addition of further characteristics, have sprung from the concepts of the former or more general sciences.

MASS AND MATTER

It has been noted above that kinetic energy depends upon another magnitude beside velocity. A conception of its nature can be obtained when we try to put different bodies in motion. In doing so the muscles of the arm perform certain quantities of work, and we feel whether the quantities are greater or smaller. In this way we obtain a clear consciousness of the fact that different bodies require quite different quantities of work for the same velocity. The property which comes into play here is called *mass*, and mass is proportional to the work which the various bodies require to attain the same velocity. Since the work and the velocity can be measured very accurately by appropriate means, mass also lends itself to a correspondingly accurate measurement.

All known ponderable bodies have mass. That means there is a regular connection between the property which makes a body tend to the earth with a certain definite force (called weight) and the property by virtue of which a body assumes certain velocities under the influences of motive causes. We can readily conceive that it is possible for us to learn only of such bodies as are heavy, that is, bodies which are *held* by the earth, since the others, if they exist at all, would naturally have left the earth long ago. That all these bodies also have mass is to be explained in a similar way. For a body of mass zero would at each impulse assume infinitely great velocity, and could therefore never be the object of our observation. Consequently, by reason of the physical conditions obtaining on the earth's surface, the bodies known to us must combine both properties, mass and weight.

The name given to this concept of the combined presence of mass and weight in space is *matter*. Experience shows that there is a law of *conservation* for these magnitudes also, according to which *whatever changes we may produce in bodies possessing weight and mass, no change will occur in the sum of their weight and mass.* According to the nomenclature previously introduced we must therefore call weight and mass substances, since they remain the same as to quantity, no matter what changes they may undergo. However, it is usual to apply the name substance to the concept of matter composed of mass and weight. In fact, scientists often go so far as to limit the name to this single instance of the various laws of conservation, and to take substance to mean exclusively the combination of mass and weight. This is connected with the conception which we are about to discuss, that all natural phenomena can ultimately be conceived as the motion of matter. Through the greater part of the nineteenth century this conception, called *scientific materialism,* was accepted almost without opposition. At present it is being more and more recognized that it was only an unproved assumption, which the development of science daily proves to be more untenable.

ENERGETIC MECHANICS

In the light of our previous observations the branch of science traditionally known as mechanics appears as the science of work and of kinetic energy. Furthermore, statics is shown to be the science of work, while dynamics, besides treating of kinetic energy in itself, also treats of the phenomena of the change of work into kinetic energy, and *vice versa.* We shall find the same relation again later, only in more manifold forms. Every branch of physics proves to be the science of a special kind of energy, and to the knowledge of each kind of energy must be added the knowledge of the relations by which it changes to the other forms of energy and *vice versa.* It is true that in the traditional division of physics this system has not been strictly carried out, since an additional and very influential motive for classification has been the regard paid to the various human sense organs.

Nevertheless this ground does not lie in the field of physics, but in that of physiology, and must, therefore, be abandoned in the interest of strict systematization.

Of the physical sciences mechanics was the first to develop in the course of historical evolution. A number of factors contributed to this end—the wide distribution of mechanical phenomena, their significance to human life, and the comparative simplicity of the principles of mechanics, which made it possible to discover them at an early date. Most to be noted is, that of all departments of physics mechanics is the first which lent itself to comprehensive *mathematical* treatment. It is true that the mathematical treatment of mechanics was possible only after idealizing assumptions had been made—perfect machines and the like—so that the results of this mathematical treatment not infrequently had very little to do with reality. The mistake of losing sight of the physical problem and of making mechanics a chapter of mathematics has not always been avoided, and it is only in most recent times that the consciousness has again arisen that the classical mechanics, in arbitrarily limiting itself to extreme idealized cases, sometimes runs the risk of losing sight of the aim of science.

THE MECHANISTIC THEORIES

Because the evolution of mechanics antedates that of the other branches of physics, mechanics has largely served as a model for the formal organization of the other physical sciences, just as geometry, which has been handed down to us from antiquity in the very elaborate form of Euclid, has largely been used as a model for scientific work in general. Such methods of analogy prove to be extremely useful at first because they serve as a guide to indicate when and where new sciences, in which all possibilities are open, can be got hold of. But later on such analogies are apt to be harmful. For each new science soon requires new methods, by reason of the peculiar manifoldness which it has to deal with, and the finding and the introduction of these new methods are easily delayed, and, as a matter of fact, often have been delayed, because scientists could not free themselves soon enough from the old analogy.

By its being based upon memory the human mind is so constructed that it cannot assimilate something entirely new. The new must in some way be connected with the known in order that it may be organically embodied in the aggregate of concepts. Therefore, it is the first involuntary impulse of our mind, in the presence of new experiences or thoughts, to look about for such points at which a linking of the unknown to the known seems possible. In the case of mechanics this necessity for finding connecting links has acted in such a way that the attempt has been made, and is still being made, to conceive and represent all physical phenomena as mechanical.

The impulse to this was first given by the extraordinary successes which mechanics has attained in the generalization and prediction of the *motions of the heavenly bodies.* The names of Copernicus, Kepler, and Newton mark the individual steps in the mechanization of astronomy. The cause of this lies in the fact that the heavenly bodies actually approximate very closely the ideal of the purely mechanical form with which classical mechanics deals. These successes encourage the attempt to apply these mental instruments that were productive of such rich results to all other natural phenomena. An old theory, according to which all physical things are composed of the most minute solid particles of matter called *atoms,* supported these tendencies and invited the attempt to regard the little world of atoms as subject to the same laws as had been found to apply so successfully to the great world of the stars.

Thus we see how this mechanistic hypothesis, the assumption that all natural phenomena can be reduced to mechanical phenomena, comes as if it were a self-understood matter, and with its claim to be a profound interpretation of nature it scarcely permits the question as to its justification to be raised at all. And the effects here have been the same as I described above in cases in which inferences from analogy are accepted too extensively or too credulously. While it is true, no doubt, that the mechanical hypothesis at first was fruitful of results in special research, because it facilitated the putting of the question—for example, we need think only of the atomic hypothesis in chemistry—later, the efforts to find further hypothetic help for the inadequacies of the hypothesis that gradually came to light, have not infrequently led scientific research to pseudo-problems, that is, to questions which are questions only in hypothesis, but to which no actual reality can be shown to correspond. Such problems, therefore, are by their very nature *insoluble,* and constitute an inexhaustible source of differences of scientific opinion.

The most flagrant of the injurious consequences of the mechanistic hypothesis appear in the scientific treatment of the mental phenomena. Ready as scientists were to represent all other life phenomena, such as digestion, assimilation, and even generation and propagation, as the consequence of an extremely complicated play of certain atoms, their courage never went so far as to apply this principle to mental life and to consider that by mechanics the last word had been said on the subject.

It is because of this hesitancy to bring mental phenomena under the same mechanistic principle as all the other phenomena that the philosophical systems had to search for some other means to connect the mental world with the mechanical, and the efforts of the philosophers to bring about this end have been most varied. Of the various doctrines that have come down to us, that of the *pre-established harmony* proposed by Leibnitz is in the ascendant in our day, and is now called the theory of the *psycho-physical parallelism*. According to this theory it is assumed that the mental world exists alongside, and quite independent of, the mechanical, but that the things have been so prearranged that mental processes take place simultaneously with certain mechanical processes (according to some, with all mechanical processes) in such a way that, although the two series do not influence each other in the least, they always correspond to each other precisely. How such a relation has come about and how it is maintained remains unsaid, or is left to future explanation.

We need only think of the content of this hypothesis with an unbiased mind to lose all relish for it at once. In fact, it has no other *raison d'être* than the presumption that the mental and the mechanical world are opposed to each other. As soon as we abandon the thesis that the non-mental world is exclusively mechanical, we acquire the possibility again of finding for the theory of mental phenomena a constant and regular connection with the theories of all other phenomena, especially with the phenomena of life. Therefore it will be found most expedient in every respect, instead of rendering scientific research one-sided and almost blind to non-conforming facts by preconceived hypotheses, such as the mechanistic hypothesis, to seek, as hitherto, from step to step, the new elements of manifoldness which must be taken account of in the progressive upbuilding of science and to limit ourselves faithfully to them in the formation of general ideas.

Chapter 17

ÉMILE MEYERSON
(1859–1933)

INTRODUCTION

EMILE MEYERSON was born in 1859 in Lublin, then a part of Russian Poland. He studied at the Universities of Dresden, Leipzig, and Berlin, after which he undertook additional work in chemistry at Göttingen, Heidelberg, and again in Berlin. In 1882 Meyerson went to France where he worked as a chemist until 1898. In that year he accepted an administrative position in Paris and held it until 1923. He continued to study, concentrating on the history of natural science in general and the history of chemistry in particular. He was not so much interested in the succession of the different theories as in the principle which seemed to guide the human mind in its attempt to penetrate beneath the somewhat superficial generalizations of our everyday knowledge of nature. He was convinced that the various forms which scientific theory has taken in the course of history can be understood only in the light of a single principle which is somehow immediately connected with the very nature of the human mind and thus, for that reason, could be called a priori as intended by Kant. Meyerson believed this to be the only way to account for something permanent and unchanging in a world of spatial and temporal change and qualitative multiplicity.

In his development of this idea Meyerson took up the tradition of the French movement in philosophy of science which began near the end of the nineteenth century as a result of Boutroux, Poincaré, Bergson, and Brunschvicg's work, collectively known as "critique of science." Meyerson did not repeat their ideas, he developed them further. His elaboration of the ideas of this tradition reveals his great familiarity with Scholastic philosophy and the philosophies of Leibniz and Kant, and also with Descartes, Lavoisier, and Carnot.

Meyerson joined Joseph Wilbois and Pierre Duhem in rejecting Comte's positivism, as well as any form of skepticism, resolutely choosing instead for a rationalist approach in philosophy of science.

In *Identity and Reality* (1908) and in *On the Explanation in the Sciences* (1921) he also objected to Poincaré's principle which held that natural science is free to choose those hypotheses which are most convenient for our understanding. In Meyerson's view natural science is a science of the real world and consequently must adapt itself to the very nature of things. This interest in the "ontological" dimension of natural science made him object to Mach's view, too, in which the work of science is seen as limited to discovering laws of mere phenomena in the interest of their practical application. He considered Mach's attempt to reduce scientific laws to generalizations that are required for practice as a form of skepticism and to be avoided. Practice for him meant only one phase of the life of man's mind; in addition to this, the mind could and must bring to light the rationality which is present in nature itself. Meyerson did not fail to see, however, that in its practical realization natural science must start with certain hypotheses; it merely presupposes that nature is framed according to an order that can be explained rationally. It also presupposes that the principle of conformity to law is universally valid in nature. Furthermore, in Meyerson's view, without concrete hypotheses scientific investigation is impossible. All this accounts for the fact that a scientific theory will always be more than a copy of reality, though this does not alter the fact that *in addition to this* science must genuinely attempt to understand the natural phenomena by reducing them to their true causes. But in so doing, Meyerson contended, science will understand nature only if it is able to penetrate to the very being of things, for it is only on that level that the original postulates and hypotheses can be justified. It is on that level, too, that it becomes clear that what originally manifested itself as a multiplicity appears to be a unity. This is why Meyerson claimed, as did Bergson, that the aim of science consists in the discovery of rationality *in* nature, and that rationality consists in the reduction of differences to identity in a way that these differences will be absorbed in that identity without remainder. Such an ideal can never be completely materialized because of the eternal presence of residual elements of irrationality in nature.

In 1925 Meyerson applied these general insights to Einstein's theory of relativity (*La déduction relativiste*). In Meyerson's opinion the theory of relativity is to be interpreted as a theory of reality; he felt that it was not too difficult to show that this theory proves the inadequacy of Kant's conception of space and time, and that space and time do not consist in a priori forms of man's sensitivity, but refer immediately to the essence of natural things. Meyerson later applied similar ideas to quantum mechanics in his last book, *Réel et déterminisme dans la physique quantique* (1933).

The selection to follow is taken from the *Conclusions* of Meyerson's first major work, *Identity and Reality*. The first part of this last chapter of the book briefly summarizes the work's main thesis: It is not true that the sole end of science is action, nor that it is solely governed by the desire for economy in this action. Science also wishes to make us *understand* the world. That is why science is not and cannot be positive in Comte's sense of this term, namely, stripped of all ontology. On the contrary ontology is of a piece with science itself and cannot be separated from it. Those who pretend to separate them are unconsciously using a quite universally accepted ontology, a sort of common-sense metaphysics more or less transformed by science of the past. Who denies the ontological character of science must, if he wants to be consistent, at the same time deny the principle of conformity to law. But in so doing science would be made impossible (pp. 385–386).

SELECTIVE BIBLIOGRAPHY

Meyerson, É., *Identity and Reality* (1908), trans. (1930) Kate Loewenberg. New York, 1962.

———, *De l'explication dans les sciences*. Paris, 1921.

———, *La déduction rélativiste*. Paris, 1925.

———, *Réel et déterminisme dans la physique quantique*. Paris, 1933.

Sée, H., *Science et philosophie d'après la théorie de M. É. Myerson*. Paris, 1932.

Mertz, A., *Meyerson; une nouvelle philosophie de la connaissance*. Paris, 1934.

Kelly, T. R., *Explanation and Reality in the Philosophy of É. Meyerson*. Princeton, N.J., 1937.

Boas, G. A., *A Critical Analysis of the Philosophy of Émile Meyerson*. Baltimore, 1930.

Stumpfer, S., *L'explication scientifique selon M. É. Meyerson*. Luxembourg, 1929.

IDENTITY OF THOUGHT AND NATURE
AS THE FINAL GOAL OF SCIENCE

*F*ROM the theoretical considerations which precede, is it possible to draw any indications concerning the methods science ought to follow? It seems to us that, on the whole, our results end in confirming the processes which scientists have applied more or less consciously up to the present time; and this is but natural, seeing that our special desire has been to extract rules from these very processes. We believe that we have shown that the methods of science hold together better than is generally supposed and that their basis is not that usually attributed to them; and in this way we have come to understand better the necessity of certain developments. Thus atomic theories must be maintained in science. Certainly the treatises of mathematical physics are full of them at the present time. But at times one detects a sort of discomfort in the authors; feeling that their spatial representations end in contradictions, they excuse themselves for using them. Van't Hoff, for example, one of the authors of the theory of the asymmetric atom of carbon, seems to regret that one is obliged to use the molecular conception, upon whose hypothetical character he lays stress. This necessity, however, appears provisional to him.[1] Maxwell, who has done so much for the development of mechanical conceptions, whose principal work—his treatise on electricity—is, as Poincaré states,[2] dominated by the desire to show in each particular case the possibility of mechanical explanations, and who has at least partially succeeded in introducing them

---

[1] van't Hoff: *Leçons de chimie physique*, trans. Corvisy, 1st Part. Paris, 1898, p. 9.
[2] H. Poincaré: *Electricité et Optique*. Paris, 1901, p. IV–VIII. Cf. *La science et l'hypothèse*, p. 249.

This article is from **Identity and Reality** by Émile Meyerson, trans. Kate Loewenberg (London, 1930), pp. 405–421. Copyright © 1930 by George Allen & Unwin, Ltd. Reprinted by permission of Dover Publications, Inc.

---

into the domain of Carnot's principle, which until then seemed closed to them, Maxwell himself sometimes yields to this tendency. In his communication to the Mathematical and Physical Section of the British Associaion, before giving a brief and brilliant exposition of the kinetic theory of gases, he warns his hearers that this is only a metaphor, an illustration, and is only useful as such. There are men who can get along without this help, and Maxwell admits implicitly that this is a mark of superiority; but the majority have need of it, and science must satisfy both.[3] Duhem, who has so clearly recognized the essential character of the theories . . . , declares also that the use of mechanical theories is a question of personal convenience.[4] These reservations seem unjustified. No doubt, it is possible, by completely neglecting the natural development of science, to give to certain of its parts the appearance of pure empiricism. Attempts of this kind will always be interesting, as is everything which proceeds logically from a single point of view. Such an exposition will also have the additional advantage of making us see clearly the results that have been reached. But it involves inconveniences also.

The principle of inertia, we have seen, can be proved experimentally; the entire branch of mechanics depending upon it can consequently be treated, conformably to the programme of Kirchhoff, as a "descriptive science." But for the conservation of energy we have found experimental demonstrations to be insufficient; if, therefore, one makes these proofs its basis, starting from the work of Joule and neglecting all prior developments, not only is a heresy committed from an historical point of view,[5] but one runs the risk of completely perverting the signification and import of the principle. Moreover, expounded without hypotheses, the experimental results appear to us as something definite, finished, without our having seen the way which has led us to them, nor the way which can lead us farther; for science is not Baconian, and experiment alone, without the help of hypothesis, cannot lead far. That is why the image of science or of a part of science, offered us thus, will be in a sense static. Whereas as a matter of fact science is in perpetual flux, is dynamic.

We have seen . . . that Comte believed that certain laws possessed

---

[3] MAXWELL: *Scientific Papers*, Vol. II, p. 219.

[4] DUHEM: *L'évolution de la mécanique*, p. 186.

[5] HELMHOLTZ (*Vortraege und Reden*. Brunswick, 1896, p. 407) insists particularly on the fact that the leading principles which had inspired him in his work of 1847 on the conservation of energy did not seem to him "new but, on the contrary, very old."

this character of definiteness, which led him to veto any research capable of modifying them; and it is clear that this conception would have been impossible except for his horror of all explanatory theory.

Without wishing consequently to proscribe the attempts of which we have been speaking, especially in the exposition of chapters of science in a very advanced phase of evolution, we believe that the scientist ought, every time that the development demands it, to use to a great extent kinetic considerations. It is certain that from the point of view even of the strictest experimental science, we have an interest in following causal deductions to the end, even though they are the most abstract in appearance. Boltzmann, in stating the speculations of which we have spoken concerning Carnot's principle, has very rightly insisted upon this fact, that they ought to be considered as unimportant, for they may suggest experiments about the limits of the divisibility of matter, the size of the spheres of action, etc.

The scientist in building up his hypotheses ought not even to be too much afraid of ultimate contradictions. The agreement between the causal image and the phenomenon, between thought and nature, cannot be complete; but there is a real and profound analogy. Every mechanical theory of a series of phenomena constitutes an immense scientific advance; it will lead surely to discoveries, for it contains a revelation of the essence of these phenomena. However perfect in many respects Fresnel's theory was, it would certainly have been wrong to take literally the affirmation of the existence of the luminiferous ether endowed with the contradictory properties we know. But that there was similarity, more than that, an identity of properties between light and the mode of spatial motion called undulation, that was certain, and this truth has remained even when Fresnel's conceptions have given place to those of Maxwell. Now these undulations could only be conceived of in a medium; it was, therefore, wise to accept the luminiferous ether in spite of its irreconcilable properties.[6] One must be resigned to these contradictions which result from the limitations of our understanding. Doubtless, one should try to reduce them to a minimum and a great progress will have been made every time an agreement has been found in the theory of two

[6] MACH, whose impartiality on this question can in no way be doubted, declares that if Young and Fresnel had abandoned the undulatory hypothesis (because of the difficulties experienced in imagining them in a fluid) the loss to science would have been as great as if Newton had abandoned the theory of gravitation (*Erkenntnis und Irrtum*. Leipzig, 1905, p. 251).

distinct parts of science. But no theory, however perfect it may be, will ever be logical or intelligible to the very end. Scientists have been very much concerned with certain fundamental problems of mechanism. Must the elementary particles be considered as infinitely elastic or as infinitely hard? Leibniz and Huygens have earnestly discussed this question. In the same way we have seen that an explanation has been sought for the elasticity of atoms by motions of all kinds. Finally, there have been great discussions on the question whether we ought to consider the atom as a corpuscle or as a centre of forces. This research and these discussions have been very useful; there would evidently be an advantage in definitely putting aside action at a distance, which is too repugnant to our understanding. By reducing all transitive action to a single type, we should have, in the first place, a minimum of the irrational, which is always satisfactory from a theoretical point of view; and besides, this hypothesis would without doubt be truer—that is, would present a greater analogy with reality, and would better than any other enable us to discover more relations between phenomena. This would be a real advance and would have its immediate repercussion even on the part of science devoted solely to law: it is probable that our calculations would be simplified by it, for it is less complicated to suppose that a particle is only influenced by what is contiguous to it than to admit that it is directly under the influence of the entire universe. An eminent theorist has shown in this respect that Maxwell's equations, which are based upon the exclusion of action at a distance, are only a particular case, a simplification, of those of Helmholtz, where such action, on the contrary, is posulated;[7] and another scientist, while declaring that he would consider the corpuscular and dynamic conceptions as equivalent, confesses nevertheless to have been surprised by the simplicity and elegance of certain formulæ of Hertz.[8]

As to the ultimate end which these theories have in view, an end essentially unattainable, it lies in the explanation of phenomena by an element differentiated to the least possible extent from space, or, if you will, it lies in the reduction of matter to space. From this point of view science can never turn aside from the path traced for it by perhaps the most powerful mind humanity can boast of—Descartes. (Descartes' mistake was in believing that deductions could produce anything but hypothetical constructions.) Never will any theory what-

---

[7] PLANCK: *Die Maxwell'sche Theorie. Wiedemann's Annalen,* 1899, Spl.
[8] L. LANGE: *Das Inertialsystem. Philosophische Studien,* 1902, p. 55.

soever be simply true. We saw that the superiority of corpuscular mechanism over dynamism resides in the fact that it corresponds more nearly to a postulate relative to space; and that the virtue, the "explicative force," of mechanism in general comes solely from the causal principle—that is, from the fact that the persistence of something is stipulated—and not from the fact that the thing of which the conservation is postulated, the dynamic or corpuscular atom, can ever be rendered comprehensible, its essential function, transitive action, being fundamentally inaccessible to our intelligence. The end of the reduction can only be irrational. It may very well be that in this respect electric theories, which presuppose an action from place to place, the mechanical nature of which is not determined, constitute the best solution. But while waiting for that solution to be firmly established and universally accepted, scientists are right in using in their theories, without too many scruples, both corpuscles and actions at a distance, neglecting momentarily the contradictions in which they end.

So scientists are wise in not wrangling too much over the unconscious "deceits" of the theories we have noted, such as the theory which considers the ether both as identical and differentiated. . . . These are simple consequences of the causal instinct which ends in the causal illusion, and they are justified by the partial agreement between our thought and reality. As Duhem has said: "No teaching of physics can (and we add *ever* could) be given under a form which leaves nothing to be desired from a logical point of view."[9] In the same line of thought one must not be too surprised if physicists are led to use, side by side, two or even several contradictory theories or incongruous mechanical models, as Maxwell has done amongst others. No doubt, this is rather inconvenient; a theory well regulated and logical would be infinitely preferable, because, being *truer*, it would have much more chance of rendering us service. Duhem is right, therefore, in protesting against the abuse of these procedures;[10] but perhaps too much rigour in this matter is not advisable. In a science like that of electricity, which is in a state of rapid development, somewhat irregular procedures of this kind may render great service. And it must also be added that since common sense can here be of no direct help to us (lacking as we do for electricity the organ of im-

[9] DUHEM: *La théorie physique*, p. 424.
[10] *Ib.*, p. 145 and ff.

mediate sensation), material images, whatever they be, are often indispensable to uphold our wavering imagination.

Yet, in using atomic theories, it would be wise for the scientist to see clearly the nature of these conceptions, and the nature of the *a priori* elements which they conceal. He will then understand better that their basis is immutable, because they rest upon what constitutes the foundations of the human mind. Very great scientists at times have been turned aside from the right path on this point. Thus Boltzmann admitted that there might be a question of changing atoms.[11] There would have been nothing to say against this hypothesis if this physicist had conceived the atoms (or what supposedly were such) as composed of smaller particles which remained immutable. But, according to all appearances, he wished to speak of ultimate particles of matter, and hence his supposition is inadmissible. Atoms which should simply change in time (nothing else could be imagined) would change without cause; they could not, therefore, explain anything—that is, they would lose their reason for being, and being only creatures of reason they would cease to exist. Nor must we ever lose sight of the fact that we do not know where the analogy between mechanism, or, better still, between the causal conception of the world and nature ceases. The only thing of which we are sure is that it must cease somewhere. If, then, we come upon a phenomenon which does not seem to conform to this conception, we doubtless have the right to try every way of making it conform; but we are not permitted to throw it aside, or reject it, if it appears unruly. It may be precisely that the future has many ideas of this kind in store for us. After having sought a long time for what especially persists, science, since the importance of Carnot's principle has been clearly recognized, is turning its attention more and more toward that which is modified, toward perpetual flux, and it is clear that on this ground causal considerations will always be more or less at fault.

Within these limits we would say with Ostwald[12] that the identity of thought and being postulated by Spinoza, Schelling, and Hegel, remains the programme of science, a programme toward the realization of which its efforts tend. This tendency shows itself through the influence of the causal considerations of which mechanical theories constitute a particular expression.

It is permissible, therefore, to state that science really tends toward

---

[11] BOLTZMANN: *Wiedemann's Annalen*, Vol. LX, 1897, p. 240.
[12] OSTWALD: *Vorlesungen*, p. 6.

the reduction of all phenomena to a universal mechanism or atomism, defining these terms so as to include electrical theories, and remembering that the causality of being, so near a relative of the causality of becoming, demands that the elementary particles be made of a single matter possessing only a minimum of qualities, in such a way that it may be, to a certain extent, identified with space or its hypostasis, ether. Not that this reduction is really possible, nor that we can believe that this atomism constitutes the essence of things, nor that it is capable of offering a system free from contradiction—but because it is, amongst all the images which our intellect is capable of conceiving, the only one, which, satisfying at least to a certain degree our tendency in the direction of identity offers at the same time real and sometimes surprising agreements with phenomena. It is, therefore, in following up this image, in rendering it more and more adequate to the facts, that we have the greatest chance of knowing these latter better. In other words, reduction to mechanism and atomism is not in itself an end but a means. It is a rule which guides the progress of science, as Lange has shown.[13]

It cannot be said that science approaches indefinitely reduction to mechanism, if by this last term is meant a logical, coherent hypothesis, free from contradictions. Science only accomplishes finite progress, and all the mechanical hypotheses which it forms being contradictory in themselves—that is, absurd at bottom—it always remains separated by an infinite distance from that logical conception toward which it seems to tend.

It is as a guide, as the directing principle, that mechanism has rendered invaluable services to science and—the past being a sure guarantee for the future—will doubtless render still more. Adopting it frankly as such, we shall have, moreover, the advantage of ridding theoretical science of certain phantoms (as Bacon would say) which haunt it, such as the "tendency to unity" or the "tendency to simplicity." In a certain sense the tendency to unity exists, since our understanding, denying all diversity in time and space, tends to reduce finally the totality of phenomena to an indistinct whole. But this tendency is not an independent principle and it is not at all mysterious. It is a direct consequence of the principle of identity. In regard to simplicity we must make distinctions. Science surely tends to simplify acquired knowledge—that is, to summarize it by formulating

[13] LANGE: *Geschichte des Materialismus,* 4th ed. Iserlohn, 1882. Preface, HERMANN COHEN, p. ix.

more and more general laws and theories; this is a consequence of the principle of the economy of effort, which is the source of empirical science. But it is not true to say that as science advances our conception of a real phenomenon gains in simplicity; for if science often discovers the simple beneath the complex, at other times, as Poincaré shows, it is, on the contrary, the complex which science discovers beneath the appearances of simplicity. Thus the simplicity of Newton's law might very well be only apparent. "Who knows whether it is not due to some mechanism, to the impact of some subtile matter . . . and whether it did not become simple only through the play of averages and large numbers."[14] And if of two formulæ, of two theoretical solutions, we ought always, the merits being equal, to adopt the simplest, it is not correct that of two eventualities the one which corresponds to the simplest theory must be realized. It was simple to suppose that the planets turn around the sun in circles, and since Copernicus had at his disposal only crude observations, he acted logically in adopting this hypothesis. An ellipse is, as a matter of fact, a more complicated line than a circle, and the fact that the sun is in one of its foci, the other remaining empty, shocks our sense of symmetry at first; but this did not prevent Kepler, in his turn, from being right in not hesitating in this case. In the same way the laws of Mariotte and Gay-Lussac on the volume of gases are very simple; does this prove that they are correct—that is, that they are laws really followed by nature itself? In the present state of science the question is no longer even asked; we know very well that no gas exactly follows the laws of which we are speaking, and in stating them or in applying them we are extremely careful to specify that we are dealing with an entirely hypothetical thing which we designate as "perfect gas." We saw, however, that the question presented itself to Comte . . . , and that he settled it in the opposite sense. Milhaud, in a remarkable analysis, has shown that the source of Comte's inspiration was a real dogma, and he has disclosed its foundations.[15]

---

[14] H. POINCARÉ: *La science et l'hypothèse*, p. 176; cf. *Thermodynamique*, p. vii.
[15] MILHAUD: *L'idée d'ordre chez Auguste Comte. Revue de métaphysique*, IX, 1901, p. 539. LÉVY-BRUHL, *La philosophie d'Auguste Comte*, 2nd ed. Paris, also remarks that in Comte "the scientific interest, however intense it may be, is subordinated to the social interest." Cf. *ib.*, pp. 5, 25. It seems, indeed, that this particular conviction of Comte's flows in no way from his conception of science, which would end rather by considering laws as the ephemeral expression of a momentary condition of the science of a period (cf. *Cours*, Vol. VI, pp. 600–601, 622, 630, 642), the restrictions (*ib.*, pp. 601, 623) appearing

They are quite foreign to the philosophy of the physical sciences, having their origin in a sociological conception. Comte was extremely preoccupied with the idea of order; he had placed it at the very centre of his system. Having put aside theological traditions, he could establish order only on the basis of experience. Experience, therefore, must needs set up definite principles and laws, which should remain immovable for all time.

Law, such as science really understands it, is an ideal construction, an image transformed by our understanding of the ordering of nature; it cannot, therefore, directly express reality or be truly adequate to it. It did not exist before we had formulated it, and it will no longer exist when we shall have merged it in a wider law. To suppose that an empirical rule conceived by us will no longer be modifiable in the future is to affirm, on the contrary, that that rule exists objectively in nature itself; for no one could pretend that in the knowledge of rules we shall ever be able to pass definite limits, since there is no possibility of tracing these limits, or even of conceiving their existence. In this respect, it is asserted that nature is in agreement with thought. The latter, in formulating laws, must allow itself to be guided by considerations of simplicity. Thus, in the last instance, we attribute these same considerations to nature. Let us observe, moreover, that every definition of simplicity can only be relative to the faculties of our mind, to the means it has at its disposal and to its habits. As Le Roy has shown,[16] the sine function which enters, for instance, into the law of refraction appears simple to us, because we are accustomed to use it, and we even possess tables for it; but if we were obliged to express it under a purely arithmetical form, by a polynomial, it would be very complicated. To-morrow, perhaps, we shall discover a new process for the mathematical calculus which will make the problems now embarrassing us seem simple. We have here most evidently an accidental element which it seems difficult to attribute to nature. Moreover, we never really do so. When an astronomer, laboriously, by successive approximations, calculates the "perturbations" which the planets cause one another, he has not the least doubt that nature solves this same problem instantaneously with an absolute exactitude and without any difficulty. As Fresnel has said: "Nature

---

as something foreign to the very substance of the doctrine. It is true nevertheless, as we have already indicated (p. 398f.), that the error could not have taken place if Comte had not proscribed all theoretical research.

[16] E. LE ROY: *Un positivisme nouveau. Revue de métaphysique*, IX, 1901, p. 146.

is not embarrassed by difficulties of analysis."[17] This is why Poincaré believes that nature "has too often given the lie" to those who proclaim that it loves simplicity,[18] and why Duhem also comes to the conclusion that simplicity, "so ardently sought for, is an unattainable chimera."[19]

Let us add, as Poincaré indicated in the passage quoted a little before, that even where nature may appear to us as simple, this may be purely an appearance. This simplicity may easily cover a real diversity of very numerous facts; it would then be *statistical*. When we watch from some distance the movement of a crowd emerging from a gateway, we discover without difficulty that the phenomenon has an entirely regular aspect; yet each one of the individuals which compose it is making very different movements. In the same way the regularity of births and deaths in important aggregates covers a group of facts peculiar to each individual.

Upon this same postulate of the simplicity of nature is based a fundamental objection which is sometimes brought against atomism, and to which notably Stallo seems to give much importance.[20] In formulating the kinetic theory of gases, one evidently seeks to explain that state of matter by the solid state, or, if you will, by the ultrasolid. Now it is easy to be convinced, that everywhere and always —whether it is a question of purely physical phenomena, such as the change of volume as a function of the pressure and the temperature, diffusion, specific heat, or even of chemical reactions (the law of Gay-Lussac)—the laws which regulate the gaseous state are simpler than the laws applicable to the solid state of matter. Therefore, it is concluded, "if there be a typical and primary state of matter, it is not the solid, but the gaseous," and consequently "the gaseous form is the true basis for the explanation of the solid form, and not, conversely, the solid for the explanation of the gas."[21]

This objection, in the light of what we have just established, appears to us as purely specious. It presupposes that the simpler phenomenon must be, by that very fact, more fundamental. Now, we have just seen, simplicity here may only by *statistical*, and as the

[17] A. FRESNEL: *Mémoire sur la diffraction de la lumière. Mémoires de l'Académie royale des sciences*, Years 1821–1822, Vol. V, p. 340.
[18] H. POINCARÉ: *Thermodynamique*, p. vii.
[19] DUHEM: *L'évolution de la mécanique*, p. 343.
[20] STALLO: *l.c.*, p. 132 and ff.
[21] *Ib.*, p. 134.

number of molecules in one cubic centimetre of gas is incomparably larger than the number of human individuals in any aggregate of which we know, or the existence of which we can even suppose in the future, and as the regularity of phenomena evidently increases as a function of this number, it is not astonishing that the laws which gases obey are much more regular than those of human statistics.

So also it is certain that if we do not adopt mechanism as a guide our analogical reasoning will wander, so to speak, without a compass on the limitless ocean of possibilities. This is clearly seen in the strange hypotheses of the *Natur-Philosophen*, and it is this which has rendered their reasonings so sterile. And yet one cannot pass too absolute a judgment. The analogies between nature and our understanding are many and profound. A vigorous mind may, therefore, by simple analogical reasoning arrive at important scientific discoveries. Indeed, there is no doubt that many great discoveries are due to reasonings of this kind.[22] It is not, therefore, astonishing that Oersted, starting from the doctrines of "the philosophy of nature," discovered electro-magnetism; and it may very well be that the relations formulated by Ostwald, which seem at present to be rather numerical combinations, may bring us to generalizations of great value.

It may, however, be predicted apparently without too great a risk of error that the theories and the hypotheses which will be at the basis of this kind of reasoning will be ephemeral. Only the foundation of mechanism, the explanation of phenomena by motion, is and will be really eternal.

As long as humanity attempts to develop science, mechanism will continue to be developed with it. The return to peripateticism, extolled with such force and learning by Duhem,[23] appears impossible

[22] DUHEM: *La théorie physique*, p. 50 and ff. Duhem's thesis goes still farther: he wishes to establish that "search for explanations was not Ariadne's clew." The services rendered by mechanism seem to us, however, very difficult to deny.
[23] It is but right to state, however, that Duhem toward the end of his life seems to have somewhat changed his opinion on this subject. Thus in his important book on the *Système du Monde* (Paris, 1913 and following years, Vol. I, pp. 194, 196, 240, 284), after having brought out that none of the notions of peripatetic mechanism "has the least analogy with the notion of mass" of modern physics, and that Aristotle's theory of motion is open to an objection which "ruins" it, he praises the University of Paris for having, in the fourteenth century, "begun to substitute a sensible dynamics for Aristotle's dynamics," and states that in the years to come "there will not rest one stone upon another" in the edifice erected by this latter. [Addition to the 3rd ed.]

to us. Indeed, it does not seem to us that the pure doctrine of Aristotle was a truly scientific doctrine. It became so, as in the case of the alchemists, only by deviation from it. Nor do we believe that the recent developments of theoretical physics and chemistry, as, for instance, in the work of Gibbs, constitute really an advance in the direction of Aristotelian physics. The analogies are thin and superficial. The similarity established between change of state and motions is probably the only real one; and even on this point the resemblance is perhaps more in the expressions than in the conceptions of the two doctrines. When we are told that the modern theory contemplates change "in itself," as Aristotle did, it seems to us that at bottom the analogy holds especially of a kind of negative fact, namely that neither regards atoms as intervening, the intrusion of which, being the expression of a strict causality, ends in re-establishing identity, that is in denying change itself. But modern theories of change pretend in no way to penetrate to the bottom of it; they are not explicative, as Aristotle's conceptions were believed to be. Duhem himself recognizes this difference. It follows that these theories do not exclude mechanism as an explanation. Gibbs himself used mechanistic conceptions, and no one pretends that the most recent development of physics, which, we have seen, seems entirely directed toward atomism, has entered into conflict at any point whatsoever with his theories.[24] The two kinds of conception seem to live together peacefully.—But as to the process itself of analogical reasoning, it must be clearly understood that it is still more indestructible, if that is possible, than mechanism, for by it alone can we approach reality. Whatever we do we are always obliged to suppose—at least momentarily—that nature proceeds, as does our reasoning. The error of Descartes and the *Natur-Philosophen*, as also of Comte, consisted solely in using analogy, not to formulate assumptions to verify, but for apodictic affirmations.

It is only just, however, to point out that something of this spirit subsists in our present physics. When we assign a peculiar place to the principles of conservation, and when in general we attribute to every proposition emanating from the principle of identity an importance surpassing its experimental basis, we evidently assume in nature a tendency analogous to that of our mind. Are we wrong? We have already replied to that question, the analogy between our intelligence

[24] BOLTZMANN: *Leçons sur la Théorie des gaz*, 2nd Part. Paris 1905, p. 206, establishes "that Gibbs had continually on his mind this idea of molecular theory, even when he did not use equations of molecular mechanics."

and nature cannot be denied. Moreover, it seems evident that the fundamental opinions of humanity have not varied much in that respect. Anaxagoras, and before him Hermotimus, Aristotle tells us, "proclaimed that it is intelligence which, in nature as well as in animate beings, is the cause of order and regularity everywhere in the world."[25] But we are forced to go farther than these ancient philosophers, since beyond order we again perceive the plasticity of nature with respect to the causal principle. Unless we suppose, as did Spir, that there is in this agreement an "organized deception,"[26] planned in advance, we are indeed forced to assume this analogy and to admit a partial harmony.

Should we wonder at it? Doubtless, if we oppose the world of consciousness to that of the noumenon. But it must not be forgotten that this, on the whole, is only a metaphysical theory; that it is we who have created the conception of the noumenon, and that we created it in order that it may act. If it remains without action it vanishes immediately, just like the gods of Epicurus, or the ether of physicists when we deprive it of mass. We therefore postulate action at the same time as the concept, and consequently also the analogy. If, therefore, in this dualistic concept, the analogy appears as a miracle, it is a miracle of the same order as sensation.

But this conception is not the only one. It is equally allowable for me to consider my ego as a part of the great whole or to judge, on the contrary, that the entire world is only my sensation.

> Are not the mountains, waves and skies a part
> Of me and of my soul, as I of them?

says Byron in verses Schopenhauer loved to cite.[27] If we adopt a conception of this kind, wonder disappears, or rather the difficulty will then consist in understanding how it happens that the analogy is not more complete, that there remain unknowable, transcendent elements.

Here again we are on the ground of metaphysics proper. We cannot, indeed, evade a reply to this question: Are the results at which theoretical science has arrived of a nature to determine our choice between the different systems which metaphysics offers us? To this

[25] ARISTOTLE: *Metaphysics*, Book I, Chap. III, § 28.
[26] SPIR: *l.c.*, pp. 9, 18, 317.
[27] BYRON: *Manfred;* cf. SCHOPENHAUER, *Die Welt als Wille und Vorstellung*, Vol. I, p. 213.

question von Hartmann, in a remarkable book,[28] appears to give an affirmative answer. He states that science, with the help of experience and of observation, and starting from common-sense notions, arrives finally at substituting for them an entirely different conception, mechanism. But, on the other hand, science, though destroying the reality of common sense, maintains time and space. It culminates, therefore, in a noumenon subject to the conditions of time and space —that is, in a determined, metaphysical system, called "transcendental realism."

In the light of the results at which we have arrived we cannot evidently recognize this conclusion as valid. Let us state in the first place that, contrary to what Hartmann seems to suppose, mechanism in its turn is not an end; it is only a stage, just like common sense, a somewhat artificial resting-place upon a road which does not admit of such. We have put aside the greater part of sensible qualities, declaring them, with Democritus, conventional; we only wish to retain what is strictly necessary to define a body. It is here that the difficulty arises. What can we logically retain? According to the reply which we give to this question, the form of the mechanical system which we adopt will vary; but in scrutinizing things profoundly we shall soon come to the conviction that we can retain no sensible quality, the atom cannot even "harbour" its impenetrability, which is manifestly an "occult quality," and the body vanishes into space, which has for a logical corollary the disappearance of space itself and of time. Consequently, in applying the reasoning of Hartmann, one would conclude that science leads us not to his "transcendental realism," but to idealism, or, if you wish, to the most absolutely negative dogmatism, since its formula would be: Nothing exists or can exist. But we know that mechanism is not a *result* of science. Science confirms it to a certain extent, just as experience of ordinary life seems to confirm the naïve realism of common sense; it is invariably the agreement between reason and reality, evidently a partial agreement. But mechanism is itself prior to science, or, at least, coeval with it. The fact that theoretical science, constructed with the aid of mechanism, conserves the notions of time and space is not at all enigmatic. It is like those numerical games made for the amusement of children, who marvel at finding their initial number after a whole series of complicated operations. Finally, as a last argument against Hartmann's reasoning, it cannot even be said that mechanism really

---

[28] ED. VON HARTMANN: *Das Grundproblem der Erkenntnistheorie.* Leipzig, s.d., p. 21 and ff.

conserves intact the concepts of space and time. For we have seen that mechanical theories notably suppose time to be reversible, which assuredly clears the way for its elimination, and enables us to see once more that mechanism is only a stage.

And yet does not the process of this elimination suggest to us that there might be something justified in the argument in question? Indeed, if our understanding postulates elimination, reality resists it, and its resistance is shown by Carnot's principle. . . . We therefore have but to change the point of departure of the reasoning and substitute Carnot's principle for mechanism: does not the generality of this formula prove that reality cannot be conceived independently of time? No doubt, in a certain measure. What we see clearly, indeed, in this is that we cannot exclude conditions of time in the consideration of a phenomenon. But we knew that from the beginning; we knew that every external phenomenon was inconceivable for us outside of the conditions of time and space. Carnot's principle simply expresses this truth in a clearer way. Science, on this point also, teaches us nothing about the noumenon; it only states precisely that there is a partial agreement between our intelligence and the external world. One may start from the fact of this agreement to arrive at the existence of the external world, as Leibniz[29] did amongst others; but one may also use, as do idealistic philosophers, the fact that this agreement is only partial, that there is also disagreement, to prove the impossibility of the external world. One may finally, like Kant, attempt a conciliation by supposing that the agreement is due to intuitive elements indissolubly mixed with our sensation. These are matters which are the exclusive property of metaphysics.

A study of the history of philosophy readily assures us of this truth. If the solutions which the moderns have proposed for these problems differ from those formulated in antiquity or during the Middle Ages, it is rather in form than in substance, and it seems that the progress of the physical sciences can only definitely influence this form of the solution. The agreement and the disagreement of which we are speaking are manifest with the advent of common sense. Our sensations really present themselves in groups, and render possible the constitution of a world of persistent objects . . .; but as soon as we seek to penetrate a little more deeply into the nature of things, we see this world dissolve. . . . Shall we say that this second evolution already belongs to science? No doubt, but we have seen that, from this point of view, there is an absolute continuity between science and common

29 Cf. COUTURAT: *La logique de Leibniz.* Paris, F. Alcan, 1901, p. 258.

sense. As a matter of fact, atomism belongs to the dawn of human thought; common sense, science, and metaphysics are, so to speak, blended in it.

Science in progressing does not abolish atomism; on the contrary, it develops it and renders it more precise. But at the same time it posits the antagonistic conception in Carnot's principle. In other words, its tendency is toward both the abolition of reality and its affirmation. In it the two opposite philosphical tendencies exist together peacefully. Therefore, from the metaphysical point of view one can draw from it no conclusion going beyond Hermotimus' statement as properly modified.

We may remark that we are here repeating, in slightly different terms, a proposition which we have already expressed. In fact, since theory is in search of a reality behind that of the world of common sense, the affirmation: "no theory will ever be simply true" . . . amounts to this: "one cannot consider the entity created by a theory as a thing in itself." A very different meaning, however, is sometimes attributed to statements of this kind. By declaring that hypotheses could not be transformed into realities, that in their nature they were neither true nor false, that they were unverifiable in their essence— propositions perfectly correct if applied to the totality of our assumptions about the external world—one has sought to affirm a fundamental difference between the world of theory and that of common sense. Now the difference, we have seen, does not exist from this point of view; and if by the term *reality* we understand not that of the thing itself, but that of the sensible object, as common sense conceives it, the proposition becomes certainly incorrect. In the first place, from the historical point of view it is easy to show that some theories must have passed into common sense. Sound certainly did not appear to our far-off ancestors as a vibration; but men have learned a long time since to see and to feel these vibrations under certain conditions, and it cannot, it would seem, be contested that, for a large part of humanity at present, the concept of these vibrations is a part of sound considered as a real object. The transformation which the sun, the moon, the sky have undergone considered as objects is of the same order. Moreover, from a philosophical point of view, we saw that as soon as we admit of the intervention of memory in our perception, we are, in fact, forced to admit also that of our *knowledge*, which leads us equally to affirm the evolution of common sense. Thus it is possible that what was an hypothesis, an assumption, agrees so well, in the consequences that we draw from it, with our sensations, that a connection, an association more and more intimate becomes established, and that finally the latter is instantaneously and automatically called forth by the former. At that moment the hypothesis will be a part of the reality of common sense; it will become in Le Roy's terminology, a brute fact. But from the logical point of view, little will be changed; brute facts being at bottom only causal hypotheses, just like scientific facts and theories.

The elaboration of scientific hypotheses is the continuation of the process which creates the realities of common sense; but the work being conscious, their authority is thereby lessened. We feel that between the hypotheses and the facts there are the laws; these latter, while not expected, like hypotheses, to penetrate into the secret of nature's work, appear to us because of this as nearer to the facts themselves. So when the law, stipulating the conservation of a concept, seems to create a veritable object, almost a thing in itself,[30] this rule exercises upon our mind the double authority of laws and hypotheses. This is evidently only a slightly different form of the explanation which we gave before . . ., but perhaps it will seem more direct under this new aspect. It seems also to point out more clearly the peculiar nature of the propositions in question, the reason why, in spite of what we have learned about the rôle of causality in science, about its incessant intervention at the moment of the latter's genesis, they constitute, nevertheless, quite evidently a well-defined class, that of *plausible* statements. This is because in this case the influence of the causal tendency attains its culminating point, shows itself in the very heart of the domain which appears to belong to pure law alone; the statement appears to be deducible directly from the causal principle and yet it is susceptible of being treated as an experimental law.

To a less degree, because less immediately derived from the causal principle, all those laws which are susceptible of a mechanical or spatial interpretation, like those which regulate the radiation of heat or like Newton's law, partake of this authority. Poincaré, speaking of the law of attraction, attributes a particular dignity to the constant 2, which is there in the quality of exponent; he declares it *essential,* whereas the greater part of the others would only be *accidental.* It seems to us that no other foundation can be found for this distinction than the fact that in the first case there exists a possible spatial interpretation.

[30] It is right to recall that Ostwald's system resulted in elevating energy to this dignity (p. 349 f.).

Chapter 18
ERNST CASSIRER
(1874–1945)

INTRODUCTION

*E*RNST CASSIRER was born in Breslau, Germany, on July 28, 1874. He received his higher education at the Universities of Berlin, Leipzig, Heidelberg, and Marburg. Cassirer spent most of his time in the study of philosophy and its history: in Marburg he was strongly attracted to Cohen's version of the neo-Kantian philosophy. Cassirer worked for a time as a *Privatdocent* at the University of Berlin. In 1919 he was appointed professor of philosophy at the University of Hamburg, where he remained until 1933, when he left Germany and the threatening National Socialist movement. He taught at Oxford between 1933 and 1935, then at Göteborg University in Sweden from 1935 to 1941, and finally at Yale (1941–1944) and Columbia (1944–1945) in the United States. Cassirer died in New York on April 13, 1945.

Cassirer was a prolific writer, with publications to his credit in at least three major fields in philosophy, history of philosophy, philosophy of science, and, in particular, philosophy of culture. His historical works include the well-known: *Leibniz' System in seinen wissenschaftlichen Grundlagen* (1902); *Das Erkenntnisproblem in der Philosophie und Wissenschaft der neueren Zeit* (3 vols., 1906–1920); *Freiheit und Form* (1916); *Kants Leben und Lehre* (1918); *Die Philosophie der Aufklärung* (1932); and *Descartes* (1939).

In philosophy of science Cassirer attempted mainly to apply Cohen's critical method to domains not covered by the latter's treatises, that is, especially to the sciences of mind. However, he also wrote some valuable studies on the philosophy of natural science, his *Substanzbegriff und Funktionsbegriff* in 1910, and the later works *Zur Einsteinschen Relativitätstheorie* in 1921 and *Determinismus und Indeterminismus in der modernen Physik* in 1937. In these works, and even more so in his works on philosophy of culture, it becomes evident that Cassirer's historical studies had led him to the insight that

Cohen's neo-Kantianism was to be transcended in many aspects. Gradually some of the basic ideas of Hegelian philosophy began to manifest themselves, but finally Cassirer moved to a philosophical view which places him above any attempt to link his thinking to a specific philosophical trend or school.

Philosophie der symbolischen Formen (3 vols., 1923–1929) is recognized as Cassirer's main work. In 1944 he wrote a further personal reflection on the work which was published in English under the title *An Essay on Man*. Cassirer's aim in these books was to try to uncover what is specifically human; in his opinion it was to be found in the distinction between the symbol and mere signs. Signs, he claimed, are univocal; higher animals are able to "understand" them. Symbols on the other hand are essentially universal and multi-dimensional; once man discovers them a new world opens up for him. He can talk, take distance from the concrete, develop science, master nature, develop a world of culture. Signs ask only for a passive response while symbols appeal to man's creativity and require a creative answer to each situation.

From this, in Cassirer's view, it follows that it does not make sense to try to understand man by an interpretation of his "substance"; on the contrary, he who wishes to understand man must interpret his functional behavior. Man then is genuinely found only in his own creations: language, science, culture, myth, religion. His creative culture as a whole must be described as a progressing self-liberation of man in society with language, science, art, and religion as the main phases of the process. In each, man shows a new kind of power, the power to continuously build up a new, and evidently ideal, world. Philosophy's task, according to Cassirer, is to try to find the unity in this ideal world, recognizing that each function has its own meaning and task but that all of them must supplement each other. Each in its own way opens a new horizon for mankind and reveals a new dimension of man's own mode of being.

In his book *Substance and Function,* from which the selection which follows is taken, Cassirer wished to show that the view of our concepts as formed by abstraction from a great number of particular instances is inadequate and unacceptable. Concepts, taken at least in their function as instruments of human knowledge, are already presupposed by any process of classification which itself, self-evidently, is a necessary condition for induction and abstraction. As Cassirer says, "What lends the theory of abstraction support is merely the circumstance that it does not presuppose the contents, out of which the concept is to develop, as disconnected particularities, but that it tacitly thinks them in the form of an ordered manifold from the first. The concept, however, is not deduced thereby, but presupposed; for, when we ascribe to a manifold an order and connection of elements, we have already presupposed the concept if not in its complete form, yet in its fundamental function" (*Ib.*, p. 17). In fact the theory of abstraction also shows another decisive weakness in that it focuses exclusive attention upon the attributes which a class of things have in common, whereas the particularities of each member of the class are left out of consideration. In Cassirer's view, however, the "genuine concept does not disregard the peculiarities and particularities which it holds under it, but seeks to show the necessity of the occurrence and connections of just these particularities . . ." (*Ib.* 19–20). Thus he comes to the conclusion that the more universal a concept is, the more rich in content will it be (*Ib.*).

Cassirer claims that before any classification presupposed by abstraction can take place, our mind must be able to establish such classification in a rational way. This is done, he says, by a process of identification on the basis of a certain criterion. By using any one particular as an instance which satisfies the conditions determined by the criterion, the process of identification collects a group of similar particulars, related to one another and bound together by this common criterion. The material given by our perceptions then can be ordered in many ways according to the criterion we have chosen; but in any case each criterion forms a special series of perceptions in which a certain relation among the elements of this series prevails. This relation again is determined by the criterion one has chosen.

Cassirer calls this structure of concepts as a successive series of terms connected with one another by a certain criterion "a functional concept." All our mathematical concepts are of this type; the same holds true for our mechanical, physical, and chemical concepts, and for all scientific concepts in general. By analyzing and describing these basic functional concepts in their genesis, one can explain the continuous process of objectivation in and through which the mind gradually brings regularity and order into disconnected, fluctuating, and chaotic sensations by making them into clear and distinct objects.

SELECTIVE BIBLIOGRAPHY

Cassirer, E., *Leibniz' System in seinen wissenschaftlichen Grundlagen.* Marburg, 1902.

————, *Das Erkenntnisproblem in der Philosophie und Wissenschaft der neueren Zeit* (3 vols.). Berlin, 1906–1920.

————, "Kant und die moderne Mathematik," *Kantst.*, **12**(1907) 1–40.

————, *Substance and Function* (1910) and *Einstein's Theory of Relativity* (1921), trans. W. C. and M. C. Swabey. La Salle, Ill., 1923.

————, "Philosophische Probleme der Relativitätstheorie," *Die Neue Rundschau*, **31**(1920) 1337–1357.

————, *Determinismus und Indeterminismus in der modernen Physik.* Gothenburg, 1937.

————, *The Problem of Knowledge*, trans. W. H. Woglom and C. W. Hendel. New Haven, Conn., 1950.

The Philosophy of Ernst Cassirer, ed. P. A. Schilpp. Evanston, Ill., 1949.

Hamburg, C. H., *Symbol and Reality. Studies in the Philosophy of Ernst Cassirer.* The Hague, 1956.

Frank, P., *Modern Science and Its Philosophy.* New York, 1961. Pp. 34–35, 173–183.

Paton, H. J. and Klibansky, R., *Philosophy and History. Essays Presented to Ernst Cassirer.* New York, 1964.

ON FUNCTIONAL CONCEPTS
IN NATURAL SCIENCE

ROBERT MAYER'S METHODOLOGY
OF NATURAL SCIENCE

*T*he discoverer of the fundamental law of modern natural science agrees entirely in his methodological views with the series of great investigators that starts with the Renaissance. Robert Mayer begins with the same theoretical definition of the problem of physics as is found in Galileo and Newton in the most diverse applications. The logical continuity appears unbroken, in all the material remodeling of physics introduced with the principle of energy. "The most important, not to say the only, rule for the true investigation of nature is this: to remain persuaded that it is our task to learn to know the phenomena before we seek explanations or ask about higher causes. Once a fact is known on all sides, it is thereby explained and the work of science is ended. This assertion may be pronounced by some as trivial, or combated by others with ever so many reasons; yet certain it is, not only that this rule has been too often neglected down to the most recent times, but that all speculative attempts, even of the most brilliant intellects, to raise themselves above facts instead of taking possession of them, have up till now borne only barren fruit."[1] This is precisely such language as Kepler used against the alchemists and mystics of his time, or as Galileo used against the Peripatetic philosophy of the school. The question as to how heat arises from diminishing motion or how heat is again changed into motion, is declined by Robert Mayer, just as Galileo avoided the question as to

[1] Robert Mayer, *Bemerkungen über das mechanische Aequivalent der Wärme*, (*Mechanik der Wärme*, ed. by von Weyrauch, 3rd ed., Stuttgart 1893, p. 236).

---

the cause of weight. "I do not know what heat, electricity *etc.* are in their inner essence—just as little as I know the inner essence of a material substance or of anything in general; I know this, however, that I see the connection of many phenomena much more clearly than has hitherto been seen, and that I can give a clear and good notion of what a force is." This, however, is all that can be required of an empirical investigation. "The sharp definition of the natural limits of human investigation is a task of practical value for science, while the attempt to penetrate the depths of the world-order by hypotheses is a counterpart of the efforts of adepts." In the light of this conception, only numbers, only the quantitative determinations of being and process ultimately remain as the firm possession of investigation. A fact is understood when it is measured: "a single number has more true and permanent value than a costly library of hypotheses."[2]

HYPOTHESES AND NATURAL LAWS

Here a new problem of permanent significance is indicated, along with the rejection of the false problem. A problem is held to be *explained*, when it is *known* (*bekannt*) perfectly and on all sides. This definition must, indeed, be accepted without limitation; but back of it arises the further question, as to under what *conditions* a phenomenon is to be taken as known in the sense of physics. The "knowledge" of a phenomenon, which exact science brings about, is obviously not the same thing as the bare sensuous cognizance of an isolated fact. A process is first known, when it is added to the totality of physical knowledge without contradiction; when its relation to cognate groups of phenomena is clearly established, and finally to the totality of facts of experience in general. Every assertorical affirmation of a reality, at the same time, implies an assertion concerning certain relations of law, *i.e.*, implies the validity of universal rules of connection. When the phenomenon is brought to a fixed *numerical expression*, this logical relativity becomes most evident. The constant numerical values, by which we characterize a physical object or a physical event, indicate nothing but its introduction into a universal serial connection. The individual constants mean nothing in themselves; their meaning is first, established by comparison with and differentiation from other

---

[2] See Mayer's letter to Griesinger (*Kleinere Schriften u. Briefe*, ed. by Weyrauch, Stuttgart 1893, p. 180, 226 *etc.*)

values. Thereby, however, reference is made to certain logical pre-suppositions which lie at the basis of all physical enumeration and measurement,—and these presuppositions form the real "hypotheses," that can no longer be contested by scientific phenomenalism. The "true hypothesis" signifies nothing but a principle and means of measurement. It is not introduced *after* the phenomena are already known and ordered as magnitudes, in order to add a conjecture as to their absolute causes by way of supplement, but it serves to make possible this very order. It does not go beyond the realm of the factual, in order to reach a transcendent beyond, but it points the way by which we advance from the sensuous manifold of sensations to the intellectual manifold of measure and number.

THE PRESUPPOSITIONS OF PHYSICAL "MEASUREMENT"

Ostwald, in his polemic against the use of hypotheses, has laid great emphasis on the difference between the hypothesis as a *formula* and the hypothesis as a *picture*. Formulae contain merely algebraic expressions; they only express relations between magnitudes, which are capable of direct measurement and thereby of immediate verification by observation. In the case of physical pictures (*Bildern*), on the contrary, all such means of verification are lacking. Often, indeed, these pictures themselves appear in the guise of mathematical exposition, so that the given criterion of differentiation seems, at first glance, insufficient. But in every case there is a simple logical procedure, which always leads to a clear discrimination. "When every magnitude appearing in the formula is itself measurable, then we are concerned with a lasting formula or with a law of nature; . . . if, on the contrary, magnitudes, which are not measurable, appear in the formula, then we are concerned with a hypothesis in mathematical form, and the worm is in the fruit."[3] While this postulate of measurability is justified, it is erroneous to regard measurement itself as a purely empirical procedure, which could be carried out by mere perception and its means. The answer given here signifies only the repetition of the real question; for the numbered and measured phenomenon is not a self-evident, immediately certain and given starting-point, but the result of certain conceptual operations, which must be traced in detail. In fact, it soon appears that the bare attempt

[3] Ostwald, *Vorlesungen über Naturphilosophie*, Leipzig 1902, p. 213 f.

to measure implies postulates that are never fulfilled in the field of sense-impressions. We never measure sensations as such, but only the *objects* to which we relate them. Even if we grant to psychophysics the measurability of sensation, this insight remains unaffected; for even granting this assumption, it is clear that the *physicist* at least never deals with colors or tones as sensuous experiences and contents, but solely with vibrations; that he has nothing to do with sensations of warmth or contact, but only with temperature and pressure. None of these concepts, however, can be understood as a simple copy of the facts of perception. If we consider the factors involved in the measurement of motion, the general solution is already given; for it is evident that the physical definition of motion cannot be established without substituting the geometrical body for the sensuous body, without substituting the "intelligible" continuous extension of the mathematician for sensuous extension. Before we can speak of motion and its exact measurement in the strict sense, we must go from the contents of perception to their conceptual limits. . . . It is no less a pure conceptual construction, when we ascribe a determinate velocity to a non-uniformly moving body at each point of its path; such a construction presupposes for its explanation nothing less than the whole logical theory of infinitesimal analysis. But even where we seem to stand closer to direct sensation, where we seem guided by no other interest than to arrange its differences as presented us, into a fixed scale, even here theoretical elements are requisite and clearly appear. It is a long way from the immediate sensation or heat to the exact concept of temperature. The indefinite stronger and weaker of the impression offers no foothold for gaining fixed numerical values. In order even to establish the schema of measurement, we are obliged to pass from the subjective perception to an objective functional correlation between heat and extension. If we give to a certain volume of mercury the value of 0 degrees, and to another volume of mercury the value of 100 degrees, then in order to divide the distance between the two points thus signified into further divisions and subdivisions, we must make the assumption that the differences of temperature are *directly proportional* to the volume of the mercury. This assumption is primarily nothing but an hypothesis suggested by empirical observation, but is in no way absolutely forced upon us by it alone. If we go from solid bodies to fluid, from the mercury thermometer to the water thermometer, then, for purposes of measurement, the simple formula of proportionality must be replaced by a more complex formula, according to which the correlation between temperature values and volume values is established.[4] In this example, we see how the simple *quantitative* determination of a physical fact draws it into a network of theoretical presuppositions, outside of which the very question as to the measurability of the process could not be raised.

THE PHYSICAL "FACT" AND THE PHYSICAL "THEORY"

This epistemological insight has been increasingly clarified by the philosophical work of the physical investigators themselves. It is Duhem above all, who has brought the mutual relation between physical fact and physical theory to its simplest and clearest expression. He gives a convincing and living portrayal of the contrast between the naïve sensuous observation, which remains merely in the field of the concrete facts of perception, and the scientifically guided and controlled experiment. Let us follow in thought the course of an experimental investigation; imagine ourselves, for instance, placed in the laboratory where Regnault carried out his well-known attempt to test the law of Mariotte; then, indeed, we see at first a sum of direct observations, which we can simply repeat. But the enumeration of these observations in no way constitutes the kernel and essential meaning of Regnault's results. What the physical investigator objectively sees before him are certain conditions and changes in his instruments of measurement. But the *judgments* he makes are not related to these instruments, but to the objects, which are measured by them. It is not the height of a certain column of mercury that is reported, but a value of "temperature" that is established; it is not a change which takes place in the manometer, but a variation in the pressure, under which the observed gas stands, that is noted. The peculiar and characteristic function of the scientific concept is found in this *transition* from what is directly offered in the perception of the individual element, to the form, which the elements gain finally in the physical statement. The value of the volume which a gas assumes, the value of the pressure it is under, and the degree of its temperature, are none of them concrete objects and properties, such as we could coördinate with colors and tones; but they are "abstract

[4] *Cf.* the pertinent exposition of G. Milhaud, *Le Rationnel*, Paris 1898, p. 47 ff.

symbols," which merely connect the physical theory again with the actually observed facts. The apparatus, by which the volume of a gas is established, presupposes not only the principles of arithmetic and geometry, but also the abstract principles of general mechanics and celestial mechanics; the exact definition of pressure requires, for its complete understanding, insight into the deepest and most difficult theories of hydrostatics, of electricity, *etc.* Between the phenomena actually observed in the course of an experiment, and the final result of this experiment as the physicist formulates it, there lies extremely complex intellectual labor; and it is through this, that a report regarding a single instance of a process is made over into a judgment concerning a law of nature. This dependence of every practical measurement on certain fundamental assumptions, which are taken as universally valid, appears still more clearly when we consider, that the real outcome of an investigation never comes directly to light, but can only be ascertained through a critical discussion directed upon the exclusion of the "error of observation." In truth, no physicist experiments and measures with the particular instrument that he has sensibly before his eyes; but he substitutes for it an ideal instrument in thought, from which all accidental defects such as necessarily belong to the particular instrument, are excluded. For example, if we measure the intensity of an electric current by a tangent-compass, then the observations, which we make first with a concrete apparatus, must be related and carried over to a general geometrical model, before they are physically applicable. We substitute for a copper wire of a definite strength a strictly geometrical circle without breadth; in place of the steel of the magnetic needle, which has a certain magnitude and form, we substitute an infinitely small, horizontal magnetic axis, which can be moved without friction around a vertical axis; and it is the totality of these transformations, which permits us to carry the observed deflection of the magnetic needle into the general theoretical formula of the strength of the current, and thus to determine the value of the latter. The corrections, which we make and must necessarily make with the use of every physical instrument, are themselves a work of mathematical theory; to exclude these latter, is to deprive the observation itself of its meaning and value.[5]

---

[5] *Cf.* the excellent exposition of Duhem, in which this connection is explained in its particulars and illuminated on all sides. (*La Théorie Physique, son objet et sa structure,* Paris 1906.)

UNITS OF MEASUREMENT

This connection appears from a new angle, when we realize that every concrete measurement requires the establishment of certain *units*, which it assumes as constant. This constancy, however, is never a property that belongs to the perceptible as such, but is first conferred upon the latter on the basis of intellectual postulates and definitions. The necessity of such postulates is seen especially in the fundamental physical problem of measurement, *viz.*, in the problem of the *measurement of time*. From the beginning, the measurement of time must forego all sensuous helps, such as seem to stand at the disposal of the measurement of space. We cannot move one stretch of time to the place of another, and compare them in direct intuition, for precisely the characteristic element of time is that two parts of it are never given at once. Thus there only remains a conceptual arrangement made possible by recourse to the phenomena of movement. For abstract mechanics, those times are said to be equal, in which a material point left to itself traverses equal distances. Here again we find the concept of the mass-point, and thus a purely ideal concept of a limit; and once more it is the hypothetical assumption of a universal *principle*, which first makes possible the unit of measure. The law of inertia enters as a conceptual element into the explanation of the unit of time. We might attempt to eliminate this dependence by going over from rational mechanics to its empirical applications, and seeking to establish a strictly uniform motion in the field of the concrete phenomena. The daily revolution of the earth, as it seems, offers the required uniformity in all the perfection that could ever be taken into account for purposes of measurement. The unit here would be directly given us by the interval, which lies between two successive culminations of the same star. More exact consideration, however, renders apparent the difference, which always remains between the ideal and empirical measure of time. The inequality of the stellar days is what is rather demanded now, on the basis of theoretical considerations, and is confirmed by empirical reasons. The friction, which arises from the continuous change of the ebb and flow of the tides, produces a gradual diminution of the velocity of the rotation of the earth, and thus a lengthening of the stellar day. The desired exact measure again eludes us, and we are forced to more remote intellectual assumptions. All of these gain their meaning only by relation to some physical law, which we tacitly assume with them. Thus the time, in which the emanation of radium loses its radio-activity, has recently been proposed as an exact unit of measurement;

in this, the law of exponents, in accordance with which the diminution of effect takes place, serves as a foundation. It is analogous to this, that the principles and theories of optics are presupposed in order to introduce the wave-lengths of certain rays of light as the foundation of the *measurement of distance*. What guides us in the choice of units is thus always the attempt to establish certain laws as universal. We *assume* the empirically entirely "equal" stellar days to be unequal, in order to maintain the principle of the conservation of energy. The *real* constants are thus fundamentally, it has been justly urged, not the material measuring-rods and units of measurement, but these very laws, to which they are related and according to whose model they are constructed.[6]

THE VERIFICATION OF PHYSICAL HYPOTHESES

The naïve view, that measurements inhere in physical things and processes like sensuous properties, and only need to be read off from them, is more and more superseded with the advance of theoretical physics. Nevertheless, the relation of law and fact is thereby altered. For the explanation, that we reach laws by comparing and measuring individual facts, is now revealed as a logical circle. The law can only arise from measurement, because we have assumed the law in hypothetical form in measurement itself. Paradoxical as this reciprocal relation may appear, it exactly expresses the central logical problem of physics. The intellectual anticipation of the law is not contradictory, because such anticipation does not occur in the form of a dogmatic assertion, but merely as an initial intellectual assumption; because it does not involve a final answer, but merely a question. The value and correctness of this assumption are first shown when the totality of experiences are connected into an unbroken unity on the basis of it. The correctness of the assumption, on the other hand, cannot indeed be assured by our verifying every hypothesis and every theoretical construction directly in an individual experience, in a particular sensuous impression. The validity of the physical concept does not rest upon its content of real *elements of existence*, such as can be directly pointed out, but upon the strictness of *connection*, which it

[6] *Cf.* Henri Poincaré, *La mesure du temps, Revue de Métaphysique et de Morale* VI, 1898. Concerning the theoretical presuppositions of the determination of units of measurement, *cf. especially* Lucien Poincaré *La Physique moderne, dtsch. v. Brahn,* Lpz. 1908; and also Wilbois, *L'Esprit positif, Revue de Métaph.* IX (1901).

makes possible. In this fundamental character, it constitutes the extension and continuation of the mathematical concept. . . . Thus the individual concept can never be measured and confirmed by experience for itself alone, but it gains this confirmation always only as a member of a theoretical complex. Its "truth" is primarily revealed in the consequences it leads to; in the connection and systematic completeness of the explanations, which it makes possible. Here each element needs the other for its support and confirmation; no element can be separated from the total organism and be represented and proved in this isolation. We do not have physical concepts and physical facts in pure separation, so that we could select a member of the first sphere and enquire whether it possessed a copy in the second; but we possess the "facts" only by virtue of the totality of concepts, just as, on the other hand, we conceive the concepts only with reference to the totality of possible experience. It is the fundamental error of Baconian empiricism that it does not grasp this correlation; that it conceives the "facta" as isolated entities existing for themselves, which our thought has only to copy as faithfully as possible. Here the function of the concept only extends to the subsequent inclusion and representation of the empirical material; but not to the testing and proving of this material.[7] Although this conception has been obstinately maintained within the epistemology of natural science, nevertheless there are many signs that physics itself in its modern form has definitely overcome it. Those thinkers, also, who urge strongly that *experience* in its totality forms the highest and ultimate authority for all physical theory, repudiate the naïve Baconian thought of the *"experimentum crucis."* "Pure" experience, in the sense of a mere inductive collection of isolated observations, can never furnish the fundamental scaffolding of physics; for it is denied the power of giving mathematical form. The intellectual work of understanding, which connects the bare fact systematically with the totality of phenomena, only begins when the fact is represented and replaced by a mathematical symbol.[8]

THE MOTIVE OF SERIAL CONSTRUCTION

Yet if we conceive the final result of the analysis of physical theory in this fashion, there remains a paradox. Of what value is all the

---

[7] *Cf.* more particularly *Erkenntnisproblem* II, 125 ff.

[8] *Cf.* in particular Duhem, *La Théorie Physique*, p. 308 ff.

intellectual labor of physics, if we must ultimately recognize that all the complication of investigation and its methods simply removes us more and more from the concrete fact of intuition in its sensuous immediacy? Is all this expenditure of scientific means worth while, when the final outcome is and can be nothing but the transformation of facts into symbols? The reproach raised by modern physics, in its inception, against scholasticism, that scholasticism replaced the consideration of *facts* by that of *names*, now threatens to fall upon physics itself. Nothing but a new nomenclature seems to be gained, by which we are more and more alienated from the true reality of sensation. In fact, this consequence has occasionally been pointed out: the *necessity*, to which physical theory tends, has been opposed to the evidence and *truth*, which comes to consciousness in the experience of individual facts. This separation, however, rests upon a false abstraction; it is based in the attempt to isolate two moments from each other, which are inseparably connected together by the very presuppositions of the formation of concepts. It has been shown, in opposition to the traditional logical doctrine, that the course of the mathematical construction of concepts is defined by the procedure of the *construction of series*. We have not been concerned with separating out the common element from a plurality of similar impressions, but with establishing a principle by which their diversity should appear. The unity of the concept has not been found in a fixed group of properties, but in the rule, which represents the mere diversity as a sequence of elements according to law. . . . A consideration of the fundamental physical concepts has confirmed and broadened this view. All these concepts appear to be so many means of grasping the "given" in series, and of assigning it a fixed place within these series. Scientific investigation accomplishes this last definitely; but in order for it to be possible, the serial principles themselves, according to which the comparison and arrangement of the elements take place, must be theoretically established. The individual thing is nothing for the physicist but a system of physical constants; outside of these constants, he possesses no means or possibility of characterizing the particularity of an object. In order to distinguish an object from other objects, and to subsume it under a fixed conceptual class, we must ascribe to it a definite volume and a definite mass, a definite specific gravity, a definite capacity for heat, a definite electricity, *etc.* The measurements, which are necessary for this, however, presuppose that the aspect, with respect to which the comparison is made, has been previously conceived with conceptual rigor and exactitude. This aspect is never given in the original impression, but has to be worked out theoretically, in order to be then applied to the manifold of perception. The physical analysis of the object into the totality of its numerical constants is thus in no sense the same as the breaking-up of a sensuous thing into the group of its sensuous properties; but new and specific categories of *judgment* must be introduced, in order to carry out this analysis. In this judgment, the concrete impression first changes into the physically determinate object. The sensuous quality of a thing becomes a physical object, when it is transformed into a serial determination. The "thing" now changes from a sum of properties into a mathematical system of values, which are established with reference to some scale of comparison. Each of the different physical concepts defines such a scale, and thereby renders possible an increasingly intimate connection and arrangement of the elements of the given. The chaos of impressions becomes a system of numbers; but these numbers first gain their denomination, and thus their specific meaning, from the system of concepts, which are theoretically established as universal standards of measurement. In this logical connection, we first see the "objective" value in the transformation of the impression into the mathematical "symbol." It is true that, in the symbolic designation, the particular property of the sensuous impression is lost; but all that distinguishes it as a *member of a system* is retained and brought out. The symbol possesses its adequate correlate in the *connection* according to law, that subsists between the individual members, and not in any constitutive part of the perception; yet it is this connection that gradually reveals itself to be the real kernel of the thought of empirical "reality."

THE PHYSICAL CONCEPTS OF SERIES

The relation, that is fundamental here, can be illumined from another standpoint by connecting it with the ordinary psychological theory of the concept. In the language of this theory, the problem of the concept resolves into the problem of "apperceptive connection." The newly appearing impression is first grasped as an individual, and first gains conceptual comprehension through the apperceptive interpretation and arrangement that it undergoes. If this reference of the individual to the totality of experience were lacking, the "unity of consciousness" itself would be destroyed,—the impression would no longer belong to "our" world of reality. In the sense of this conception, we can characterize the various physical concepts of measurement evolved by scientific theory as the real and necessary *apperceptive concepts* for all empirical knowledge in general. Without them, indeed, it has been shown there is no arrangement of the factual in series, and thus no thorough reciprocal determination among its individual members. We would only possess the fact as an individual *subject*, without being able to give any *predicate* for determining it more closely. It is only when we bring the given under some norm of measurement, that it gains fixed shape and form, and assumes clearly defined physical "properties." Even before its individual value has been empirically established within each of the possible comparative series, the fact is recognized, that it necessarily belongs to some of these series, and an anticipatory schema is therewith produced for its closer determination. The preliminary deductive work furnishes a survey of the possible kinds of exact correlation; while experience determines which of the possible types of connection is applicable to the case in hand. Scientific experiment always finds several ways before it, which theory has prepared, and between which a selection must be made. Thus no content of experience can ever appear as something absolutely strange; for even in making it a content of our thought, in setting it in spatial and temporal relations with other contents, we have thereby impressed it with the seal of our universal concepts of connection, in particular those of mathematical relations. The material of perception is not merely subsequently moulded into some conceptual form; but the thought of this form constitutes the necessary presupposition of being able to predicate any character of the matter itself, indeed, of being able to assert any concrete determination and predicates of it. Now it can no longer seem strange, that scientific physics, also, the further it seeks to penetrate into the "being" of its object only strikes new strata of numbers, as it were. It discovers no absolute metaphysical qualities; but it seeks to express the properties of the body or of the process it is investigating by taking up into its determination new "parameters." Such a parameter is the *mass*, which we ascribe to an individual body in order to render rationally intelligible the totality of its possible changes and its relations with respect to external impulses to motion, or the *amount of energy*, which we regard as characteristic of the momentary condition of a given physical system. The same holds of all the different magnitudes, by which physics and chemistry progressively determine the bodies of the real world.[9] The more deeply we enter into this procedure, the more clear becomes the character of the scientific concept of the thing and its difference from the metaphysical concept of substance. Natural science in its development has everywhere used the *form* of this latter; yet in its progress it has filled this form with a new content, and raised it to a new level of confirmation.

[9] *Cf.* the striking developments of G. F. Lipps, *Mythenbildung und Erkenntnis,* Lpz. 1907, p. 211 ff.

Part IV
TOWARD
CONTEMPORARY
PHILOSOPHY OF
SCIENCE: 1910-1927

Chapter 19
CHARLIE DUNBAR BROAD
(1887–1971)

CHARLIE DUNBAR BROAD studied at Cambridge University, then taught at Cambridge, Dundee, Bristol, and again at Cambridge. He is a representative of the neo-realistic movement and, with Moore and Russell, one of the founders of English analytic philosophy. Unlike Moore, who moved to philosophy from philology, Broad gravitated to philosophy from science. As a result of his careful analysis of the fundamental concepts of natural science, he gradually moved in the direction of neo-realism. His earlier publications deal mainly with philosophy of science, but his later works focus on ethical, religious, and psychological themes. In this context the following publications are relevant: *Perception, Physics, and Reality* (1914), *Scientific Thought* (1923), and "Critical and Speculative Philosophy" (1924).

At first Broad believed that philosophy should be absolutely independent from religion and all world views. He was not personally interested in the development of a philosophical system, and in fact even questioned whether it was philosophy's task to develop systems and world views. In his opinion philosophy's task should consist in carefully analyzing and determining the meaning of basic concepts, not only in the realms of epistemology and philosophy of science, but for ethics as well. In the latter domain a philosopher is to apply himself to an accurate analysis and description of fundamental ethical concepts such as good, evil, right, wrong, duty, obligation, and responsibility without attempting to develop an "objective" set of "intersubjectively valid" norms and values. Here, too, philosophy is to remain no more than a highly critical and descriptive science. Broad held speculative philosophy, which for him was identical with English Hegelianism and dealt with metaphysics, world views, theology, and even mysticism, to be of minor importance.

In his own philosophy Broad began from the fact that in our everyday life we make use of a number of basic concepts such as thing, substance, cause, space, time, and such like, which are taken over by science from common sense with only minor modifications. These concepts are used by common sense and science without any clear idea of their meaning and mutual relationship. But, according to Broad, when people use these concepts consistently the former must mean something, and therefore they must have a certain understanding of their meaning, albeit often vague and confused. Consequently the most important task of philosophy is to analyze the concepts used in our common life and in science, determine their precise meanings and mutual relations, and describe them as accurately as possible.

It is possible to conclude here that the task of philosophy has thus been reduced to a mere discussion concerning the meaning of words, but Broad rejected such a view as absolutely wide of the mark, acknowledging that any analysis of fundamental concepts, once it has been made, is to be expressed in words: yet insisting that this is true for every other discovery as well. The meaning of critical analysis as used by philosophy is not to determine the meaning of words, but to ascertain exactly what properties are present in objects when a certain concept of them is predicated, and what properties are absent when it is not. Here philosophy is not interested in a matter of words, but rather in questions concerning things and their properties.

Broad believed also that in addition to the critical analysis of basic concepts philosophy has another task closely connected to the first. People not only use vague concepts, but also hold a great number of uncritical beliefs and convictions used in ordinary life and in science too. We constantly assume, for instance, that every event must have a cause; that the events of nature are governed in their course by natural laws; that we live in a world of things the existence of which is independent of our knowledge of them. These beliefs and convictions, which common sense and science accept without any form of radical criticism, can (as experience teaches us in many particular cases) be mere prejudices. Therefore, Broad contends, they call for criticism; this can be materialized only after they have been clearly formulated, or only after the concepts they necessarily imply have been analyzed and defined. Then will come the task of critically testing these assumptions, to be done by resolutely and honestly exposing them to any and all reasonable objections. For these two branches of the philosophical enterprise, the analysis and definition of fundamental concepts and the clear statement and resolute criticism of our fundamental beliefs, Broad reserves the designation Critical Philosophy.

In Broad's view, however, there was nevertheless another type of philosophy possible: Speculative Philosophy. It has a different subject matter, uses a different method, and leads to results of a different degree of certainty from those of critical philosophy. Speculative philosophy takes its starting point in the results of the sciences, adds to them the results of the religious and ethical experiences of mankind and then, by reflecting on the whole, tries to reach some general conclusion as to the nature of the universe. For Broad, speculative philosophy is meaningful only if it presupposes critical philosophy as its necessary basis and always keeps in mind the relative certainty of its conclusions. But if speculative philosophy takes its starting point in critically verified concepts and conceptions and does remember its limitations, then according to Broad, it is of great value. For example, in all sciences, with the exception of psychology, one deals with things and their changes, but leaves out of account the human mind which observes them. In psychology, on the other hand, one deals with the human mind and its processes, leaving out of consideration the objects with which we are in contact by means of them. A scientist who limits himself to either of these subjects is likely, therefore, to get a very one-sided view of the world. In fact it is almost impossible not to fall into one of these one-sided views unless a resolute attempt is made to think synoptically of all the facts.

The selection which follows is from Broad's first publication. It deals with Mach's phenomenalism, a subject to which Broad returned in an article (1914–1915) published in the *Proceedings of the Aristotelian Society.*

SELECTIVE BIBLIOGRAPHY

Broad, C. D., *Perception, Physics, and Reality.* London, 1914.

———, *Scientific Thought.* London, 1923.

———, "Critical and Speculative Philosophy," in *Contemporary British Philosophy* (First Series), ed. J. H. Muirhead. London, 1924.

———, *The Mind and Its Place in Nature.* London, 1925.

Braithwaite, R. B., *Scientific Explanation.* New York, 1960. Pp. 293–318 (*passim*).

Lean, M., *Sense, Perception, and Matter.* London, 1953.

The Philosophy of C. D. Broad. Ed. by A. Schilpp. New York, 1959.

ON PHENOMENALISM[1]

*B*EFORE coming to . . . the Causal Theory of Perception and its effects on our beliefs in the reality of the objects of our perceptions, I propose to devote a short space to the discussion of Phenomenalism.

No apology is needed for discussing this theory somewhere in any essay which deals with the question of the information, if any, that perception can give us about reality. And when this question is raised with particular reference to the philosophical position of the truths of natural science, such a discussion is essential in view of the fact that Mach and his school are phenomenalists, and hold that phenomenalism is the philosophic theory that is best suited to be the basis for physics. The only preliminary point that does call for some explanation is why I should discuss phenomenalism here rather than after the chapter on the Causal Theory of Perception. It will be said that it is arguments based on the causal theory that chiefly undermine naïf realism, and that it would be a more reasonable order to discuss phenomenalism as an alternative when the difficulties of naïf realism were becoming insuperable than here, where they have hardly fairly begun.

We may anticipate the results of the next chapter so far as to agree

---

[1]It must be understood that in this chapter I am not discussing arguments for phenomenalism—so far as I am aware—that any phenomenalist uses. The average phenomenalist bases his position on the kind of considerations that Mr Moore overthrew in his *Refutation of Idealism*. For the purposes of the philosophy of science Mach is the most important phenomenalist. But he has no positive arguments for his position that are worth discussing. He is vitiated by the fallacy that Mr Moore overthrew, and it seems that his main reason for holding the doctrine is that he supposes—for some inscrutable reason—that it is scientific and 'economic' to a preëminent degree. Anyhow he has now retired behind the formula "Es giebt keine Machschen Philosophie." I have therefore tried to invent the argument which a philosophical phenomenalist might reasonably use, and to criticise it.

---

that the main difficulties of naïf realism do spring from the causal account of perception; but we may still defend the order that we have adopted. For we shall try to show in the present chapter that causal arguments that refute naïf realism cannot be used to support phenomenalism. Moreover, the main problem about phenomenalism from the point of view of the philosophy of natural science is to be found in the question of its relation to causal laws, and so there is good reason for discussing it directly after we have finished with Causality. I shall therefore proceed to discuss phenomenalism without further apology.

First of all, what precisely is meant by phenomenalism? In the sense in which we propose to discuss it it is the theory about the reality of the world with which we come in contact in perception which is diametrically opposite to that of naïf realism. It holds, not merely that the objects of all our perceptions exist only when they are perceived, but also that there are no permanent real things with laws of their own that cause these perceptions and in some measure resemble their objects. The laws of science, stated in terms of such supposed realities and their states, are for it mere transcriptions of laws connecting the perceptions that people actually have, and these perceptions and their laws are all that we can hope to make the objects of science.

Phenomenalism, unlike naïf realism, is a position that needs proof. Every man is a realist except in so far as experience and reflexion force him away from that position. But nobody becomes a phenomenalist except by argument. Nor is phenomenalism the position at which we naturally arrive on leaving naïf realism.[2] As soon as the average man is forced away from naïf realism at any point he always assumes that, in the case of every characteristic that does not raise some special difficulty, he perceives the real, and that events in that real cause the perception of that characteristic which he now has to believe to be appearance. What, then, are the arguments for phenomenalism?

Clearly it has to refute both naïf realism and the modified form of realism that is put in its place. We have already discussed the arguments against naïf realism that are independent of causality, and seen that most of them have very little weight. And the arguments that remain to be discussed in the next chapter from the relativity of perceptions to an organ will not really prove phenomenalism. The argument is a little complicated, and we had better put it formally, in

---

[2] As with naïf realism so with phenomenalism there is no room for appearance, and the plain man does soon distinguish appearance and reality.

order to avoid all chance of error. Let p be the proposition 'phenomenalism is true.' Let q be the proposition that the objects of our perceptions depend on the structure of our organs. Can we prove p from this, i.e, can we at the same time assert q *and* $q \supset p$? In the first place, if phenomenalism be true, our perceptions cannot depend on the permanent structure of our organs, for they will have no permanent structure. They exist when somebody perceives them, but not otherwise. Hence, unless you can be sure that, e.g., somebody always perceives your eye when you perceive a colour, there is no truth in the statement that what you perceive by sight depends on your eye and its structure. Every time you perceive a colour when no one does perceive your eye, you have a perception of colour which does not depend on the existence of an organ and its structure. Hence we can assert the proposition $p \supset \sim q$. Taking this together with $q \supset p$, we get $q \supset \sim q$. But $q \supset \sim q \,.\, \supset .\sim q$. Hence we reach the conclusion that $q \supset p \,.\, \supset .\sim q$. Thus to assert *both* $q \supset p$ and q would involve the assertion of both q and $\sim q$. You cannot do this, and therefore you cannot prove p from the argument. It is true that Berkeley, whose argument is properly phenomenalistic, is so shocked at this result that he introduces God either—for the point is uncertain—to perceive your eye when no one else does, or to be a permanent cause which can make people perceive an eye whenever someone else perceives a colour. But either alternative is a departure from pure phenomenalism. The first alternative is ridiculous, unless there be other grounds for believing in the existence of God. Taken as an argument for God, the position might be stated as follows: 'I have produced a theory about the unreality of the objects of our perceptions which is intrinsically so contrary to what people generally believe that it needs powerful proofs. What would be a strong proof, if it were consistent both with known facts and with my theory, is unfortunately inconsistent with them. It would cease to be so if I introduced a new fact, viz. a percipient God. Therefore it is obvious that such a God must exist.' The second alternative takes us away from phenomenalism to a form of idealism, for it now holds that our perceptions have permanent causes common to all of us under like circumstances; but it goes beyond this by supposing that these causes are to be found in the volition of a single person whom it identifies with the God of theology. The first step takes us from phenomenalism to a form of realism, the second to a form of idealism.

We may then, I think, agree that no arguments based on relativity of perception to an organ can by themselves legitimately lead to the proposition that all that we ever perceive is appearance, and that there corresponds nothing permanent in the real to the objects that we time and again perceive, and that common-sense takes to be relatively permanent realities. I do not, of course, wish to deny at this stage that the propositions about relativity to an organ could be stated in a roundabout way in terms of phenomenalism. I only want to show that they could not when so stated be consistently made into an argument to prove the truth of phenomenalism in general.

There is, in fact, no direct argument for phenomenalism that can make any claim on us. The doctrine, if held at all, can only be held reasonably on some such grounds as the following: Suppose it were found that naïf realism could not be maintained, and that the causal view that is substituted for it leaves us in complete agnosticism about the real, then the cry might well arise: Why not drop all reference to the real and state everything in terms of perceptions and the laws of their connexion? To this question the answer must be: Either you do not intend to attempt to find any causal laws that will tell us what perceptions to expect or you do. In the former case you must remember that your theory is less well off than the one that boldly assumes real causes of our perceptions rather like their objects, and assumes that they obey certain laws. For these assumptions do account for a good many of the perceptions that we have, and there is no reason to suppose that there is any à priori objection to making them. In the latter case, if you are to keep to pure phenomenalism you will have to account for present perceptions by causal laws that bring in no data beside other perceptions. And it may certainly be questioned whether you will be able to do this. If you bring into your laws anything like 'possible perceptions,' or 'perceptions beneath the threshold of consciousness,' you have, however, left pure phenomenalism. For you are now assuming the existence of something that is not, and never has been, the object of a direct awareness, and that is precisely what the ordinary scientist does in his assumptions about the real world. The question will then merely be (a) whether his assumption that the real world is on the whole very much like the objects that he perceives is à priori less probable than yours, and (b) whether your theory can explain the occurrence of the perceptions that we actually have as well as the rival theory.

I think it is perfectly clear that an absolutely pure phenomenalism that wishes to explain and anticipate our perceptions can be ruled out of court. We will suppose that it is allowed to assume present perceptions and those that it can remember. It is quite clear that with these alone there are no causal laws possible that will account for the perceptions that we may expect to have anything like as well as the assumptions which science makes will do. To make such laws possible we shall certainly have to take into account the perceptions that other people have, that we and they might have, and those that we have had but have forgotten. The question is whether the processes by which the phenomenalist—who has now ceased to be a pure phenomenalist—arrives at his beliefs in all these other perceptions would not equally justify the plain man's assumption of a real world more or less like what he perceives. We will consider, then, the phenomenalist's answer to the two questions proposed to him on the last page.

(a) Why should it be held to be à priori more probable that what is real is perceptions than that it is something like the objects of our perceptions? To this the phenomenalist would answer: Everyone has perceptions, and perceptions at least must certainly exist whether they themselves are objects of direct awareness or not. On the other hand, the objects of our perceptions are clearly only known to exist in the relation of being perceived. What right, then, have you to suppose that they could exist out of this relation? Thus the argument is that in assuming other perceptions as the real causes of our present ones we are only assuming that the real consists of what we know on other grounds must be capable of existing unperceived; whilst in assuming that it is like the objects of our perceptions, we are assuming that what exists unperceived is like that which is clearly only experienced as perceived, and cannot be proved independently to exist in any other state.

We must consider both sides of this argument carefully. We must keep separate the two distinctions of conscious and unconscious perceptions, and of perceptions which are, and those which are not, reflected upon. These two differences are often confused. Let us begin with that between those that are and those that are not reflected upon. It is quite clear that when I perceive a tree or any of the objects of ordinary life I do not generally reflect upon this perception and say that I know that I perceive a tree.[3] And it is generally considered that the opposite view would involve an infinite regress in knowledge that is psychologically impossible. On the other hand, my awareness of the tree, whether reflected upon or not, would be called a 'conscious

---

[3] This does not of course prove that I am not as a matter of fact directly aware of my perception when I have it, but merely that I *do* not make a certain judgment which I *could* not make unless I were directly aware of the perception.

awareness,' as against the so-called unconscious perceptions that I am supposed to have by Leibniz, or believers in the sub-liminal self, when I hear the roaring of the waves. These are supposed to consist of perceptions of the numberless little noises due to the rolling of the separate stones, and they are supposed to differ from any perception like that of the tree in a definite way which would perhaps best be described as a difference of intensive magnitude. Now, it is clear that the distinction between perceptions that are and those that are not reflected upon is a valid one, and can be witnessed by introspection, but there is much more doubt whether the distinction between conscious and unconscious perceptions is valid. Unconscious perceptions have often been introduced where either there was no need to assume any event in the brain or the mind, or where all that was needed was a persistent state of the brain or the mind, or both, which alone does not produce or constitute a perception, but in company with other such states, or under new bodily or mental conditions is capable of giving rise to a perception. But there is no more reason for calling a psychical state of this kind an unconscious perception than for calling a match-box an unlighted bonfire. Similarly if we grant that at a given moment there may be unanalysed detail in the object of a perception which attention can discover we do not assume any but conscious perceptions. Before it was discovered it was part of the object of a total conscious perception; afterwards it forms the objects of several new conscious perceptions. Perceptions, then, are either conscious and like our perceptions of trees and chairs, or there is no reason for calling them perceptions at all. And it is certainly not essential to the existence of perceptions that they should be reflected upon.

Hence, to be plausible, the phenomenalist position has to be stated as follows. The assumption that the real causes of our perceptions are perceptions that we have had and have forgotten (i.e. on which we have ceased to be able to reflect) or are perceptions in other people (i.e. perceptions on which we never could have reflected) is so far more probable than any other alternative à priori in that we know that perceptions can and do exist unreflected upon, whilst we do not know that the objects of those perceptions, or anything like them, exists unperceived.

Before discussing this argument in its present form we must say a word in explanation of the second half of it. This is the argument that, since the objects of our perceptions are clearly only known to exist when perceived, it is a greater assumption to suppose that the unknown real causes of our perceptions are like their objects than that they are other perceptions, since we have now seen that other perceptions are able to exist unreflected upon. This does not, of course, mean that when I perceive a tree I perceive it as something perceived by me, for this would be just to deny the true assertion of the other part of the argument that perceptions can and do exist unreflected upon. What it means is that a thing is only directly known to exist while it is actually an object of perception, whether, as a matter of fact, we did or did not reflect at the moment of perceiving it that this was the case. The point, then, is that, whilst there is no ground independent of the success with which the assumption meets in accounting for the occurrence of our perceptions which makes it necessary to suppose that the objects of perceptions, or anything like them, exist unperceived, it is certain that perceptions can and do exist unreflected upon. The assumption that the unknown real causes of our perceptions are other perceptions that are unreflected upon by us is therefore *à priori* a more probable or a less improbable assumption than that they are like the objects of our perceptions.

Now that we have stated this argument as fairly as possible, there are two criticisms to be made upon it, one from each side. From the side of the argument that perceptions can certainly exist unreflected upon, we must ask whether the sense in which this is true is the sense in which it will make the phenomenalist assumption of forgotten perceptions and perceptions in other people as the real causes of our perceptions more probable *à priori* than the rival assumption. The phenomenalist wants to be able to assume forgotten perceptions and perceptions in other people. He argues that these are just perceptions that are not reflected upon by the person who assumes them, and that, since it is known that such can exist, the assumption of them is *à priori* the most probable one that can be made about the nature of the real. I think that this argument loses most of its weight when we examine a little more carefully the meaning of reflexion. We shall see, in fact, that, if the assumption of forgotten states of mind in oneself is *à priori* slightly more probable than that of a real world like the objects of our perception, the assumption of perceptions in other people —which is certainly necessary for phenomenalism even more than for other views—is not more probable.

When we say that it is certain that perceptions can and do exist unreflected upon, we mean that we can now remember to have had many perceptions of which we can also remember that at the time at which we had them we did not think or say to ourselves: 'I have now a perception of X.' This, it is to be noted, involves memory. I am directly aware now of the perception that I had some time ago, and I am directly aware that I did not at the moment at which I had the perception have the sort of experience that enables me to say, 'I have such and such a perception.' Such a proposition would, of course, have been true; but I did not then have, as I do now, that direct awareness of my perception at that time which would have enabled me to assert it. You can only show that you have perceptions of which you are not directly aware when you have them by becoming directly aware of them later, and not by the alleged infinite regress that accompanies the opposite view. That regress only applies to the assertion that you cannot know without knowing that you know, and knowing that you know that you know, and so on. But it certainly does not disprove the possibility that every perception might be accompanied by a coexistent awareness of it. We only discover that this is not the case by becoming directly aware through memory of a past perception, and also of the fact that there was no awareness of that perception contemporary with its occurrence.[4]

We can now see that phenomenalism has just as much and just as little right to assume other minds as common-sense and science have to assume that the causes of our perceptions are like their objects in general character. It is true that we know directly that perceptions have existed at moments when we were not directly aware of them. But it is also clear that the only perceptions of which we know this are those of which we have been directly aware at some time. Now, the perceptions of other people just are perceptions of which we never can be directly aware. Hence the passage from perceptions in us of which we were not aware when they existed to the assumption of perceptions in other people is not a mere assumption of more of the same kind, but an assumption that perceptions of which we can never be directly aware exist, whilst all that we know directly is that perceptions of which we are sometimes, but not always, aware exist. For all that direct experience can tell us, it might be the case that the only perceptions that can exist are those of which we are sometimes directly aware. Of course, I do not use this as an argument to show that we make a mistake in assuming perceptions in ourselves and others that are never the objects of direct awareness, but merely to show that the phenomenalist who assumes perceptions of which he can never *ex hypothesi* be directly aware, makes a jump just as much as the man

---

[4] How far memory can be taken to prove a negative on such a point I should hesitate to dogmatise.

who assumes realities like the objects of his perceptions but which are not perceived. To be strictly fair, however, we must grant that the phenomenalist's jump from solipsism is not as great as that of the man who holds that the real in its general character resembles the objects of his perceptions. To assume that there exist perceptions of which I am never directly aware when I know that there are perceptions of which I am very rarely aware, and which exist when I am not aware of them, is undoubtedly a less assumption than to suppose that something like that which I can never know to exist except as an object of perception exists unperceived.

And in the matter of assuming forgotten perceptions of his own the phenomenalist's position is still less open to cavil. A forgotten perception is one of which I believe that I could at one time have been directly aware, but of which I can not at the present be directly aware. Now, of course, it is true that the only perceptions of which the phenomenalist can be immediately certain that they existed when he was not directly aware of them are those of which he can be directly aware when he makes the judgment, and *ex hypothesi* perceptions that he has forgotten are not in that position. So that there is a jump into a slightly different territory when I assume perceptions of which I am no longer able to be directly aware. But the jump is less in this case than that which takes the phenomenalist out of solipsism. There he started with perceptions of which he was immediately certain that they existed when he was not aware of them, but could only be thus certain because he could be aware of them when he made the judgment, and assumed perceptions of which he never could be aware. Here he starts from the same set of perceptions and assumes others which only differ from them by the fact that he can no longer be aware of them at will, although he believes that he could have been aware of them at some time.

We must now consider from the other side what is to be said about the phenomenalistic argument that its assumption about the nature of the real is *à priori* more probable than that of the scientific realist. Is it true that there are no other considerations that ought to be noticed in trying to find the relative *à priori* probabilities of the rival assumptions about the real beside those mentioned in the phenomenalistic argument? It is clear that in considering the *à priori* probabilities we ought to take into account *all* relevant knowledge and belief as well as that about the comparative success with which the two assumptions account for our perceptions. Now, as we have constantly pointed out, people invariably begin with the conviction that what they perceive is real. It is only by arguments that they can be led out of this position, and when they pass out of it to the belief that what they perceive is appearances, the perception of which is caused by a reality, they still do not maintain a perfectly open mind as to the nature of that reality. It is quite true that purely on the grounds of the law of causation there is no reason why a cause should resemble its effect; some causes do so and others do not. But people do not decide here solely on the grounds of the law of causality. They suppose that in this particular case the cause does resemble— not indeed the effect, which is a psychical event, and is therefore more like the cause on the phenomenalistic theory—but the object of the perceptions that are the effects. If we started with perfectly open minds from the position that our perceptions as psychical events have causes which exist but cannot be perceived, then the phenomenalistic argument that it is *à priori* more probable that those real causes should be of the same general nature as perceptions than that they should be of the same general nature as the objects of perceptions would, as we have seen, have a certain though not a great weight. But when we are considering the *à priori* probabilities in this subject we must take into account, not merely the law of causality and the amount of knowledge and ignorance that it allows us, but also any other relevant judgments about probability that may be made on this point. Now it seems to me clear that people do think that it is more likely that the real shall be like the objects of their perceptions than like their perceptions as psychical events. This is not a belief, as we have seen and shall see, to which the standard arguments for subjective idealism[5] can bring any valid objection. Of course there are many beliefs which are very generally held, and the opposite of which it is very difficult to believe, which are as a matter of fact false. But I am not suggesting that it is more probable *à priori* that the real world should be like the objects of our perception than like the perceptions themselves, *because* most or all people think so. I merely wish to suggest that when a man recognises that there is no reason why the real world should not resemble in character the objects of his perceptions he will undoubtedly hold that it is more

---

[5] It should be noted, however, that these do not exhaust the arguments on which Idealism has been based. The arguments for *esse* = *percipere* do not depend, as Mr Moore seemed to hold in his *Refutation of Idealism,* on the doctrine that *esse* = *percipi*. He forgot such arguments as the Hegelian Dialectic and Lotze's doctrine that substances must be selves. I do not think that these arguments prove idealism, but this is not the place to discuss them.

probable that it does than that it does not, and that this belief must be taken into account by every man who has it when he considers the relative probability *à priori* of phenomenalism and scientific realism. Our beliefs may be wrong, but so long as we have them and no valid argument can be found against them, it is our duty to judge in accordance with them, and not to pretend that they are non-existent.

I think it is fair then to conclude that the phenomenalistic argument, when superposed on the belief that our perceptions have causes, does not succeed in making it *à priori* more probable that the nature of those realities should be like that of our perceptions rather than that it should be like that of the objects of our perceptions. We will pass then to the relative final probabilities of the two theories, i.e. the probabilities that they gain by the agreement of the perceptions that they predict with those which we actually have. This takes us to the question which formerly I called (*b*). If two hypotheses be equally probable *à priori* and explain known facts, their relative final probabilities are dependent on how well they explain the facts. By explaining the facts is meant making that probable which is actually found in most cases to be true. In the present reference there are two points to be noted: (1) Whether it is necessary at all to go outside the perceptions that we now have and can remember; and (2) whether, if we do so, laws entirely in terms of perceptions will explain better than laws in terms of realities whose general nature is like that of the objects of our perceptions.

It is quite clear that if we keep entirely to what we can remember we shall be able to find but few causal laws among our perceptions. We must therefore give up all hope of being able to predict what perceptions we are likely to have in given circumstances or else assume something beside those perceptions that we can at a given moment reflect upon directly. The assumption that we make will have its initial probability increased in proportion to the success which it has in accounting for what is found to be true. The question that remains therefore is what assumption does this best.

Phenomenalism has a plausible argument to prove that the assumption of real causes of our perceptions like their objects in general character cannot explain better than the assumption of perceptions, because the hypotheses are really equivalent. This argument runs as follows: Suppose you do assume that your perceptions are *caused* by realities that obey certain laws among themselves. The only evidence for the particular kinds of realities and for the laws that they obey are the regularities that can be observed among your perceptions.

Surely, then, it would be just as well to say that these regularities justify causal laws among perceptions of the form 'the occurrence of the perceptions, p, q, r at t makes that of s, t, u at $t + T$ probable,' as to say that the first set were caused by X, that X makes Y probable and that Y causes the second set. The argument then is that you do not directly need to assume anything fresh. You say that the observed regularities among perceptions are evidence for certain causal laws among them. Then no doubt you will find that these sometimes inexplicably break down unless you assume that you had perceptions on which you cannot now reflect, i.e. forgotten perceptions. The first point to notice about this argument is that, as we have already seen, the amount of regularity among the perceptions that can be remembered at any given time is very small indeed. As a matter of fact, we all assume on no grounds at all that what we perceive is more or less the same as what is real, and *when* this assumption is made as it is, it enables us to discover and make probable very many regularities which without it would not have been noticed or judged probable. The suggested laws of the form mentioned above connecting the occurrence of perceptions with each other could certainly never have been discovered if it had not been the case that by nature we make the assumption that the real world resembles the objects that we perceive. The phenomenalist then will have to regard our incorrigible tendency to make this assumption as a *'felix culpa,'* like that which led to the Redemption by way of the Fall. Still I do not think that we have any right to hold that because we should not have discovered certain propositions if we had not made certain assumptions, therefore the assumptions must be held to be more probable than the propositions which would not have been discovered without them. The essence of the phenomenalist argument is that, in as far as the assumption of the real world with its laws could ever give a verifiable result, it could be replaced by hypothetical propositions about the occurrence of perceptions, and that as far as it could not give a verifiable result it was useless.

I think, however, that the position of the phenomenalist is open to criticism. It unquestionably follows from the laws of probability that if a proposition or set of propositions p strengthen the probability of a proposition q, then if q be found to be true the probability of p is strengthened. Now it is quite clear that the assumption of a real world with certain laws is a set of propositions that strengthen the probability of what is actually found to happen. And it is also clear that it is a different assumption from that of laws connecting perceptions, even though it always has to be verified, if at all, by noting the occurrence of certain perceptions under certain circumstances which must also appear as perceptions. Hence it is quite certain that this particular assumption, like any other that makes probable what is as a matter of fact true, has its probability increased. Of course the phenomenalist would be perfectly right if he said that the discoverable order in our perceptions does not by itself prove the existence of a regular world of reals that cause them. For if any regular system led us to another regular system as probably existing, we should have no right to stop at the first external world. Its regularity would make it very probable that there was another that caused it, and so on. The fact is that the strengthening of probability takes place multiplicatively, and, if the *à priori* probability of the assumption be small, the final probability of the existence of such an assumed world will be negligible. Hence you cannot prove to a man who thinks it very unlikely that anything exists except perceptions, that it is probable that a real world like the objects of perceptions exists on the ground that it explains so well what actually takes place.

But, granted that no one really is in the position that he thinks it positively improbable that there should be a real world like the objects of his perceptions, except through listening to erroneous arguments, we must compare the strengthening of the probability of such an external world with the strengthening of the probability of laws connecting our perceptions, which are equivalent to the former in all cases capable of verification. Let us take some simple scientific law and consider it—say the proposition that 'heat causes metals to expand.' The ordinary theory would be that the metals, the temperature, and the length exist whether we perceive them or not; that, under suitable circumstances, we can perceive them; and that, when we do so, we are actually able to perceive that connexion which we believe to hold between change of temperature and change of length, whether we perceive it or not. The phenomenalist position is that it is absurd to assume any more than the law about perceptions, that whenever we have a perception of a metal with a certain degree of temperature we can perceive a certain length, and that this will increase as the felt temperature increases. This, the phenomenalist would say, is all that ever could be verified in the realist's law, and therefore all in his assumption that any verification can render probable. Now if all laws were as simple as this the phenomenalist might be right; but this is not the case. The actual position that we have to face is the following one: In general it is not a case of finding regular conjunctions of perceptions and arguing from this that there is a causal law of the form that the occurrence of p_1 strengthens the probability of the occurrence of p_2 and makes it practically certain. There are some such regularities among our perceptions, and if we confined ourselves to them, I think it would be true to say that such hypothetical propositions about perceptions were better justified than the assumption of real connexions of events that take place alike when we do and when we do not perceive them. But the propositions at which we arrive as probable when we assume a real world like the objects of our perceptions, are not propositions at which we could have arrived from any actual recurrent series of connected perceptions, for the excellent reason that such a recurrent series will not in general have taken place. Thus we must not merely say that the assumption of real causes of our perceptions obeying certain laws enables us to discover regularities in our perceptions that we should otherwise miss. What we must say is that the assumption actually renders probable propositions about the connexion of our perceptions that would not be rendered probable by our actual perceptions alone, and which are verified. Propositions of the form 'if p occurs it is practically certain that q will occur, where p and q are perceptions,' can only be rendered probable by our perceptions when p and q actually do constantly occur together. But in general both p and q scarcely ever occur at all, and therefore it is true to say that our perceptions alone do not render such propositions probable. But this just means that unless the assumption is probably true these propositions are not probable. Now the less probable a proposition is antecedently, i.e. without the assumption of a given proposition, the more probable does that assumption become if it makes the antecedently improbable proposition soluble, and the latter is as a matter of fact found to be true.

Let us take an example from the wave-theory of light. It can be shown to follow from this that there will be a bright spot in the middle of the shadow cast by a small circular object like a coin. Now the phenomenalist would say: 'Why suppose that this verifies the wave-theory in as far as that involves what cannot be perceived? Why not say at once that it makes probable the law that I shall always perceive a bright spot in the middle when I perceive a circular shadow. This is the only part of the law that can be verified. True I should never have thought of looking for a bright spot there if I had not made the assumption of the wave theory, but this merely shows that the assumption was a fortunate aid to the discovery of a hypothetical proposition connecting my perceptions.' In accordance with what we have said, the following is the right answer to this contention. Certainly it is perfectly true that what has been discovered involves a hypothetical law about perceptions. But if you keep to perceptions the law was not probable unless there was an actual coexistence in all or most cases of the shadow and the bright spot. Now, as a matter of fact, on your theory there was not such a coexistence. It is useless for you to tell us that there was a bright spot, but that you did not notice it, because you did not expect it to be there. If phenomenalism be correct it was not there when you did not notice it. Hence unless the assumption of the wave theory be true, or at least probable, your law about perceptions and their connexion does not merely remain undiscovered, it actually has practically no probability. On the other hand, when you have made the assumption of the wave-theory, and learnt from it that such a spot ought to be found, the fact that you find it and that antecedently to the assumption there was practically no probability of its being there strengthens the probability of the assumption enormously. In fact, the whole criticism of the phenomenalistic argument under discussion might be put as follows: Propositions asserting constant conjunctions hypothetically can only be rendered probable either by the actual observable recurrence of both terms in conjunction or by the fact that they can be inferred from probable hypotheses. In most of the hypothetical laws about the conjunctions of perceptions a constant recurrence of both terms in conjunction is not as a matter of fact experienced. It follows that such laws are not probable unless they follow from some probable hypothesis. But if such an hypothesis can be found, and the conjunction can be verified, even in a single case the very fact that, apart from the hypothesis, it had no particular probability, makes the probability of the hypothesis stronger.

We cannot therefore accept the phenomenalistic position that we may just as well assume hypothetical laws about our perceptions as causes of them obeying among themselves the laws laid down by physics. The laws of physics will indeed in general be able to be stated in terms of connexions between possible perceptions; but the laws so stated will have little or no probability apart from the truth or probability of the assumption about the real world and its laws. And the mere statement of the laws of physics in terms of possible perceptions would not be without difficulty. Atoms, molecules, and ether-waves could not be put directly into phenomenalistic laws. For they are supposed to be of such a nature that they could not possibly be perceived. Hence a phenomenalistic law about them would take the form: If I had the perception p which, *ex-hypothesi*, I cannot have, I should always have also the perception q which, *ex-hypothesi*, I cannot have. If, then, we are to be able to carry science to any pitch it is essential that we should be able to allow the existence of real causes of our perceptions as at least possible, and when this is done the most probable laws for them to obey among themselves are those which science finds it necessary to assume in order to account for what is perceived. Now these laws are not in the least like those which perceptions obey among themselves, although they are of course connected with the latter. They are in fact laws about the kind of changes that we can observe in the object of a single continuous perception; and the only common characteristics of the objects of our perceptions and the perceptions themselves is that both have temporal relations and can enter into causal laws. Hence, until anyone can make up a theory in terms of laws like those that hold between perceptions which will explain our perceptions better than the theory of science, we shall be justified in holding that if there be a real world at all it probably resembles the objects of our perceptions.

Chapter 20
BERTRAND RUSSELL
(1872-1970)

INTRODUCTION

BERTRAND ARTHUR WILLIAM RUSSELL was born on May 18, 1872. His father, who died when Russell was only three years old, had wished him to be brought up an agnostic, but his grandmother found a subtle way of getting around this plan by arranging for Russell to have a private education. By 1890 Russell had prepared to enter Cambridge to study mathematics and philosophy. Four years later he left that university and spent some months as attaché at the British embassy in Paris, then he went on to Berlin, where he studied social democracy, finally settling down in Haslemere, where he devoted himself to the study of philosophy and the preparation of a book entitled *A Critical Exposition of the Philosophy of Leibniz* (1900). He published *The Principles of Mathematics* in 1903, and from then on worked with his friend Whitehead in further developing the mathematical logic of Frege and Peano. Between 1910 and 1913 Russell and Whitehead published the *Principia Mathematica* in three volumes.

Russell was appointed lecturer at Trinity College, Cambridge, in 1910, the same year he published *Philosophical Essays*. In 1914 *Our Knowledge of the External World* appeared. About this time Russell's pacifist activities in connection with World War I began to cause trouble for him; eventually he received a prison sentence of six months, which he used for the writing of *Introduction to Mathematical Philosophy*, subsequently published in 1919. Russell then visited Russia and China before returning to England to earn his livelihood through lecturing, writing popular books, and journalism. During this difficult period he also published such important studies as the Introduction to the second edition of *Principia Mathematica* (1925), and *The Analysis of Matter* (1927). In 1931 Russell succeeded to the peerage vacated by the death of his brother, and since that time has lectured all over the world on philosophy, social affairs, and politics, and has written many books on philosophy, mathematics, science, politics, morals, religion, and education. He has been a fellow of the Royal Society since 1908; he received the Order of Merit in 1949 and was awarded the Nobel Prize for literature in 1950.

Between 1896 and 1960 Russell's philosophical views changed considerably, particularly his views on metaphysics, ethics, and religious problems, but, in all, his basic outlook toward philosophy has not changed substantially since 1914. From then on he has continued to maintain that the subject matter of philosophy is different from that of science. Unlike the latter, which deals with limited realms of objects, philosophy is concerned with the nature of the universe and man's place in it. Russell believes, however, that in studying the problems connected with this subject matter philosophy must try to keep in close contact with the latest findings of the sciences, which is not to say that philosophy should found its insights on the findings of the sciences—for Russell an impossible notion. Instead he holds that philosophy must be consistent with the present state of science, and in addition must try to substitute constructions out of known entities for inferences to unknown entities, whenever and wherever this is possible. He also maintains that philosophy must attempt to solve its problems by strictly rational, even scientific, methods.

In delineating his view concerning the relation between philosophy and science, Russell carefully specifies that it is not so much the results of the sciences which are of prime importance for philosophy as the methods they use. In this connection he says, "There are two different ways in which a philosophy may seek to base itself upon science. It may emphasize the most general *results* of science and seek to give even greater generality and unity to these results. Or it may study the *methods* of science, and seek to apply these methods with the necessary adaptations to its own particular province. Much philosophy inspired by science has gone astray through preoccupation with the *results* momentarily supposed to have been achieved. It is not results but *methods* that can be transferred with profit from the sphere of the special sciences to the sphere of philosophy" (*Mysticism and Logic,* pp. 105–106).

Russell does not primarily understand by *methods* those of the empirical sciences which are often referred to using the terms experiment, measurement, formulations of hypotheses and laws, since these methods primarily refer to concrete and particular findings about the world. The methods which philosophy is to apply in the study of its subject matter are the methods of the formal sciences, as described in

Russell's logical investigations. On the basis of these Russell claims that all problems, insofar as they are genuinely philosophical problems, are problems of logic. This does not mean for him, however, that mathematical logic is directly of philosophical importance; as he expresses it, after the beginnings, logic belongs rather to mathematics. It is only logic's beginnings which are relevant to philosophy; its later developments will only be of indirect use in philosophy in that they enable us to deal easily with abstract conceptions, suggest fruitful hypotheses, and make it possible for us to see quickly what is the smallest store of materials necessary for the construction of a logical edifice (*Knowledge*, pp. 33, 39-40). As far as the beginnings of logic are concerned, Russell claims that logic showed at that point that analysis must be *the* method of philosophy; that what is to be analyzed does not consist in things, but in facts, and, finally, that logic provides us with an infinite number of possible hypotheses to be applied in the analysis of any complex fact (*Ib.*, 40-53).

In the third lecture of *Our Knowledge of the External World* Russell applies this method of logic to the problem of our knowledge of the external world (*Ib.*, pp. 55-56). He turns his attention first to our prescientific knowledge, our common sense knowledge of facts, which in his view (following the empiricist line) takes its starting point in sense data. All other common-sense objects, he maintains, are built up from these data and so are constructions from them. Such a view inevitably leads to the crucial question of whether the existence of anything other than our own immediate sense data can be inferred from the existence of those data. It could be said that there must be real material things because they have caused our sense impressions. Whereas in *The Problems of Philosophy* (1912) Russell held the view that a certain belief in the reality of material things is a belief which simplifies and systematizes our account of our experiences, and that, therefore there seems to be no good reason for rejecting it (p. 38), later in *Our Knowledge of the External World* he comes to the conclusion that a belief in an external cause of our sense impressions serves no useful purpose. In *The Analysis of Matter* (1927) Russell returned to his original view and, instead of accepting the causal theory as an "instinctive belief" reasonably tried to defend the view as a valid thesis.

The following selection was the fourth lecture in the series *Our Knowledge of the External World*, in which Russell applies his logical-analytic method to our scientific knowledge of the external world.

SELECTIVE BIBLIOGRAPHY

Russell, B., *Our Knowledge of the External World*. London, 1914.

———, *Mysticism and Logic*. London, 1918.

———, *Introduction to Mathematical Philosophy*. London, 1919.

———, *Analysis of Matter*. London, 1927.

———, *An Outline of Philosophy*. London, 1927.

———, *An Inquiry into Meaning and Truth*. London, 1940.

———, *Human Knowledge*. London, 1948.

———, *Logic and Knowledge* (Essays 1901–1950), ed. R. C. Marsh. London, 1956.

———, *My Philosophical Development*. London, 1959.

———, *On the Philosophy of Science*, ed. C. A. Fritz. Indianapolis, 1965.

Wood, A., *Bertrand Russell: The Passionate Sceptic*. New York, 1958.

The Philosophy of Bertrand Russell, ed. P. A. Schilpp. Chicago, 1944.

Fritz, C. A., *Russell's Construction of the External World*. London, 1952.

Wedberg A., *Bertrand Russell's Empiricism*. Uppsala, 1937.

Götland, E., *Bertrand Russell's Theories of Causation*. Uppsala, 1952.

Warnock, G. J., *English Philosophy Since 1900*. London, 1958.

O'Connor, D. J., "Betrand Russell," in *A Critical History of Western Philosophy*, ed. D. J. O'Connor. New York, 1965. Pp. 473–491.

THE WORLD OF PHYSICS AND
THE WORLD OF SENSE

*A*MONG the objections to the reality of objects of sense, there is one which is derived from the apparent difference between matter as it appears in physics and things as they appear in sensation. Men of science, for the most part, are willing to condemn immediate data as "merely subjective," while yet maintaining the truth of the physics inferred from those data. But such an attitude, though it may be *capable* of justification, obviously stands in need of it; and the only justification possible must be one which exhibits matter as a logical construction from sense-data—unless, indeed, there were some wholly *a priori* principle by which unknown entities could be inferred from such as are known. It is therefore necessary to find some way of bridging the gulf between the world of physics and the world of sense, and it is this problem which will occupy us in the present lecture. Physicists appear to be unconscious of the gulf, while psychologists, who are conscious of it, have not the mathematical knowledge required for spanning it. The problem is difficult, and I do not know its solution in detail. All that I can hope to do is to make the problem felt, and to indicate the kind of methods by which a solution is to be sought.

Let us begin by a brief description of the two contrasted worlds. We will take first the world of physics, for, though the other world is given while the physical world is inferred, to us now the world of physics is the more familiar, the world of pure sense having become strange and difficult to rediscover. Physics started from the common-sense belief in fairly permanent and fairly rigid bodies—tables and chairs, stones, mountains, the earth and moon and sun. This common-sense belief, it should be noticed, is a piece of audacious metaphysical theorising; objects are not continually present to sensation, and it may be doubted whether they are there when they are not seen or felt. This problem, which has been acute since the time of Berkeley, is ignored by common sense, and has therefore hitherto been ignored by physicists. We have thus here a first departure from the immediate data of sensation, though it is a departure merely by way of extension, and was probably made by our savage ancestors in some very remote prehistoric epoch.

But tables and chairs, stones and mountains, are not *quite* permanent or *quite* rigid. Tables and chairs lose their legs, stones are split by frost, and mountains are cleft by earthquakes and eruptions. Then there are other things, which seem material, and yet present almost no permanence or rigidity. Breath, smoke, clouds are examples of such things—so, in a lesser degree, are ice and snow; and rivers and seas, though fairly permanent, are not in any degree rigid. Breath, smoke, clouds, and generally things that can be seen but not touched, were thought to be hardly real; to this day the usual mark of a ghost is that it can be seen but not touched. Such objects were peculiar in the fact that they seemed to disappear completely, not merely to be transformed into something else. Ice and snow, when they disappear, are replaced by water; and it required no great theoretical effort to invent the hypothesis that the water was the same thing as the ice and snow, but in a new form. Solid bodies, when they break, break into parts which are practically the same in shape and size as they were before. A stone can be hammered into a powder, but the powder consists of grains which retain the character they had before the pounding. Thus the ideal of absolutely rigid and absolutely permanent bodies, which early physicists pursued throughout the changing appearances, seemed attainable by supposing ordinary bodies to be composed of a vast number of tiny atoms. This billiard-ball view of matter dominated the imagination of physicists until quite modern times, until, in fact, it was replaced by the electromagnetic theory, which in its turn developed into a new atomism. Apart from the special form of the atomic theory which was invented for the needs of chemistry, some kind of atomism dominated the whole of traditional dynamics, and was implied in every statement of its laws and axioms.

The pictorial accounts which physicists give of the material world as they conceive it undergo violent changes under the influence of modifications in theory which are much slighter than the layman might suppose from the alterations of the description. Certain features, however, remained fairly stable until the last few years. It was always assumed that there is *something* indestructible which is capable of motion in space; what is indestructible was always very small, but did not always occupy a mere point in space. This view still dominated the Rutherford-Bohr theory of the structure of the atom. Since 1925, however, under the influence of De Broglie, Heisenberg, and Schrödinger, physicists have been led to dissolve the atom into systems of wave-motions, or radiations coming from the place where the atom was supposed to be. This change has brought physics much nearer to psychology, since the supposed permanent material units are now merely logical constructions. In regard to space and time, relativity has introduced a fundamental structural change by merging them in the one four-dimensional space-time. Both these changes have made physics easier to reconcile with psychology than was formerly the case. Both sciences now demand certain departures from common-sense metaphysics, and fortunately the departures they demand harmonize with each other.

Common sense, and physics before the twentieth century, demanded a set of indestructible entities, moving relatively to each other in a single space and a single time. The world of immediate data is quite different from this. Nothing is permanent; even the things that we think are fairly permanent, such as mountains, only become data when we see them, and are not immediately given as existing at other moments. So far from one all-embracing space being given, there are several spaces for each person, according to the different senses which give relations that may be called spatial. Experience teaches us to obtain one space from these by correlation, and experience, together with instinctive theorising, teaches us to correlate our spaces with those which we believe to exist in the sensible worlds of other people. The construction of a single time offers less difficulty so long as we confine ourselves to one person's private world, but relativity has shown that, when we pass beyond one private world, it is space-time, not space and time, that we need. Thus, apart from any of the fluctuating hypotheses of physics, two main problems arise in connecting the world of physics with the world of sense, namely (1) the construction of permanent "things," and (2) the construction of a single space-time. We will consider these two problems in succession.

(1) The belief in indestructible "things" very early took the form of atomism. The underlying motive in atomism was not, I think, any empirical success in interpreting phenomena, but rather an instinctive belief that beneath all the changes of the sensible world there must be something permanent and unchanging. This belief was, no doubt,

fostered and nourished by its practical successes, culminating in the conservation of mass; but it was not produced by these successes. On the contrary, they were produced by it. Philosophical writers on physics sometimes speak as though the conservation of something or other were essential to the possibility of science, but this, I believe, is an entirely erroneous opinion. If the *a priori* belief in permanence had not existed, the same laws which are now formulated in terms of this belief might just as well have been formulated without it. Why should we suppose that, when ice melts, the water which replaces it is the same thing in a new form? Merely because this supposition enables us to state the phenomena in a way which is consonant with our prejudices. What we really know is that, under certain conditions of temperature, the appearance we call ice is replaced by the appearance we call water. We can give laws according to which the one appearance will be succeeded by the other, but there is no reason except prejudice for regarding both as appearances of the same substance.

One task, if what has just been said is correct, which confronts us in trying to connect the world of sense with the world of physics, is the task of reconstructing the conception of matter without the *a priori* beliefs which historically gave rise to it. In spite of the revolutionary results of modern physics, the empirical successes of the conception of matter show that there must be some legitimate conception which fulfils roughly the same functions. The time has hardly come when we can state precisely what this legitimate conception is, but we can see in a general way what it must be like. For this purpose, it is only necessary to take our ordinary common-sense statements and reword them without the assumption of permanent substance. We say, for example, that things change gradually—sometimes very quickly, but not without passing through a continuous or nearly continuous series of intermediate states. What this means is that, given any sensible appearance, there will usually be, *if we watch*, a continuous series of appearances connected with the given one, leading on by imperceptible gradations to the new appearances which common sense regards as those of the same thing. Thus a thing may be defined as a certain series of appearances, connected with each other by continuity and by certain causal laws. In the case of slowly changing things, this is easily seen. Consider, say, a wall-paper which fades in the course of years. It is an effort not to conceive of it as one "thing" whose colour is slightly different at one time from what it is at another. But what do we really *know* about it? We know that under suitable circumstances—i.e. when we are, as is said, "in the room"—we perceive certain colours in a certain pattern: not always precisely the same colours, but sufficiently similar to feel familiar. If we can state the laws according to which the colour varies, we can state all that is empirically verifiable; the assumption that there is a constant entity, the wall-paper, which "has" these various colours at various times, is a piece of gratuitous metaphysics. We may, if we like, *define* the wall-paper as the series of its aspects. These are collected together by the same motives which led us to regard the wall-paper as one thing, namely a combination of sensible continuity and causal connection. More generally, a "thing" will be defined as a certain series of aspects, namely those which would commonly be said to be *of* the thing. To say that a certain aspect is an aspect *of* a certain thing will merely mean that it is one of those which, taken serially, *are* the thing. Everything will then proceed as before: whatever was verifiable is unchanged, but our language is so interpreted as to avoid an unnecessary meta-physical assumption of permanence.

The above extrusion of permanent things affords an example of the maxim which inspires all scientific philosophising, namely "Occam's razor": *Entities are not to be multiplied without necessity.* In other words, in dealing with any subject-matter, find out what entities are undeniably involved, and state everything in terms of these entities. Very often the resulting statement is more complicated and difficult than one which, like common sense and most philosophy, assumes hypothetical entities whose existence there is no good reason to believe in. We find it easier to imagine a wall-paper with changing colours than to think merely of the series of colours; but it is a mistake to sup-pose that what is easy and natural in thought is what is more free from unwarrantable assumptions, as the case of "things" very aptly illus-trates.

The above summary account of the genesis of "things," though it may be correct in outline, has omitted some serious difficulties which it is necessary briefly to consider. Starting from a world of helter-skelter sense-data, we wish to collect them into series, each of which can be regarded as consisting of the successive appearances of one "thing." There is, to begin with, some conflict between what common sense regards as one thing, and what physics regards as an unchanging collection of particles. To common sense, a human body is one thing, but to science the matter composing it is continually changing. This conflict, however, is not very serious, and may, for our rough prelim-inary purpose, be largely ignored. The problem is: by what principles shall we select certain data from the chaos, and call them all appearances of the same thing?

A rough and approximate answer to this question is not very difficult. There are certain fairly stable collections of appearances, such as landscapes, the furniture of rooms, the faces of acquaintances. In these cases, we have little hesitation in regarding them on successive occasions as appearances of one thing or collection of things. But, as the *Comedy of Errors* illustrates, we may be led astray if we judge by mere resemblance. This shows that something more is involved, for two different things may have any degree of likeness up to exact similarity.

Another insufficient criterion of one thing is *continuity*. As we have already seen, if we watch what we regard as one changing thing, we usually find its changes to be continuous so far as our senses can perceive. We are thus led to assume that, if we see two finitely different appearances at two different times, and if we have reason to regard them as belonging to the same thing, then there was a continuous series of intermediate states of that thing during the time when we were not observing it. And so it comes to be thought that continuity of change is necessary and sufficient to constitute one thing. But in fact it is neither. It is not *necessary*, because the unobserved states, in the case where our attention has not been concentrated on the thing throughout, are purely hypothetical, and cannot possibly be our ground for supposing the earlier and later appearances to belong to the same thing; on the contrary, it is because we suppose this that we assume intermediate unobserved states. Continuity is also not sufficient, since we can, for example, pass by sensibly continuous gradations from any one drop of the sea to any other drop. The utmost we can say is that discontinuity during uninterrupted observation is as a rule a mark of difference between things, though even this cannot be said in such cases as sudden explosions.

The assumption of continuity is, however, successfully made in physics (apart, perhaps, from quantum phenomena). This proves something, though not anything of very obvious utility to our present problem: it proves that nothing in the known world is inconsistent with the hypothesis that all changes are really continuous, though from too great rapidity or from our lack of observation they may not always appear continuous. In this hypothetical sense, continuity may be allowed to be a *necessary* condition if two appearances are to be classed as appearances of the same thing. But it is not a *sufficient* condition, as appears from the instance of the drops of the sea. Thus something more must be sought before we can give even the roughest definition of a "thing."

What is wanted further seems to be something in the nature of fulfilment of causal laws. This statement, as it stands, is very vague, but we will endeavour to give it precision. When I speak of "causal laws," I mean any laws which connect events at different times, or even, as a limiting case, events at the same time provided the connection is not logically demonstrable. In this very general sense, the laws of dynamics are causal laws, and so are the laws correlating the simultaneous appearances of one "thing" to different senses. The question is: How do such laws help in the definition of a "thing"?

To answer this question, we must consider what it is that is proved by the empirical success of physics. What is proved is that its hypotheses, though unverifiable where they go beyond sense-data, are at no point in contradiction with sense-data, but, on the contrary, are ideally such as to render all sense-data calculable from a sufficient collection of data all belonging to a given period of time. Now physics has found it empirically possible (apart from certain problems in quantum theory, as to which there is still uncertainty) to collect sense-data into series, each series being regarded as belonging to one "thing," and behaving, with regard to the laws of physics, in a way in which series not belonging to one thing would in general not behave. If it is to be unambiguous whether two appearances belong to the same thing or not, there must be only one way of grouping appearances so that the resulting things obey the laws of physics. It would be very difficult to prove that this is the case, but for our present purposes we may let this point pass, and assume that there is only one way. We must include in our definition of a "thing" those of its aspects, if any, which are not observed. Thus we may lay down the following definition: *Things are those series of aspects which obey the laws of physics.* That such series exist is an empirical fact, which constitutes the verifiability of physics.

It may still be objected that the "matter" of physics is something other than series of sense-data. Sense-data, it may be said, belong to psychology and are, at any rate in some sense, subjective, whereas physics is quite independent of psychological considerations, and does not assume that its matter only exists when it is perceived.

To this objection there are two answers, both of some importance.

(a) We have been considering, in the above account, the question of the *verifiability* of physics. Now verifiability is by no means the same thing as truth; it is, in fact, something far more subjective and psychological. For a proposition to be verifiable, it is not enough that it should be true, but it must also be such as we can *discover* to be true. Thus verifiability depends upon our capacity for acquiring knowledge, and not only upon the objective truth. In physics, as ordinarily set forth, there is much that is unverifiable: there are hypotheses as to (*a*) how things would appear to a spectator in a place where, as it happens, there is no spectator; (*β*) how things would appear at times when, in fact, they are not appearing to anyone; (*γ*) things which never appear at all. All these are introduced to simplify the statement of the causal laws, but none of them forms an integral part of what is *known* to be true physics. This brings us to our second answer.

(b) If physics is to consist wholly of propositions known to be true, or at least capable of being proved or disproved, the three kinds of hypothetical entities we have just enumerated must all be capable of being exhibited as logical functions of sense-data. In order to show how this might possibly be done, let us recall the hypothetical Leibnizian universe. . . .

In that universe we had a number of perspectives, two of which never had any entity in common, but often contained entities which could be sufficiently correlated to be regarded as belonging to the same thing. We will call one of these an "actual" private world when there is an actual spectator to which it appears, and "ideal" when it is merely constructed on principles of continuity. A physical thing consists, at each instant, of the whole set of its aspects at that instant, in all the different worlds; thus a momentary state of a thing is a whole set of aspects. An "ideal" appearance will be an aspect merely calculated, but not actually perceived by any spectator. An "ideal" state of a thing will be a state at a moment when all its appearances are ideal. An ideal thing will be one whose states at all times are ideal. Ideal appearances, states, and things, since they are calculated, must be functions of actual appearances, states, and things; in fact, ultimately, they must be functions of actual appearances. Thus it is unnecessary, for the enunciation of the laws of physics, to assign any reality to ideal elements: it is enough to accept them as logical constructions, provided we have means of knowing how to determine when they become actual. This, in fact, we have with some degree of approximation; the starry heaven, for instance, becomes actual whenever we choose to look at it. It is open to us to believe that the ideal elements exist, and there can be no reason for *disbeliev-*

ing this; but unless in virtue of some *a priori* law we cannot *know* it, for empirical knowledge is confined to what we actually observe.

Permanent things, even as a logical construction, are no longer quite adequate to the needs of physics. It seems that the persistence of electrons and protons is only an approximate fact. After a quantum change in an atom, according to Heisenberg, we can no longer identify a given electron with a definite one of those existing before the change. Moreover, it is thought that, in the stars, an electron and a proton sometimes destroy each other, with the result that their energy (or mass) is converted into a non-material form. In fact, the difference between matter and other forms of energy has become much less than it used to be before energy and mass were found to be the same thing.

(2) The three main conceptions of physics are space, time, and matter. Some of the problems raised by the conception of matter have been indicated in the above discussion of "things." But space and time also raise difficult problems of much the same kind, namely, difficulties in reducing the haphazard untidy world of immediate sensation to the smooth orderly world of geometry and kinematics. Let us begin with the consideration of space.

People who have never read any psychology seldom realise how much mental labour has gone into the construction of the one all-embracing space into which all one man's sensible objects are supposed to fit. Kant, who was unusually ignorant of psychology, described space as "an infinite given whole," whereas a moment's psychological reflection shows that a space which is infinite is not given, while a space which can be called given is not infinite. What the nature of "given" space really is, is a difficult question, upon which psychologists are by no means agreed. But some general remarks may be made, which will suffice to show the problems, without taking sides on any psychological issue still in debate.

The first thing to notice is that different senses have different spaces. The space of sight is quite different from the space of touch: it is only by experience in infancy that we learn to correlate them. In later life, when we see an object within reach, we know how to touch it, and more or less what it will feel like; if we touch an object with our eyes shut, we know where we should have to look for it, and more or less what it would look like. But this knowledge is derived from early experience of the correlation of certain kinds of touch-sensations with certain kinds of sight-sensations. The one space into which both kinds of sensations fit is a construction, not a datum.

And besides touch and sight, there are other kinds of sensation which give other, though less important, spaces: these also have to be fitted into the one space by means of experienced correlations. And as in the case of things, so here: the one all-embracing space, though convenient as a way of speaking, need not be supposed really to exist. All that experience makes certain is the several spaces of the several senses, correlated by empirically discovered laws. The one space may turn out to be valid as a logical construction, compounded of the several spaces, but there is no good reason to assume its independent metaphysical reality.

Another respect in which the spaces of immediate experience differ from the space of geometry and physics is in regard to *points*. The space of geometry and physics consists of an infinite number of points, but no one has ever seen or touched a point. If there are points in a sensible space, they must be an inference. It is not easy to see any way in which, as independent entities, they could be validly inferred from the data; thus here again, we shall have, if possible, to find some logical construction, some complex assemblage of immediately given objects, which will have the geometrical properties required of points. It is customary to think of points as simple and infinitely small, but geometry in no way demands that we should think of them in this way. All that is necessary for geometry is that they should have mutual relations possessing certain enumerated abstract properties, and it may be that an assemblage of data of sensation will serve this purpose. Exactly how this is to be done, I do not yet know, but it seems fairly certain that it can be done.

The following illustrative method, simplified so as to be easily manipulated, has been invented by Dr. Whitehead for the purpose of showing how points might be manufactured from sense-data. We have first of all to observe that there are no infinitesimal sense-data; any surface we can see, for example, must be of some finite extent. But what at first appears as one undivided whole is often found, under the influence of attention, to split up into parts contained within the whole. Thus one spatial object may be contained within another, and entirely enclosed by the other. This relation of enclosure, by the help of some very natural hypotheses, will enable us to define a "point" as a certain class of spatial objects, namely all those (as it will turn out in the end) which would naturally be said to contain the point. In order to obtain a definition of a "point" in this way, we proceed as follows:

Given any set of volumes, or surfaces, they will not in general converge into one point. But if they get smaller and smaller, while of any two of the set there is always one that encloses the other, then we begin to have the kind of conditions which would enable us to treat them as having a point for their limit. The hypotheses required for the relation of enclosure are that (1) it must be transitive; (2) of two *different* spatial objects, it is impossible for each to enclose the other, but a single spatial object always encloses itself; (3) any set of spatial objects such that there is at least one spatial object enclosed by them all has a lower limit or minimum, i.e. an object enclosed by all of them and enclosing all objects which are enclosed by all of them; (4) to prevent trivial exceptions, we must add that there are to be instances of enclosure, i.e. there are really to be objects of which one encloses the other. When an enclosure-relation has these properties, we will call it a "point-producer." Given any relation of enclosure, we will call a set of objects an "enclosure-series" if, of any two of them, one is contained in the other. We require a condition which shall secure that an enclosure-series converges to a point, and this is obtained as follows: Let our enclosure-series be such that, given any other enclosure-series of which there are members enclosed in any arbitrarily chosen member of our first series, then there are members of our first series enclosed in any arbitrarily chosen member of our second series. In this case, our first enclosure-series may be called a "punctual enclosure-series." Then a "point" is all the objects which enclose members of a given punctual enclosure-series. In order to ensure infinite divisibility, we require one further property to be added to those defining point-producers, namely that any object which encloses itself also encloses an object other than itself. The "points" generated by point-producers with this property will be found to be such as geometry requires.[1]

The question of time, so long as we confine ourselves to one private world, is rather less complicated than that of space, and we can see pretty clearly how it might be dealt with by such methods as we have been considering. Events of which we are conscious do not last merely for a mathematical instant, but always for some finite time, however short. Even if there be a physical world such as the mathematical theory of motion supposes, impressions on our sense-organs produuce sensations which are not merely and strictly instantaneous, and therefore the objects of sense of which we are immediately conscious are

---

[1] The above method requires certain improvements, which will be found in Whitehead, *Concept of Nature*.

not strictly instantaneous. Instants, therefore, are not among the data of experience, and, if legitimate, must be either inferred or constructed. It is difficult to see how they can be validly inferred; thus we are left with the alternative that they must be constructed. How is this to be done?

Immediate experience provides us with two time-relations among events: they may be simultaneous, or one may be earlier and the other later. These two are both part of the crude data; it is not the case that only the events are given, and their time-order is added by our subjective activity. The time-order, within certain limits, is as much given as the events. In any story of adventure you will find such passages as the following: "With a cynical smile he pointed the revolver at the breast of the dauntless youth. 'At the word *three* I shall fire,' he said. The words one and two had already been spoken with a cool and deliberate distinctness. The word *three* was forming on his lips. At this moment a blinding flash of lightning rent the air." Here we have simultaneity—not due, as Kant would have us believe, to the subjective mental apparatus of the dauntless youth, but given as objectively as the revolver and the lightning. And it is equally given in immediate experience that the words *one* and *two* come earlier than the flash. These time-relations hold between events which are not strictly instantaneous. Thus one event may begin sooner than another, and therefore be before it, but may continue after the other has begun, and therefore be also simultaneous with it. If it persists after the other is over, it will also be later than the other. Earlier, simultaneous, and later, are not inconsistent with each other when we are concerned with events which last for a finite time, however short; they only become inconsistent when we are dealing with something instantaneous.

It is to be observed that we cannot give what may be called *absolute* dates, but only dates determined by events. We cannot point to a time itself, but only to some event occurring at that time. There is therefore no reason in experience to suppose that there are times as opposed to events: the events, ordered by the relations of simultaneity and succession, are all that experience provides. Hence, unless we are to introduce superfluous metaphysical entities, we must, in defining what mathematical physics can regard as an instant, proceed by means of some construction which assumes nothing beyond events and their temporal relations.

If we wish to assign a date exactly by means of events, how shall we proceed? If we take any one event, we cannot assign our date exactly, because the event is not instantaneous, that is to say, it may be simultaneous with two events which are not simultaneous with each other. In order to assign a date exactly, we must be able, theoretically, to determine whether any given event is before, at, or after this date, and we must know that any other date is either before or after this date, but not simultaneous with it. Suppose, now, instead of taking one event A, we take two events A and B, and suppose A and B partly overlap, but B ends before A ends. Then an event which is simultaneous with both A and B must exist during the time when A and B overlap; thus we have come rather nearer to a precise date than when we considered A and B alone. Let C be an event which is simultaneous with both A and B, but which ends before either A or B has ended. Then an event which is simultaneous with A and B and C must exist during the time when all three overlap,

A
_____

B
_____

C
_____

which is a still shorter time. Proceeding in this way, by taking more and more events, a new event which is dated as simultaneous with all of them becomes gradually more and more accurately dated. This suggests a way by which a completely accurate date can be defined.

Let us take a group of events of which any two overlap, so that there is some time, however short, when they all exist. If there is any other event which is simultaneous with all of these, let us add it to the group; let us go on until we have constructed a group such that no event outside the group is simultaneous with all of them, but all the events inside the group are simultaneous with each other. Let us define this whole group as an instant of time. It remains to show that it has the properties we expect of an instant.

What are the properties we expect of instants? First, they must form a series: of any two, one must be before the other, and the other must be not before the one; if one is before another, and the other before a third, the first must be before the third. Secondly, every event must be at a certain number of instants; two events are simultaneous if they are at the same instant, and one is before the other if there is an instant, at which the one is, which is earlier than some instant at which the other is. Thirdly, if we assume that there is always some change going on somewhere during the time when any given event persists, the series of instants ought to be compact, i.e. given any two instants, there ought to be other instants between them. Do instants, as we have defined them, have these properties?

We shall say that an event is "at" an instant when it is a member of the group by which the instant is constituted; and we shall say that one instant is before another if the group which is the one instant contains an event which is earlier than, but not simultaneous with, some event in the group which is the other instant. When one event is earlier than, but not simultaneous with another, we shall say that it "wholly precedes" the other. Now we know that of two events which are not simultaneous, there must be one which wholly precedes the other, and in that case the other cannot also wholly precede the one; we also know that, if one event wholly precedes another, and the other wholly precedes a third, then the first wholly precedes the third. From these facts it is easy to deduce that the instants as we have defined them form a series.

We have next to show that every event is "at" at least one instant, i.e. that, given any event, there is at least one class, such as we used in defining instants, of which it is a member. For this purpose, consider all the events which are simultaneous with a given event, and do not begin later, i.e. are not wholly after anything simultaneous with it. We will call these the "initial contemporaries" of the given event. It will be found that this class of events is the first instant at which the given event exists, provided every event wholly after some contemporary of the given event is wholly after some *initial* contemporary of it.

Finally, the series of instants will be compact if, given any two events of which one wholly precedes the other, there are events wholly after the one and simultaneous with something wholly before the other. Whether this is the case or not, is an empirical question; but if it is not, there is no reason to expect the time-series to be compact.[2]

---

[2] The assumptions made concerning time-relations in the above are as follows:—

I. In order to secure that instants form a series, we assume:

    (a) No event wholly precedes itself. (An "event" is defined as whatever is simultaneous with something or other.)

    (b) If one event wholly precedes another, and the other wholly precedes a third, then the first wholly precedes the third.

    (c) If one event wholly precedes another, it is not simultanous with it.

Thus our definition of instants secures all that mathematics requires, without having to assume the existence of any disputable metaphysical entities.

Instants may also be defined by means of the enclosure-relation, exactly as was done in the case of points. One object will be temporally enclosed by another when it is simultaneous with the other, but not before or after it. Whatever encloses temporally or is enclosed temporally we shall call an "event." In order that the relation of temporal enclosure may be a "point-producer," we require (1) that it should be transitive, i.e. that if one event encloses another, and the other a third, then the first encloses the third; (2) that every event encloses itself, but if one event encloses another different event, then the other does not enclose the one; (3) that given any set of events such that there is at least one event enclosed by all of them, then there is an event enclosing all that they all enclose, and itself enclosed by all of them; (4) that there is at least one event. To ensure infinite divisibility, we require also that every event should enclose events other than itself. Assuming these characteristics, temporal enclosure is an infinitely divisible point-producer. We can now form an "enclosure-series" of events, by choosing a group of events such that of any two there is one which encloses the other; this will be a "punctual enclosure-series" if, given any other enclosure-series such that every member of our first series encloses some member of our second, then every member of our second series encloses some member of our first. Then an "instant" is the class of all events which enclose members of a given punctual enclosure-series.

(d) Of two events which are not simultaneous, one must wholly precede the other.

II. In order to secure that the initial contemporaries of a given event should form an instant, we assume:

(e) An event wholly after some contemporary of a given event is wholly after some *initial* contemporary of the given event.

III. In order to secure that the series of instants shall be compact, we assume:

(f) If one event wholly precedes another, there is an event wholly after the one and simultaneous with something wholly before the other.

This assumption entails the consequence that if one event covers the whole of a stretch of time immediately preceding another event, then it must have at least one instant in common with the other event; i.e., it is impossible for one event to cease just before another begins. I do not know whether this should be regarded as inadmissible. For a mathematicological treatment of the above topics, cf. N. Wiener, "A Contribution to the Theory of Relative Position," *Proc. Camb. Phil. Soc.*, xvii, 5, pp. 441–449.

The correlation of the times of different private worlds so as to produce the one all-embracing time of traditional physics is only approximately possible, for the reasons which led to the theory of relativity. So long as we confine ourselves to the surfaces of the earth, the approximation is very nearly exact. We saw, in Lecture III, that different private worlds often contain correlated appearances, such as common sense would regard as appearances of the same "thing." When two appearances in different worlds are so correlated as to belong to one momentary "state" of a thing, it would be natural to regard them as simultaneous, and as thus affording a simple means of correlating different private times. But this can only be regarded as a first approximation. What we call one sound will be heard sooner by people near the source of sound than by people farther from it, and the same applies, though in a less degree, to light. Thus two correlated appearances in different worlds are not necessarily to be regarded as occurring at the same date in physical time, though they will be parts of one momentary state of a thing. The correlation of different private times is regulated by the desire to secure the simplest possible statement of the laws of physics, and thus raises rather complicated technical problems; but from the point of view of philosophical theory, there is no very serious difficulty of principle involved, within the limits imposed by the theory of relativity.

The above brief outline must not be regarded as more than tentative and suggestive. It is intended merely to show the kind of way in which, given a world with the kind of properties that psychologists find in the world of sense, it may be possible, by means of purely logical constructions, to make it amenable to mathematical treatment by defining series or classes of sense-data which can be called respectively particles, points, and instants. If such constructions are possible, then mathematical physics is applicable to the real world, in spite of the fact that its particles, points, and instants are not to be found among actually existing entities.

The space-time of physics has not a very close relation to the space and time of the world of one person's experience. Everything that occurs in one person's experience must, from the standpoint of physics, be located within that person's body; this is evident from considerations of causal continuity. What occurs when I see a star occurs as the result of light-waves impinging on the retina, and causing a process in the optic nerve and brain; therefore the occurrence called "seeing a star" must be in the brain. If we define a piece of matter as a set of events (as was suggested above), the sensation of seeing a star will be one of the events which *are* the brain of the percipient at the time of the perception. Thus every event that I experience will be one of the events that constitute some part of my body. The space of (say) my visual perceptions is only *correlated* with physical space, more or less approximately: from the physical point of view, whatever I see is inside my head. I do not see physical objects; I see effects which they produce in the region where my brain is. The correlation of visual and physical space is rendered approximate by the fact that my visual sensations are not *wholly* due each to some physical object, but also partly to the intervening medium. Further, the relation of visual sensation to physical object is one-many, not one-one, because our senses are more or less vague: things which look different under the microscope may be indistinguishable to the naked eye. The inferences from perceptions to physical facts depend always upon causal laws, which enable us to bring past history to bear: e.g. if we have just examined an object under a microscope, we assume that it is still very similar to what we then saw it to be, or rather, to what we inferred it to be from what we then saw. It is through history and testimony, together with causal laws, that we arrive at physical knowledge which is much more precise than anything inferable from the perceptions of one moment. History, testimony, and causal laws are, of course, in their various degress, open to question. But we are not now considering whether physics is true, but how, if it is true, its world is related to that of the senses.

With regard to time, the relation of psychology to physics is surprisingly simple. The time of our experience is the time which results, in physics, from taking our own body as the origin. Seeing that all the events in my experience are, for physics, in my body, the time-interval between them is what relativity theory calls the "interval" (in space-time) between them. Thus the time-interval between two events in one person's experience retains a direct physical significance in the theory of relativity. But the merging of physical space and time into space-time does not correspond to anything in psychology. Two events which are simultaneous in my experience may be spatially separate in physical space, e.g. when I see two stars at once. But in physical space these two events are not separated, and indeed they occur at the same place in space-time. Thus in this respect relativity theory has complicated the relation between perception and physics.

The problem which the above considerations are intended to elucidate is one whose importance and even existence has been concealed by the unfortunate separation of different studies which prevails throughout the civilised world. Physicists, ignorant and contemptuous of philosophy, have been content to assume their particles, points, and instants in practice, while conceding, with ironical politeness, that their concepts laid no claim to metaphysical validity. Metaphysicians, obsessed by the idealistic opinion that only mind is real, and the Parmenidean belief that the real is unchanging, repeated one after another the supposed contradictions in the notions of matter, space, and time, and therefore naturally made no endeavour to invent a tenable theory of particles, points, and instants. Psychologists, who have done invaluable work in bringing to light the chaotic nature of the crude materials supplied by unmanipulated sensation, have been ignorant of mathematics and modern logic, and have therefore been content to say that matter, space, and time are "intellectual constructions," without making any attempt to show in detail either how the intellect can construct them, or what secures the practical validity which physics shows them to possess. Philosophers, it is to be hoped, will come to recognise that they cannot achieve any solid success in such problems without some slight knowledge of logic, mathematics, and physics; meanwhile, for want of students with the necessary equipment, this vital problem remains unattempted and unknown.

There are, it is true, two authors, both physicists, who have done something, though not much, to bring about a recognition of the problem as one demanding study. These two authors are Poincaré and Mach, Poincaré especially in his *Science and Hypothesis*, Mach especially in his *Analysis of Sensations*. Both of them, however, admirable as their work is, seem to me to suffer from a general philosophical bias. Poincaré is Kantian, while Mach is ultra-empiricist; with Poincaré almost all the mathematical part of physics is merely conventional, while with Mach the difficulties of a purely sensational physics are somewhat underestimated. Nevertheless, both these authors, and especially Mach, deserve mention as having made serious contributions to the consideration of our problem.[3]

When a point or an instant is defined as a class of sensible qualities, the first impression produced is likely to be one of wild and wilful paradox. Certain considerations apply here, however, which will again be relevant when we come to the definition of numbers. There is a whole type of problems which can be solved by such definitions, and

---

[3] Since 1914, when the above was written, much admirable work has been done on the above topics. I should mention especially, as of first-class importance, the work of Whitehead, Eddington, and Heisenberg.

almost always there will be at first an effect of paradox. Given a set of objects any two of which have a relation of the sort called "symmetrical and transitive," it is almost certain that we shall come to regard them as all having some common quality, or as all having the same relation to some one object outside the set. This kind of case is important, and I shall therefore try to make it clear even at the cost of some repetition of previous definitions.

A relation is said to be "symmetrical" when, if one term has this relation to another, then the other also has it to the one. Thus "brother or sister" is a "symmetrical" relation: if one person is a brother or a sister of another, then the other is a brother or sister of the one. Simultaneity, again, is a symmetrical relation; so is equality in size. A relation is said to be "transitive" when, if one term has this relation to another, and the other to a third, then the one has it to the third. The symmetrical relations mentioned just now are also transitive—provided, in the case of "brother or sister," we allow a person to be counted as his or her own brother or sister, and provided, in the case of simultaneity, we mean complete simultaneity, i.e. beginning and ending together.

But many relations are transitive without being symmetrical—for instance, such relations as "greater," "earlier," "to the right of," "ancestor of," in fact all such relations as give rise to series. Other relations are symmetrical without being transitive—for example, difference in any respect. If A is of a different age from B, and B of a different age from C, it does not follow that A is of a different age from C. Simultaneity, again, in the case of events which last for a finite time, will not necessarily be transitive if it only means that the times of the two events overlap. If A ends just after B has begun, and B ends just after C has begun, A and B will be simultaneous in this sense, and so will B and C, but A and C may well not be simultaneous.

All the relations which can naturally be represented as equality in any respect, or as possession of a common property, are transitive and symmetrical—this applies, for example, to such relations as being of the same height or weight or colour. Owing to the fact that possession of a common property gives rise to a transitive symmetrical relation, we come to imagine that wherever such a relation occurs it must be due to a common property. "Being equally numerous" is a transitive symmetrical relation of two collections; hence we imagine that both have a common property, called their number. "Existing at a given instant" (in the sense in which we defined an instant) is a transitive symmetrical relation; hence we come to think that there really is an instant which confers a common property on all the things existing at that instant. "Being states of a given thing" is a transitive symmetrical relation; hence we come to imagine that there really is a thing, other than the series of states, which accounts for the transitive symmetrical relation. In all such cases, the class of terms that have the given transitive symmetrical relation to a given term will fulfil all the formal requisites of a common property of all the members of the class. Since there certainly is the class, while any other common property may be illusory, it is prudent, in order to avoid needless assumptions, to substitute the class for the common property which would be ordinarily assumed. This is the reason for the definitions we have adopted, and this is the source of the apparent paradoxes. No harm is done if there are such common properties as language assumes, since we do not deny them, but merely abstain from asserting them. But if there are not such common properties in any given case, then our method has secured us against error. In the absence of special knowledge, therefore, the method we have adopted is the only one which is safe, and which avoids the risk of introducing fictitious metaphysical entities.

Chapter 21
ALFRED NORTH WHITEHEAD
(1861–1947)

INTRODUCTION

*A*LFRED NORTH WHITEHEAD was born at Ramsgate, Kent, England, on February 15, 1861. He studied mathematics at Cambridge, became a fellow at Trinity College, and later taught mathematics at University College in London, until he left for the United States in 1924 to become professor of philosophy at Harvard University. It was between 1910 and 1913 that Whitehead collaborated with Russell in writing the *Principia Mathematica;* in its three volumes the latter undertakes to deliver a fundamental study of the structure of logical and mathematical thinking. This was a preferred area for Whitehead at the beginning of his career; in his earlier publications philosophy of mathematics and logical studies clearly occupied the predominant place and, together with his later contributions in the same areas, have exerted a decisive influence on logic and philosophy of mathematics. In his second period Whitehead gave much attention to the natural sciences, writing not only *An Enquiry Concerning the Principles of Natural Knowledge* (1919), *The Concept of Nature* (1920), and *The Principle of Relativity* (1922), but also several important articles and essays on topics immediately related to the philosophy of natural science. The selection chosen for presentation here was Whitehead's contribution to a symposium later published in *Problems of Science and Philosophy.* Subsequent works such as *Religion in the Making* (1926), *The Aims of Education* (1929), and *The Function of Reason* (1929) show beyond a doubt that his interest had shifted in the direction of purely metaphysical questions, in addition to his continuing great interest in logic and the philosophy of science.

At Harvard Whitehead developed a first outline of his so-called organic philosophy in which he attempted to provide a set of general ideas designed to describe and unify all the complex components of the world: data of mathematics and the natural sciences, but also data of the social sciences as well as the data of our aesthetic, moral, and religious experiences. With this he hoped to compensate for the overemphasis of natural science and determinism in our modern world view. His ideas here became the main subject matter of three publications: *Science and the Modern World* (1925), *Process and Reality* (1929), and *Adventures of Ideas* (1933). For a time the value of Whitehead's philosophical system, the groundwork for which was laid in his earlier works, was the subject of much debate, but for the past fifteen years the interest of philosophers in this aspect of his philosophy has been slowly diminishing. His contributions to logic, mathematics, and philosophy of science, however, continue to maintain their importance. Whitehead died on December 30, 1947, at Cambridge, Massachusetts.

Whitehead explicitly excludes all metaphysical considerations in his earlier works on philosophy of science (1919–1922). For him the subject matter of philosophy of science is nature, taken as that complex of entities and relations which is revealed in sense perception and systematically examined in the natural sciences. In his later works, however, metaphysical discussions come gradually to occupy an increasingly important place, with the first indication in the second edition of *An Enquiry Concerning the Principles of Natural Knowledge* (1924). His main interest was no longer going out to nature, but "the endeavor to frame a coherent, logical, necessary system of general ideas in terms of which every element of our experience can be interpreted" (*Process*, p. 4). While it is clear that natural science is one of the elements of our experience and so has a place in the system, and even a central one, it is certainly not the only element which asks for the attention of the philosopher. Whitehead for the greater part maintained his earlier views on science within the context of his later philosophical system, but there are notable and important modifications stemming from the new context in which the views had to appear. These modifications, one of which is for instance a revision of his earlier version of the method of extensive abstraction, are all somehow connected with his growing interest in the function of the perceiver and percipient events in cognitive activities, and other epistemological questions. Here, however, the introductory remarks which follow pertain only to Whitehead's earlier philosophy of science.

At that point, Whitehead maintained that the subject matter of science consists in nature as the "panorama" given in sense experience. Science's task is to try to formulate basic concepts applicable to nature and laws of nature which state the relations between the data of nature. However, basic concepts and laws of nature as such do not constitute genuine science; these data must be interconnected and this is to be done by (deductive) logic. "Science is essentially logical. The nexus between its concepts is a logical nexus, and the grounds for its detailed assertions are logical grounds." (*The Interpretation,* p. 25) Whitehead here reveals that he is most familiar with the empiricist criticism of deductive logic, which is why he takes great pains to defend modern deductive logic against all forms of criticism.

In Whitehead's earlier view philosophy of natural science deals with nature as given in immediate experience and science. Its job is to develop the general notions which apply to nature, and in so doing philosophy of science must exclusively concern itself with perceived nature, not with the process of perceiving or the percipient (except insofar as they are part of nature). Omitting from consideration all epistemological and metaphysical questions, philosophy of science must clarify "the concept of nature considered as one complex fact for knowledge, exhibit the fundamental entities and the fundamental relations between entities in terms of which all laws have to be stated, and secure that the entities and relations thus exhibited are adequate for the expression of all the relations between entities which occur in nature" (*Concept,* p. 17).

However, Whitehead makes clear, before one can materialize this difficult task, he must try to come to grips with some basic misconceptions of classical science and philosophy of science. To these misconceptions belongs for example the classical bifurcation of nature as immediately given in sense experience, and as the cause of our awareness (molecules, electrons, and so on). In Whitehead's view nature, is what is experienced; as such it is colorful, soundful, odorful and the like. There is no justification for trying to break up the experienced unity and assign the status of real physical properties to primary qualities, relegating secondary qualities to a "byplay of the mind." Whitehead points out that there are other misconceptions connected with Newton's view of absolute space and time which do not find support in our sense experience, and he remarks similarly in regard to the physical concept of matter as the totality of isolated particles existing independently of each other.

After this criticism of the traditional interpretation of nature in terms of primary qualities—space, time, and matter—Whitehead attempts to offer an alternate philosophy of science in terms of concepts founded in immediate experience. As such he proposes the concepts event and object. The essay which follows is a brief survey of his view; a more detailed account is to be found in *An Enquiry Concerning the Principles of Natural Knowledge.*

SELECTIVE BIBLIOGRAPHY

Whitehead, A. N., *An Enquiry Concerning the Principles of Natural Knowledge.* Cambridge, Eng., 1919.

———, *The Concept of Nature.* Cambridge, Eng., 1920.

———, *The Principle of Relativity.* Cambridge, Eng., 1922.

———, *Science and the Modern World.* New York, 1925.

———, *Process and Reality.* New York. 1929.

———, *Essays in Science and Philosophy.* New York, 1947.

———, *The Interpretation of Science* (Selected Essays), ed. A. H. Johnson. Indianapolis, 1961.

Palter, R. M., *Whitehead's Philosophy of Science.* Chicago, 1947.

The Philosophy of Alfred North Whitehead, ed. P. A. Schilpp. Evanston, Ill., 1941.

Philosophical Essays for A. N. Whitehead, ed. F. S. C. Northrop, New York and London, 1936.

Shanan, E. P., *Whitehead's Theory of Experience.* New York, 1950.

Mack, R. D., *The Appeal to Immediate Experience. Philosophic Method in Bradley, Whitehead, and Dewey.* New York, 1945.

Das, R., *The Philosophy of Whitehead.* London, 1938.

Miller, D. L. and Gentry, G. V., *The Philosophy of A. N. Whitehead.* Minneapolis, 1938.

Blyth, J. W., *Whitehead's Theory of Knowledge.* Providence, 1941.

Hammerschmidt, W. W., *Whitehead's Philosophy of Time.* New York, 1947.

TIME, SPACE, AND MATERIAL:
ARE THEY, AND IF SO IN WHAT SENSE,
THE ULTIMATE DATA OF SCIENCE?

*T*HE concepts of modern science are founded on naïve common sense as modified by Greek thought, medieval scholasticism, and Renaissance and seventeenth-century philosophy. In practice, every scientific treatise assumes as ultimate the concepts of material—here used as a more general term than matter—and of time and of space.

Force, velocity, kinetic energy, potential energy, and life are properties expressive of many-termed relations between materials and times and spaces. Namely, force, velocity, energy, and life are (in the sense in which they enter into physical science) somewhere in space at some time, and express relations of materials *inter se,* and also to various times and various spaces.

Time may be conceived either as a succession of instants of time or as a passage of periods of time. Periods of time overlap and contain one another, and thus have complicated relationships. Accordingly, the simple mathematical concept of time, as a simple linear series of durationless instants with certain mathematical properties of serial continuity, has tacitly crept from books on mathematical physics into general scientific thought as expressive of the ultimate structure of time.

The difficulty of this view is that velocity cannot be defined by simple reference to one instant. Its definition essentially involves a neighborhood of instants. Nature at an instant is simply a definite configuration of material in space, with certain space relations. Velocity and kinetic energy have evaporated in this ultimate concept of the instant. A full description of nature has (on this view) an infinite number of chapters, each chapter being the full description of the configuration of material at some one instant.

This article is from A. N. Whitehead, **Problems of Science and Philosophy.** Aristotelian Society Supplementary Vol. II (London, 1919), pp. 44–57. Reprinted by permission of A. A. Kassman, Esq., Editor of the Aristotelian Society.

But there is an appendix to this book of nature, written either by God's foresight or by man's subsequent reflection. This appendix contains the comparison of chapter with chapter. In this appendix velocity, kinetic energy, acceleration, force, and mass make their appearance. In fact, the appendix, when completed, is a treatise on mathematical physics, with a preface on causation written by the philosophers.

Unfortunately, in this book of nature the biologists fare badly. Every expression of life takes time. Nothing that is characteristic of life can manifest itself at an instant. Murder is a prerequisite for the absorption of biology into physics as expressed in these traditional concepts.

This account of nature and of physical science has, in my opinion, every vice of a hasty systematization based on a false simplicity; it does not fit the facts. Its fundamental vice is that it allows of no physical relation between nature at one instant and nature at another instant. Causation might be such a relation, but causation has emerged from its treatment by Hume like the parrot after its contest with the monkey. The fact is that this account has ruled out in advance any physical relationships between nature at different instants, and all that is left to connect nature at one instant with nature at another instant is the identity of material and the comparisons of the similarities and differences made by observant minds. Also time as a succession of instants corresponds to nothing which falls within my own direct knowledge. I can only think of it metaphorically either as a succession of dots on a line or as a set of values of an independent variable in certain differential equations. I cannot dissociate time from concrete nature and then know nature as at an instant of time; nor am I aware of any fact which is instantaneous nature. Again, in this account also the fundamental physical quantities, such as velocity, energy, etc., are excluded from nature and become merely expressive of the spectators' comparisons. There are also difficulties connected with the concept of space, which are omitted for the sake of brevity.

I believe that the account of nature which has been outlined and criticized above is the fundamental view which has pervaded scientific thought. It has not been held consistently, because, owing to its inadequacy, it is impossible to talk of nature consistently in terms of this concept. But in the case of the majority of scientists, so far as their ideas are clear they seem to mean that.[1]

[1] I have stated these criticisms of traditional concepts at greater length in *An Enquiry into the Principles of National Knowledge,* shortly to be published by the Cambridge University Press. In subsequent references it is cited as *Enquiry.*

Let us try and find our way to another account. The chief strength of the belief in the instant of time as an ultimate fact is acceptance of the present as being a present instant which we directly know. But the psychological doctrine of the specious present warns us that this is a case of warping observation by theory. What we are immediately aware of is a duration of nature with temporal extension. We are not aware of two facts, namely, a period of time and also of things existing within that period. We are aware of nature enduring, or—in other words—of the passage of nature. Thus the present contains within it antecedents and subsequents, and the antecedents and the subsequents are themselves endurances with temporal extensions. Nature at an instant is a complex abstract conception which is useful for the simple expression of certain natural relations.

Thus, awareness of nature begins in awareness of a whole which is present. Call this present whole of nature a "duration." A "duration" is a temporal slab of nature; and is all that there is, subject to the temporal limitation inherent in the awareness. This awareness of the whole is directly sensed, and is not a detailed discrimination of its parts. This sense for the being of nature is accompanied by a diversification of the duration into parts, which are more or less clearly discriminated. The awareness of nature essentially requires both factors, namely, the sense for the whole and the discrimination of parts. The parts are known as "there," namely, there in the whole. These parts of a duration are the finite events, which have indistinct demarcations simply owing to lack of perceptive vividness and of discriminative force. An example of such an event is all nature within the Roman Senate house during the death of Julius Caesar. Durations are events with a quality of unboundedness. They form the sole class of infinite events; other events, such as the death of Caesar, are finite events.

Events, infinite and finite, are the primary type of physical facts. The two fundamental relations which they can have to each other are here called "extension" and "cogredience."

One event, e, may extend over another event, e'. If this relation holds, e' is said to be part of e, and e is a whole of which e' is a part. Thus, extension is the relation of whole event to part event, and is a relation which is special to events. For example, all nature within the Senate house during Caesar's death extends over all nature within Pompey's statue during that death. This relation of extension is the common root from which extension in time and extension in space both spring. It is the essence of externality.

Two durations which belong to the same time-system are *either* completely separate, *or* one extends over the other, *or* there is a third duration which both extend over and which is their common part. A moment of a time-system is a route of approximation to the non-existent ideal of a duration without temporal extension. This route is composed of an infinite series of durations, extending over each other, the earlier in the series over the later, and so that there is no duration which they all cover. Such a series defines an instant of time, and will be here called a "moment." All observation which endeavors to gain accuracy by instantaneousness is comprised within a duration as far down a momental series as possible, and is dated at that moment. Thus "nature at a moment" is the ideal simplicity of natural relations to which we approximate as we proceed along that momental series.

Nature at a moment exhibits (among other things) the relations of a three-dimensional space: This is instantaneous space. The instantaneous points of such a space are routes of approximation constructed on the same general principle as moments; namely, a point-series is an infinite series of events, each event extending over all the events subsequent to it in the series; the whole series converges toward an ideal of an event of no extension. The details of the definition can be omitted here. An instantaneous point is better named an "event-particle." Event-particles form a four-dimensional manifold which is divided into three-dimensional instantaneous spaces which lie within the several moments.

According to the concept of time, which until recently was unquestioned, there should be but one time-series of moments, and any two moments should be "parallel" in the sense that no event-particle could lie in two distinct moments. But the more recent electromagnetic theory of electricity requires us to assume an infinite number of distinct time-series. Each time-series would (on this theory) consist of parallel moments, exactly like the single time-series, of the older theory. But if a and a' be distinct time-series, and M be a moment of a and M' of a', then M and M' intersect in the sense of containing some event-particles common to both.

This conception of distinct time-series is paradoxical and is not yet fully accepted. It explains some perplexing observations in physics, and also enables intelligible accounts to be given of the nature of flatness and straightness in instantaneous space, and of the meaning of the timeless space of physics, and of the reason for "Newtonian" (or "Galilean") axes, for reference to which Newton's Laws of Motion hold (cf. *Enquiry*). But it necessitates the paradox that two event-particles which are simultaneous (i.e., comomental) for time-system a will not in general be simultaneous for another time-system a'; though for ordinary observation the lack of simultaneity will be inappreciable. This requires that we distinguish between the creative advance (or passage) of nature and any special time-system a. The group of all time-systems embodies the physical properties of this creative advance. But any one time-system is not itself this advance; it is simply an embodiment of some of the physical properties of nature resulting from this creative advance. On the old theory of time, the time-system is the creative advance. When once the distinction is made between the creative advance and the separate time-systems, the paradox of a multiplicity of time-systems is less acute.

Some pairs of event-particles are necessarily sequential in all time-systems, but some pairs are simultaneous in some time-systems and sequential in others, and also in these other time-systems may have their sequence inverted. For ordinary perception, pairs of the later type appear as simultaneous.

I shall call an element of nature "completely concrete" where, existing as it does exist, it could be all nature. For example, on the old "single-time-system" theory a duration is completely concrete. Its first moment might be the moment of creation, and its last moment the day of judgment. But on the theory of the multiplicity of time-series a duration is not completely concrete; the creative advance of nature is fatally excluded if we assign two moments of one time-series as a beginning and an end, since this advance requires the whole group of time-series for its expression. For example, the last moment of a finite event is different for each time-system, since no moment belongs to two time-systems. Thus, on the newer theory, a beginning or an end of nature within time is excluded. I regard this conclusion as a merit in the new theory.

Cogredience is the other relation which two events can have to each other. Cogredience directly holds between a finite event and a duration, and only mediately between two events which are cogredient to the same event. Cogredience is the relation of absolute spatial position within a duration; namely, an event is cogredient with a duration when all sections of it made by durations, which are parallel parts of the given duration, exhibit the same meaning of "here within the duration." Cogredience is immediately known to us in perception. Our awareness of nature is not from a survey from without, but an awareness of a special event, the "percipient event," in respect to its own internal relations and qualities of its parts *inter se*, and also in respect to its external relations to other events. This percipient event is, roughly speaking, the bodily life of the perceiver. Throughout a sufficiently small duration the percipient event is unequivocally "here." Unless that be the case, there is no meaning to the idea "here," which, after all, is the most insistent of all concepts. The essence of the new theory is that the hereness of the percipient event is in respect to a selected duration in some special time-system, namely, that duration with which it is cogredient. This explains the palpable fact that though we are moving we are always "here." In such a case (when we conceive ourselves as moving), there is a dual perception of cogredience, namely, the cogredience proper to the percipient event and the cogredience proper to certain other events.

From cogredience and extension the whole metrical theory of time and space can be deduced (cf. *Enquiry*). We should speak more accurately in the plural, namely, of "times" and "spaces" and not of time and space. For each system of parallel durations corresponds to one definite time-system of parallel moments, each such moment being a three-dimensional instantaneous space; and each such system allows a timeless space, the space of physical science, to be constructed, with its own timeless points, timeless straight lines and timeless planes proper to it alone.

In diversifying nature into entities, events are not the only type of entities thus disclosed. There are entities which we recognize. Such entities we will call *objects*. They might have been named *recognita*, but we choose the term *object* as being simpler. An event is essentially not an object. We live through events, and they pass; but whatever is repeated is necessarily an entity of another type. In order to recognize an object it is not necessary to have perceived it before. An object is recognized within the present duration of its perception. For this present duration includes antecedent and subsequent durations; and the recognition of the object in the present is essentially a comparison of the object in the antecedents and subsequents within the present, though memory may also be a factor in the recognition.

Events may be looked upon as relations between objects, and objects may be looked upon as qualities of events. But both these points of view lead us into difficulties, and (at any rate for physical science) it is simpler to look on objects and events as fundamentally different sorts of entities disclosed in nature with certain determinate relations to each other.

Also there are an indefinite number of distinct sorts of objects. It is only necessary here to consider a few important types. Objects of any one type have relations to events which are radically distinct from those which objects of any other type bear to events. Furthermore, the concrete whole disclosed in awareness is nature, and nature exhibits both events and objects. To think of nature as the mere passage of events without objects, or as a mere collection of objects unrelated to events, is an abstraction.

Objects of the primary type will be called "sense-objects." All other objects in nature presuppose the perception of sense-objects. A sense-object is a specific sensation of feeling perceived as situated in an event. For example, the color red of a definite shade, and the peculiar feel of velvet, and the sensation of heat are all sense-objects. We recognize them and locate them in events which we will call their situations.

But this account of the relation of a sense-object to events is much too simple. The relation is a complicated many-termed relation involving in some way all events. The terms in this relation are (i) the sense-object, (ii) the percipient events, (iii) the situations, (iv) the active conditioning events (subsequently called the "active conditions"), (v) the passive conditioning events (subsequently called the "passive conditions"). In this analysis, one event may sustain many roles. We take a definite sense-object, say redness of a particular shade, and one definite percipient event, which is the standpoint of the awareness of a definite percipient. Then, granting certain active conditions (e.g., a translucent material and a red-hot poker), and the implied existence of all nature (comprising the passive conditions), redness is situated in certain events in reference to that percipient event. Thus the relation can be summarized as—Sense-object O, with situation σ, for percipient event π, with active conditions $\gamma_1 . \gamma_2 \dots$. In this summary, reference to the passive conditions may be dropped as being a necessary presupposition in perception, and for simplicity we consider only one situation. The influence of the passive conditions is shown by the provision of time and space for the general setting of the occurrence. The situation of a sense-object is necessarily within the duration with which the percipient event is cogredient; i.e., speaking loosely, the events are simultaneous. The necessary reference of the situation to a particular percipient event with assigned conditions is understood by remembering that in the perception of redness a mirror may be among the conditions. Thus the redness of the poker will be situated behind the mirror for that percipient event. This possibility brings out the fact that being the situation of a sense-object may be a very trivial property of the event which is the situation—at least from the point of view of the event, if we can imagine it thinking. But from the point of view of the percipient whose awareness derived from that percipient event the fact is important, since his knowledge of nature is entirely derived from an analysis of such situations. Furthermore, there may be no poker and no mirror, an abnormal percipient event or abnormal conditions may generate the perception without any of the usual conditions. Redness has still that situation for that percipient event; the delusion is merely the false judgement that there are certain conditions which are in fact nonexistent; namely, all perception is what it is, and the sense-object has the situation as perceived. The only error which can arise is a misjudgment as to the conditions; these conditions may be antecedent potations of alcohol and not the almost simultaneous events comprising the existence of the hot poker. The foundation of science is the careful analysis of the types of conditions and the types of percipient events which lead to perceptions of specific sense-objects in definite situations.

The preceding discussion shows us that there is a sort of sliding scale of normality in the conditioning events necessary for the perception of a given sense-object in a situation with a definite relation to the percipient event. For example, in the case of the redness of the poker, there are the conditions for direct vision, the conditions for vision by reflection in a mirror, the conditions for alcoholic delusion. The greater the abnormality, the greater is the difficulty in formulating conditions which are both necessary and sufficient. The chief danger in the philosophy of science is the concentration of interest on the most normal conditions for perception.

The next type of objects is that of "perceptual objects." A perceptual object is the determinate association of sense-objects in a series of situations which are strung together into a continuity by mutual overlapping and can thus be synthesized as one prolonged event.

The perception of a perceptual object is radically different from that of a sense-object. It involves the perception of sense-objects and something more. This additional element in the perception will be called the "conveyance" of one sense-object by the perception of another sense-object. For example, you see a horse. Primarily you have seen the color of the horse in a certain situation. But it is the horse you have perceived and not merely his color. A set of faintly discriminated sense-objects have been "conveyed" to you by the sight of the color. Perhaps, if your perceptions are very vivid, you may feel yourself patting him. But in general the sight of the color merely conveys a nameless complex of feeling which, combined with the sight of the color, is the perception of the horse. The function of judgment is to foment or to inhibit or to divert this conveyance. You do not perceive a horse because you judge it to be a horse, but because you feel it to be a horse. Judgment helps or obstructs this feeling, and adds to the immediate perception the recollection of the many qualities assigned to horses in natural history books. In our subsequent reflections we usually adorn the crudely perceived perceptual object with some such pride of knowledge.

Perceptual objects are divisible into two classes, delusive perceptual objects and physical objects which are nondelusive objects. A delusive perceptual object is an object of which the perception and situation are essentially referred to one percipient event with its actual conditions of perception. A physical object is perceivable in the same situation from the standpoint of an indefinite number of percipient events in suitable relations to the situation and with normal conditioning events. There is an element of delusiveness in all perception, illustrated by the time-lags in the perception of stars. Our knowledge of nature rests on the assumption that ordinarily this delusiveness is negligible.

Physical objects form the bridge between nature as appearance and nature as a complex of conditions for appearance. The objects whose relations to events make those events to be conditions for the appearance of sense-objects are called "scientific objects." Physical objects are both apparent objects and scientific objects. Thus the mirror is an apparent object which causes the event which is its situation (i.e., the fact that it is where it is during the time when it is) to be a condition for the sight of the redness of the poker in a situation behind it.

According to scientific theory and to common sense, the event which is the situation of a physical object is the chief conditioning event for the appearance of that physical object. To express these conditions, primarily observed in terms of sense-objects and physical objects, science has conjectured other types of scientific objects, in particular, molecules and electrons. In the present stage of science, electrons are the ultimate scientific objects, and for brevity we consider them only.

Any single electron has relations to all events. In relation to this electron there are two classes of events: (1) the occupied events, and (2) the unoccupied field. The occupation is expressed by a certain quantitative character of the occupied event which is the charge forming the nucleus of the electron; and the character of the unoccupied field, due to its relation to the electron, is expressed by the electric and magnetic forces. The electric and magnetic field expresses the electron as representing events in their character of active agents in the creative advance; the charge expresses the electron as representing events in their character of being receptive of modification in their passage to other events.

In this account of scientific principles the material ether has disappeared. It is replaced by an ether of events, which is formed of events whose character is expressed by the properties of the electromagnetic field. The continuity of nature arises purely from the extensional properties of events. Only events have parts; and only events are directly in time and space, in the sense that time and space are expressions of certain extensional and cogredient properties of events. The atomic properties of nature arise entirely from objects. Objects have no parts, though their situations have parts. Objects are only mediately in time and space through their relations to events.

An object which (in some sense) is located in an event extending throughout a duration is not necessarily located in any slice of that event contained in a duration which is a part of the original duration. For example, a molecule of iron and a tune both require a minimum of time in which to express themselves. We may call such objects "nonuniform." On the other hand, perceptual objects appear as uniform objects; namely, if the object is situated in an event which extends throughout a duration, then any duration, however small, which is contained within that duration, cuts the event in a slice which is a situation of the same perceptual object. It is always assumed that the ultimate scientific objects (at present electrons) are uniform objects. There is thus a hierarchy of stages, from the electrons which are uniform objects, to the molecules which are nonuniform objects with time-minima dependent on the periodic times of their orbital swarms of electrons, to physical objects which are uniform objects, regaining uniformity by exhibiting the average effects of billions of molecules. This hierarchy is represented in mathematical physics by Lorentz's hierarchy of microscopic and macroscopic equations, where the macroscopic equation is due to the employment of coarser mathematical machinery than that employed in the formation of equations which are microscopic relatively to it. Thus a macroscopic equation gains its macroscopic character by averaging effects which are dealt with individually in relatively microscopic equations. The macroscopic treatment of nature is not purely loss in the mere averaging out of differences. It makes evident objects (e.g., molecules, physical objects, and living beings) which would be missed by an observer who insisted on contemplating nature exclusively through the high-powered microscope required for its microscopic treatment.

I cannot see any basis for the habitual assumption that the ultimate scientific objects should be uniform. Whenever nonuniform objects emerge, then time-minima become important in physics (i.e., *quanta* of time, in the modern nomenclature). The atomic property of objects and the nonuniformity of some types of objects are obviously the basis of the quantum properties of nature which are assuming such an important position in modern physics.

The theory of material is the theory of uniform objects which endow the events in which they are located with a quantitative character. For example, an electron gives such a character to the occupied events, namely, the electric charge. Thus the concept of a quantity of material with a definite spatial configuration at an instant of time is a very complex abstraction, and is by no means a fundamental datum for science.

Chapter 22
NORMAN ROBERT CAMPBELL
(1880–1949)

INTRODUCTION

NORMAN ROBERT CAMPBELL was born in 1880, the third son of William Middleton Campbell, of Colgrain, Dumbartonshire, England. He was educated at Eton, and Trinity College, Cambridge, and then worked in the Cavendish Laboratory under the leadership of J. J. Thomson for several years; he became a fellow at Trinity College in 1904. During this period he mainly studied the "spontaneous" ionization of gases in closed vessels, but also had great interest in radioactivity and penetrating radiation. He was appointed to the Cavendish Research Fellowship at Leeds at the time when W. H. Bragg was professor there; he became an honorary fellow at Leeds University in 1913. During that time he continued research in the same direction and wrote his *Modern Electrical Theory.*

Although Campbell allotted most of his time to research, he nonetheless managed to devote some also to the study of philosophical problems connected with the natural sciences. In 1904 he decided to write a book on philosophy of science; at first he limited himself to writing separate essays on various problems which arose in the course of his scientific work. Some of these essays were published as articles in various reviews; others appeared in a small book entitled *The Principles of Electricity* (1912). By that time a first draft of *Physics: The Elements* was ready, but World War I and the preparation of other scientific publications interrupted the work until 1919. When Campbell finally came to prepare the work for publication, he decided to publish only the first two parts of it; Part III was to be briefly summarized in an Appendix and Parts IV and V omitted completely. A second edition of this monumental work appeared posthumously in 1957 under a new title, *Foundations of Science: The Philosophy of Theory and Experiment.* He apparently did not publish Parts III, IV, and V of his original work, but in 1921 he did publish a short study, *What Is Science?,* in which he summarized some of his most important insights, followed by another small book entitled *An Account of the Principles of Measurement and Calculation* in 1928.

During World War I Campbell devoted himself to applied physics; in 1914 he joined the staff of the National Physical Laboratory where he became associated with Sir Clifford Paterson in investigating the mechanism of the spark discharge produced by plugs for internal combustion engines. In 1919 Campbell joined the staff of the research laboratories of the General Electric Company where he worked until he retired in 1944. During these years he wrote many articles and several books on topics immediately connected with his research. He died on May 18, 1949, at the age of sixty-nine.

Since its appearance in 1919 Campbell's *Foundations* has been accepted as a classic in its field and has drawn the attention of scientists and philosophers to three important issues: (1) the original and detailed treatise on measurement, (2) his view on the meaning of scientific theories, and (3) his view on the relationship between philosophy and science. Some remarks about the last two points are relevant here, in view of the selection to come. His idea of the relationship between philosophy and science, will be considered first.

Campbell described the *Foundations* as the introductory volume to a larger work; this was supposed to consist of a critical analysis of contemporary physics. In contradistinction to most similar works which have been written by scientists whose main interest has been the mathematical aspect of physics, he wanted to approach physics from the experimental point of view, which explains why his view on measurement occupies the central place in his "introductory" investigation. Campbell believed that a critical analysis of science might be developed without any philosophy at all, but on the other hand, he maintained that there were certain connections between science and philosophy which ask for a more careful investigation. By analyzing the basic metaphysical concepts of reality and truth and comparing them with the scientific conception of these concepts, Campbell comes to the conclusion that metaphysics and science can never agree because they concern entirely incompatible ideas. From this it follows that science does not imply any criticism of metaphysics, but also that metaphysics is of no help whatsoever in science and philosophy of science. Campbell maintains that the kind of truth which is applicable to the one realm of investigation is just not applicable to the other (*Foundations*, pp. 230–264).

As far as scientific theories are concerned, Campbell considered it their fundamental function to provide a deductive explanation of laws,

laws being understood as propositions which assert uniformities discovered by observation and experiment. A theory consists of two sets of propositions which can be indicated by the terms "hypothesis" and "dictionary." A hypothesis is to be formulated in terms of ideas which are characteristic of a certain theory. The dictionary provides a physical interpretation of the hypothesis by translating some, but not necessarily all, of its propositions into others which do not contain any "purely theoretical concepts" and thus can be verified by suitable experiments or observations, without any further reference to the theory. The ideas which the dictionary connects with the ideas of the hypothesis are those found in our scientific laws, whereas the "hypothetical ideas" of most of the theories of physics are mathematical constants or variables. According to Campbell, a theory is said to be true if propositions concerning the hypothetical ideas which can be deduced from the hypothesis are found, according to the dictionary, to lead to laws whose truth is known or can be verified by experience.

Thus in Campbell's view a scientific theory must be capable of explaining empirically scientific laws; this explanation will consist in deducing the laws from the hypothesis with the aid of the dictionary. However, he maintained, for a theory to be valuable it must have a second characteristic: it must display an analogy, which is to say the propositions of the hypothesis must be analogous to some known laws. These analogies are not mere "aids" to the establishment of theories, but constitute an essential part of the theory (*Ib.*, pp. 113–158).

In the following selection, taken from *Foundations*, Campbell explains his view as to the truth and meaning of scientific propositions.

SELECTIVE BIBLIOGRAPHY

Campbell, N. R., *Foundations of Science. The Philosophy of Theory and Experiment* (formerly *Physics: The Elements*, 1919). New York, 1957.

———, *What Is Science?* (1921) New York, 1952.

———, *An Account of the Principles of Measurement and Calculation.* New York, 1928.

Hempel, C. G., *Aspects of Scientific Explanation and Other Essays in the Philosophy of Science.* New York, 1965. Pp. 207–209, 442–445.

Braithwaite, R. B., *Scientific Explanation.* New York, 1960. Pp. 88–114 (*passim*).

THE MEANING OF SCIENCE

Summary. Two criteria for the value of scientific propositions have been recognised, universal assent and intellectual satisfaction. A summary of the conclusions reached already is undertaken in order to show exactly how and where the two criteria are applied.

The value which a scientific proposition derives from the first criterion is called its truth, that which it derives from the second its meaning. Meaning is more important in theories than in laws, but it is insisted that laws have also meaning as well as truth.

The reasons why there has been a danger that the meaning of science will be overlooked are mainly historical. The chief changes that have taken place in the attitude towards scientific knowledge are briefly sketched, and special attention paid to the very important views of Mach. Much of those views is accepted, but it is maintained that the object of science is not "economy of thought"

The neglect of the meaning of science has led to the stifling of the scientific imagination in education and wherever else science is brought to the notice of the laity. It is maintained that science should be regarded as an art.

TWO CRITERIA IN SCIENCE

*T*HROUGHOUT our discussion when we have had to determine why a scientific proposition has value and what degree of value it has we have referred to one of two principles. First, a proposition is valuable if truly universal assent can be obtained for it; second it is valuable

This article is from **Foundations of Science** by N. R. Campbell (New York, 1957), pp. 215–229. Copyright © 1957 by Dover Publications, Inc. Reprinted by permission of the publisher and of Cambridge University Press.

if its contemplation causes intellectual satisfaction to students of science. These two principles are to some extent contrary and, if the test provided by each of them is applied to the same proposition, one might sometimes determine that the proposition is valuable and the other that it is not. For a student of science is a student of science in virtue of some difference between his intellectual constitution and that of the rest of mankind; if he finds intellectual satisfaction in a proposition it is almost certain that persons with different training and different interests can be found to whom it will give none; and on the other hand the mere fact that a proposition is approved by everyone, however different their modes of thought, will deprive it for him, not of course of all its value, but of that very special value which is the basis of the second principle. It is necessary therefore to examine the two principles rather more nearly and to determine exactly what part each of them plays in the establishment of scientific propositions.

The general nature of the solution of the apparent inconsistency is sufficiently obvious. Every scientific proposition must satisfy in some manner the first principle. Unless in some manner and to some extent universal agreement can be obtained concerning it, then it is scientifically valueless; and again if there is anything about it which directly conflicts with any proposition for which universal assent can be obtained, then it is not worthy of a moment's consideration. But many propositions can sometimes be found, all of which satisfy equally the principle of universal assent in some measure and, indeed, in the same measure, and which are yet mutually inconsistent. Among such sets of propositions some, but not all, satisfy the second principle. These propositions which satisfy the second principle are then preferred to those which do not. But if it should ever appear that one of these propositions, satisfying the second principle, does not really satisfy the first, then without any hesitation it is to be rejected.

Let us then recapitulate our previous discussion in order to see exactly at what stages in the development of scientific propositions the two principles are applied. We start with the selection of material for study, and in this task we rely wholly on the principle of universal agreement. There is a certain class of judgements concerning which universal agreement can be obtained as it can be obtained about no other class. These judgements are those about the association of sensations; not about the occurrence of sensations, for I cannot be certain that anyone else has the same sensations as I have (whatever that may mean), but about their uniform association which determines action. These judgements if they were expressible explicitly at all would be expressed as laws, but since the ideas involved in them are more fundamental than any language they are not so expressible. These ideas, which are related to the laws as concepts are to the more complex laws, are so involved in the very structure of our thought that the bare conception of their invalidity is impossible to most people. There can be here no application of the second principle because no alternative to the acceptance of these ideas could possibly be intellectually satisfactory.

The study of these fundamental concepts leads us to the formulation of more complex laws into which the fundamental concepts enter as terms between which the relation of uniformity is asserted. In the formulation of these laws we are not under the compelling necessity which excluded before the consideration of any alternative. In the first place we have more liberty of choice of the judgements that we shall consider. Most of us could not live our daily life or think at all unless we accepted all the concepts which form the basis of our conception of a material and external world; but when we turn our attention to the relations between constituent parts of the material world, though our comfort may be greatly advanced by studying some of those relations rather than others, there is nothing that makes it impossible for us to neglect intellectually any portions that we please. And as a matter of fact we do pick and choose. We deliberately exclude from the subject matter of science, out of which its propositions are developed (although not necessarily from the matter which can be "explained" by those propositions once they are established), all those judgements of the material world which are individual and personal and concerning which universal agreement is not obtainable; we include only definite "experiments" which can be repeated and can be shown to other observers. Here again we apply the first principle in determining the value of a proposition for scientific purposes, but it is not certain that we do not also apply the second. It is not certain that, even if there were no other observers available whom we could use to apply the first principle, we should not be able to select the matter which is actually selected by means of the second; for the laws which we can base on this matter are "simpler" and intellectually more satisfactory than those which could be based on matter that included judgements rejected by the test of universal agreement. This is possibly the first application of the second principle, but there is no need to decide whether it is actually employed, for actually other observers are available and actually there does not appear to be any conflict between the results of the two principles.

But in the establishment of laws we employ the second principle in a much more definite and more important way. We use it to determine, not what judgements we shall try to arrange in laws, but in what laws we shall arrange them. It is characteristic of a law that it asserts something more general than the evidence on which it is based; for that is what we mean when we say that a law predicts and that is what gives rise to the central problem of induction. We observe that, in certain circumstances, a certain potential difference at the ends of a wire is associated with a certain current through it; we assert that the same potential difference will be associated with the same current in other circumstances, differing from those in which it has been observed by the position of the earth with reference to the stars, for instance. We observe that the magnitudes of the potential difference and the magnitudes of the current which are thus associated are related by mathematical relations of which simple proportionality is one; we assert that the magnitudes of the potential differences and the currents which we have not observed will be related by the relation of proportionality, and not by one of the indefinite class of other mathematical relations which would equally well represent the relations of those which we have observed. It has been maintained, and indeed it is generally recognised, that in choosing, among the indefinite number of alternative propositions which would represent equally well the observations concerning which universal agreement can be obtained, that proposition which will be asserted and which will be expected to cover the observations which have not yet been made, our choice is determined by an application of the second principle. The simplest law is chosen; and simplicity here, however it is analysed, always reduces to some form of intellectual satisfactoriness. Moreover the satisfactoriness is not one which is necessarily apparent to all mankind or for which it is certain that universal agreement could be obtained. If it were, no reason could be given why all men should not be equal in discovering laws, given equal training in the previous results of science. It is true that judgements in this matter do not usually differ greatly; it would usually happen that all men to whom the observed facts were presented would concur in selecting a law to represent them; but they clearly might differ, and if they did it would be found that the judgement of some persons was more satisfactory than that of others. For it must be always remembered that, though there is no certain method of telling *a priori* which of the possible laws is the more satisfactory, there is a certain method of telling *a posteriori;* when new observations are made, universal assent will be obtainable for the assertion that they accord with some of the alternative laws and do not accord with others. A law is satisfactory if it thus accords with universal agreement concerning judgements made after it is enunciated, or, in other words, if it predicts truly. While therefore laws are enunciated in the first instance by the second principle of value, they are always liable to revision by the first; it is only because and so long as the indications of the first are ambiguous that the second may be admitted at all.

From laws we proceed to theories; we attempt to find propositions from which laws may be deduced. And here again we are met with ambiguity. Even if it were true—and, as we have just seen, it is not wholly true—that laws were determined entirely by the principle of universal agreement, there would still be an indefinite number of propositions from all of which the same laws might be deduced. Universal agreement will not enable us to decide between them; we must have recourse to the second principle. A theory must not only "explain" the law in the sense that the law can be deduced from it; it must also give intellectual satisfaction. And now it is much more obvious that this intellectual satisfaction is not a matter of universal agremeent, for many cases occur immediately to memory in which there has been an obvious and notorious failure to agree; and again it is more obvious that certain persons are much more likely than others to produce a satisfactory theory by an application of the second principle, and that difference in the power to do so is something which does not depend merely upon training and knowledge. A satisfactory theory is one which predicts new laws which are true, just as a satisfactory law is one which predicts observations which occur; theories, like laws, can be distinguished *a posteriori*, though not *a priori*, by the first principle of value. Just as in the establishment of laws, so in the establishment of theories, the second principle is admissible only because and in so far as the indications of the first are ambiguous.

TRUTH AND MEANING

It is important to observe that both the principle of universal agreement and that of the intellectual satisfaction of a minority are used in the establishment of laws as well as of theories and in the determination of the value of scientific propositions of all kinds. In virtue of this use of both principles, such propositions always have values of two distinct kinds which will be called (the implications of the terms must be considered further) their truth and their meaning. The truth of a scientific proposition, in this sense, is its value determined by the first principle, its meaning that determined by the second; a proposition is true in so far as it states something for which the universal assent of all mankind can be obtained; it has meaning in so far as it gives rise to ideas which cause intellectual satisfaction. Truth is here used to mean a quality which has the same significance for others as for ourselves; it is something of which we can convince others and of which we shall not be sure unless we do convince others. Meaning on the other hand is something individual and personal; it is something which depends on the qualities of my mind and is present in my mind whether or no it is present in the minds of others; a proposition may have meaning for me even if it has meaning for nobody else; and it is not certain ever that its meaning for me is the same as its meaning for anyone else.

But though all scientific propositions are valuable only if they have both truth and meaning, the relative importance of the two features may seem to vary widely. The value of a law lies chiefly in its truth, that of a theory chiefly in its meaning. It is unnecessary to insist that a theory must have truth as well as meaning, but it may be well once more to insist that a law must have meaning as well as truth. There is something about a law which gives it value and yet is not determined wholly by universal agreement. This something consists partly in its form and is of the nature of "simplicity," but it consists also in what the law asserts. A law asserts that certain observations will be made, not only in this instance and in that, but in *all* instances; it asserts that there is something necessary about the sequence which it predicts; and though men actually agree about the necessity asserted by actual laws, it is plainly possible that they should differ, and that, even if they did differ, laws would not necessarily lose their value.

The meaning of a law (as distinct from its truth) is less obvious than that of a theory because the application of the second principle is less obvious in the establishment of laws. We are conscious of much less scope in the choice of what law shall be used to describe given observations than in the choice what theory shall be used to explain given laws. Moreover men differ far less in their power of discovering laws than in their power of discovering theories; a great deal of valuable scientific work is done and many important laws discovered by persons who bring to their task little more than the facility in handling apparatus acquired by suitable training. But the absence of choice in recording his results which is usually felt by the experimenter is very often due to the fact that the choice has already been made by considerations derived from some theory; a theory indicates that there must be some constant relation between two measurable magnitudes and only leaves to the decision of experiment precisely what is that relation; it may even indicate what the relation is and leave only the decision whether a theory which gives such an indication is true. Even when a merely empirical law is sought and no guidance is derived from theory, experiments usually consist in the search for a permanent uniform relation between two or more of a quite limited class of concepts, the use of which necessarily decides many of the questions which might arise concerning which of the attendant conditions will influence the observations; we no longer think it necessary to inquire whether our experiments will be influenced by the position of the satellites of Jupiter or the progress of the latest revolution in Central America; the concepts we employ in stating our results preclude such influence. But it must be remembered that the simplicity of this sort which we now habitually assume is precarious and that events might always occur to make us investigate much more deeply and thoroughly the extent to which laws can be made to fall in with our usual conceptions of simplicity. Again in so far as the simplicity which is selected is that of a mathematical relation the choice is again determined by something that is not strictly characteristic of the law. Functions are simple according as they are amenable to mathematical analysis; there does not seem to be anything intrinsically simple about a sine or a logarithm which would be apparent to anyone unfamiliar with analytical operations. The judgement of scientific simplicity which is made is that physical laws can usually be represented by mathematically tractable functions; and this judgement, like those just mentioned, is so habitual and familiar that we have lost the consciousness of making it. The distinction between mathematically tractable and intractable functions is not made on the basis of either of the two principles of scientific value. It is doubtless something concerning which universal agreement can actually be obtained, but it is a judgement of the internal and not of the external world; and though there is concerned in it a judgement of simplicity, it is of mathematical and not of physical simplicity; between these there appears to be a difference, for the instincts of the pure mathematician and of the physicist are by no means always associated.

On this account the application of the second principle to the determination of laws is apt to be overlooked; but it is essential, and without it the problem of induction would be insoluble. Moreover the circumstances which enable us to dispense with conscious judgements of intellectual satisfaction in the establishment of laws also sometimes make them unnecessary in the formulation of theories. Here again, habit based on long experience often suggests that a theory must be of a given form and gives little scope for selection within the limits of that form; it enables those who are endowed with but little of the true scientific insight to frame theories which are found to be satisfactory.[1] But here again habit is a dangerous guide, as is shown, in the opinion of many, by the history of dynamical theories of atomic and molecular constitution. The difference in this matter between laws and theories is one of degree rather than of kind.

SOME HISTORICAL CONSIDERATIONS

Perhaps in all this discussion I have been insisting on the obvious. Everybody recognises to-day that what I have called truth is an essential element of a scientific proposition and few, if any, will deny explicitly that what I have called meaning is also important. But it does not seem to me that facts which are universally admitted openly, or their implications, are always remembered when the most general and fundamental questions concerning science are raised. In such discussions attention is apt to be concentrated on the truth and the meaning is apt to be left out of sight.

The tendency is natural. The great advance or, more accurately, the first beginnings of scientific knowledge which took place in the 16th and 17th centuries was a consequence of the recognition of the possibility of scientific truth. To say that science must be based on experiment and observation is simply to say that it must satisfy the first principle of value, for it is only concerning the results of such experiment and observation that universal agreement of the kind which is characteristic of science can be obtained. It is the neglect of truth, the failure to test evidence according to the canons of modern science, the acceptance of well-attested fact, vague rumour, and the product of riotous imagination as equally valuable—it is the attitude of mind

---

[1] Lest I should cause offence by an appearance of superiority I would add that I include myself in this category.

to which such things were possible which raises an insurmountable barrier between ourselves and the most enlightened of the ancients. That science should have meaning, they would have agreed readily; it was the doctrine that it should have truth which was strange to them. The ghost of Greek learning still stalks ruins not yet abandoned; it still disturbs timid minds and has still to be exorcised; the weapon of Galileo cannot be allowed to rust in its sheath, and while it has still to be used other dangers may be neglected.

However there is a more cogent reason why truth rather than meaning receives emphasis whenever any question is raised of the value of science or of its relation to other studies. Truth, it has been said, is a quality of which we may hope to convince others; it is a valuable quality because it is appreciated by everyone. And there is actually no doubt that scientific propositions have the kind of truth that is here attributed to them and that this truth has some value. Nobody disputes that truth, if they once agree to use that word in our sense; what they may dispute is whether or no it is misleading to call this quality truth and what is its value in comparison with that of other qualities. When therefore there has been any discussion of the value of science as a whole, its supporters have tended naturally, but, as I think, mistakenly, to insist on that element of science which everyone agrees is valuable and to attempt almost to conceal other elements concerning the value of which a difference of opinion is possible. They have tried to maintain that science is nothing but truth and indeed that it differs from other systems of thought in consisting of nothing but truth. Such were the motives which led, for instance, to Huxley's famous definition of science as "organised common sense," a phrase admirably suited for polemics and the obscuring of clear thinking. The word "organised" begs the question completely. It is not disputed that science, like all other forms of knowledge,[2] has its basis in common sense, the agreed judgements of mankind, among which are those (relating to the external world) which have the truth characteristic of science. But the problem is how and by whom this common sense is to be organised, and whether the organisation

---

[2] Perhaps this statement is not quite true. The system of thought against which Huxley was specially concerned to defend science was theology, and some theological systems rest, not on common sense, but on immediate and fundamental judgements or revelations which are definitely stated to be confined to the elect. Such systems have not the truth characteristic of science. The value of their propositions is determined wholly by principles analogous to the second and not at all by those analogous to the first.

adopted by science introduces anything which does not share the truth of the things organised.

The confusion introduced by this method of meeting attacks was all the more serious because those attacks were mainly directed against theories. The doctrines in dispute, those for example concerning the structure of the solar system or the origin of species, were theories and not laws. Though it is necessary to recognise that laws have meaning as well as truth if we are to understand how they are formulated, it is not so necessary to insist on that meaning if we merely wish to show that they are valuable; to the unscientific mass of mankind a law is almost as useful if it is applicable to each of any assigned collection of instances as if it has the necessity which is implied by its application to all instances. But the main value of theories consists in their meaning; to conceal their meaning is to render them worthless and unintelligible. The attempt to show that science was all truth led directly to an attempt to show that all valuable scientific propositions were laws. The controversialists succeeded well in bamboozling their opponents; they made them talk glibly about the "law" of gravitation and even the "law" of evolution; they succeeded in convincing a pious gentleman that there was "natural law in the spiritual world", the laws which he discovered being almost exclusively theories. But unfortunately, as so often happens, they succeeded equally well in bamboozling themselves, their supporters and their colleagues; for a generation "theory" was almost abolished from the scientific vocabulary or, when used, had always a connotation of distrust.

The writings of Mach, and perhaps even more of Poincaré, led to a saner view. Mach, in his doctrine that science sought the "simplest" interpretation of observations, recognised the meaning of scientific propositions and the application of the second principle; Poincaré, though often using the same word, allows to simplicity a much more generous scope. But even Mach, I venture to think, failed to reach the core of the matter. He justified the adoption of the "simplest" view on the ground that it is the object of science to attain "economy of thought." That doctrine, if the words are to be interpreted in any natural sense, is to me utterly intolerable. The best way to attain economy of thought, a way only too successfully followed by the vast majority of mankind, is not to think at all. Science is a branch of pure learning; thought is its object. To engage in science in order not to think would be as sensible as to engage in commerce in order not to make money. Even Mach himself in one passage suggests that his remarks may be valuable if they suggest ideas which have not previously occurred to the reader.

But of course it will be said that I misjudge Mach, and that he did not mean that the object of science was to render all thought unnecessary, but only special kinds of tiresome thought that have no intrinsic value. However, though that idea may have been in his mind, I think that its explicit recognition and statement deprives his theory of science of all foundation. He may have meant merely what Professor Whitehead said lately in a phrase that has already become classical, that operations of thought are like cavalry charges, to be reserved for critical occasions. That view, I can readily believe, is perfectly applicable to the pure mathematics of which Professor Whitehead was speaking; the formalities of mathematics, its elaborate symbolism and all the rest of it, do enable thought to be reserved and attention concentrated for vital efforts. But I do not think that it is applicable to physics; there seems to be here an example of the dangers which arise when men whose interests are primarily mathematical expound the principles of physical and experimental science. For what is there in science which corresponds at all accurately to the formal rules of addition or differentiation which enable complicated operations to be performed mechanically? I can think of nothing. The only forms of scientific thought which seem to me disagreeable, and to be avoided whenever possible, are the efforts of memory required to retain numerical constants; and these forms we avoid, not by the establishment of any scientifically valuable propositions, but simply by the mechanical device of printed tables. There is no economy or avoidance of physical thought (though there may be of mathematical thought) in stating the relation between current and resistance by the function of proportionality rather than by a more complex function; there is simply substitution of a more agreeable for a less agreeable form of thought. It is only in the applications of science that true economy is achieved by science. In practical life we may be forced to consider certain observations or possible observations, and the ordering of these observations may save time and trouble in the consideration. But in pure science we are not forced to consider any observations at all if we do not care to do so, and we can economise thought in a manner far more successful than any provided by science by simply not thinking at all.

Again, even if we admit that the process which consists of the preference of a simpler for a more complicated relation in the statement of laws is an economy of thought, what are we to say of theories? A theory is almost a pure addition to thought. Can anyone pretend that we are saved any thought about Boyle's and Gay-Lussac's laws by a knowledge that they can be explained by the dynamical theory of gases? Surely not; for those persons who are truly concerned to save thought, because they are interested only in the applications, are inclined to be utterly contemptuous of theories. If explanation were of the first kind of Chapter V and consisted in the substitution of the more general for the less general idea, then theories might achieve the same kind of "economy" as laws. But it is not this kind of explanation that is characteristic of theories; it is the second kind which substitutes the more for the less familiar, often or usually at the expense of greater complication. Only those who, like Mach, rate the value of theories very lowly and desire that they should be excluded as far as possible from science (or at least should be confined to mathematical theories which may economise in the same manner as laws), only such persons could entertain for a moment the idea that it is the object of science to economise thought.

Moreover the phrase "economy of thought" by suggesting one very definite and precise reason for selecting one of the many possible alternative laws rather than another tends to conceal the fact that there is another and much more important reason. We cannot select any law that may happen best to economise thought; we have (if we are to be successful) to select one and one only; it may and indeed does happen that this law is also one which economises thought, but those which do not economise thought are not all equally objectionable. Suppose, for example, that we have observed the relation between the current and the potential difference for a long series of integral values of the former. Then we can describe our results (1) by the function of proportionality, (2) by some function much more complicated than proportionality, but differing from it so little over the range of any observations we are likely to make, that even for fractional values of the current its values are experimentally indistinguishable from those of proportionality, (3) by a function agreeing with proportionality at integral values but differing from it very widely at fractional values. Now on the ground of economy of thought there may well be no reason for preferring a function (2) to a function (3); either may be equally complicated and much less simple than (1). But, even if there were no economy of thought, (2) would always be preferable to (3), because it will remain satisfactory when fractional values are investigated. Even if it is true that we actually select laws by the principle of economy of thought (and it is admitted that we select them by some principle not very different) there still remains the very important fact that what we achieve by our selection is not mere economy, but a truth of the law which is determined by the first principle of value and not by the second.

SCIENCE AND IMAGINATION

But the fundamental reason why the meaning of science has been unduly neglected lies in the unwillingness of men of science themselves to recognise it. Their training in the methods necessary to attain truth has impressed on them so firmly that, in this part of their work, everything that can possibly be a matter of personal opinion must be excluded, that they are afraid to admit that anything can properly form part of their study which involves deliberate, though often unconscious, choice. In the early days both of the individual and of the study such caution is both desirable and necessary; there is an undoubted temptation to relax the criterion which must be applied before truth can be firmly established. But when the individual and the study have come to full maturity the danger has passed away; there is now no need to insist at every turn that science must have a firm experimental basis. The time has come to face the facts boldly. The search for truth alone never has and never will lead to any science of value. The spirit which must be so carefully curbed in the search for truth must be given free rein when truth is attained. Our passionate desire that truth will be found in one form rather than in another must never be allowed to influence our decision in what form it actually appears; but once that matter is settled, we not only may but must choose in accordance with our desire in which of the innumerable alternative forms that truth must be expressed; the more freely we choose the more likely it is that a renewed search for truth will confirm our choice.

The attempt to conceal from ourselves that choice is necessary may stifle the imagination on which the choice depends; and therefore I make no apology for insisting, even at the risk of irrelevance, on the necessity of proclaiming openly the imaginative element in science. Our view of the meaning of science must influence our methods of teaching and training ourselves and our pupils; 19th century philosophy, with its anxiety to conceal the essential part played by imagination in scientific discovery, is largely responsible for the ineptitude of modern scientific education. The very man who laughs to scorn the doctrine that a love of literature or an imaginative appreciation of its value is to be obtained by the grinding out of Latin hexameters will proceed gravely to assert that science is to be taught only by the determination of the nodal points of lens systems. Of course for the student who means to take science seriously and hopes in his turn to take his share in its advancement, a thorough training in the experimental art is as essential as is a complete understanding of the intricacies of metre and construction to the classical scholar. But neither the scholar nor the man of science will have a living grasp of his study if he buries himself in these pedantries. The scientific imagination can be developed by tedious laboratory practice no more than the artistic imagination by the laborious study of Greek particles, by the day-long practice of a musical instrument or by unceasing copying in the galleries. It must come from direct and intimate contact over the widest possible range with the great original works which represent its noblest expression. It is doubtless difficult to introduce a student to the latest modern theories of solid structure, of atomic constitution, or of relative motion, before he has an entire understanding of what are commonly regarded as the elements of physics; but it is no more difficult than to teach a boy to read the Odyssey before he can parse and interpret any word; and the failure to overcome the difficulty is equally disastrous.

Doubtless "there is nothing like leather," but I cannot refrain from suggesting that it would have been better if chemistry had not come to be regarded as the standard and natural "elementary" science. For chemistry, so rich in laws (though not often explicitly named as such) and so powerful in the ordering of facts, is poor in theory. And since it is in theory that the highest meaning of science is expressed, chemists are more apt than the students of other sciences to overlook its vital importance. The absurdities of the "heuristic" school, fortunately short-lived, could hardly have taken root at all in any other soil; no physicist could imagine that there was any similarity between the "discovery" of a law by the elementary pupil under the eye of his teacher and the true discovery of that law when it was unknown. Chemistry has but one noteworthy theory and but one set of hypothetical ideas, the theory of the combination of atoms into molecules with its fundamental idea of valency. It is a most beautiful theory, surpassed by none other in the intellectual satisfaction it affords, but unfortunately it is not easily or certainly applicable to the compounds on which the attention of the elementary student is concentrated; we know far less about the constitution of water than about that of some organic compound with a name a yard long—long simply because the theory is so strictly applicable to it. If chemistry is to be the vehicle of elementary instruction in science, we should begin with stereo-isomers and proceed (if we have time) to the simple compounds of oxygen, hydrogen, and nitrogen.

Nor should we think only of the effect of our repression on those who are serious students of science. The opinion of our fellows, even if they are not our colleagues, cannot fail to react, directly or indirectly, on our own studies. If scientific education to-day is unsuited for those who are to make science their life work, it is even less suited for those to whom it is merely to be part of a general education. Men of science complain of the lack of a wide appreciation of scientific knowledge; what else can they expect if they offer to the world only the dry bones of knowledge from which the breath has departed? Nothing could be better adapted than the ordinary school course, with its tedious insistence on bare and uninspiring facts, to kill any rising enthusiasm. It is important certainly to impress the student with the nature of scientific truth and with the possibility of definite positive knowledge concerning the material world. No doubt it is the failure to realise that there is such knowledge, the mistaken notion that everything is a matter of opinion on which two sides should be heard, that produces, so ludicrous if it were not so lamentable, the familiar chaos in the administration of the affairs of state and industry by the half-educated persons who pride themselves on their ignorance of science. But to insist on the truth of science and to neglect its meaning is to aggravate the evil which we seek to cure; those who are endowed with any measure of creative imagination can never hold in anything but contempt a study from which such imagination appears to be wholly banished.

Such attempts as are made to exhibit the imaginative element in science are almost more disastrous than the attempts to conceal it. The "romance of science" is usually associated with childish books and popular lectures on speculative geology and "spherical" astronomy.[3] Now both geology and astronomy are magnificent sciences,

---

[3] I have heard this term used wickedly to denote the form of science which, at the end of the 19th century, was closely associated with the name of Sir Robert *Ball*.

offering superb examples of the highest meaning of science; but they also contain elements, of no importance to their earnest students, which possess a specious and flashy interest which makes a passing appeal to shallow minds. An audience of children of all ages gapes amazedly while the lecturer discourses glibly of times reckoned in millions of years and distances in thousands of millions of miles. But science has something better to offer than sensational journalism; nothing could be less characteristic of its spirit. The mere fact that the interest of the uninitiated can thus be easily stimulated with serious training suggests doubts of the value of the stimulus; nothing worth having in this world is to be had without effort.

SCIENCE AND ART

When we so often hide what is best in science and display only its less admirable features, it is not surprising that in the outside world there is suspicion of its ultimate intellectual value. There has been in recent years a great improvement in the general appreciation of the meaning of science; but open antagonism has been in part replaced by an armed neutrality which indicates no better understanding, but merely greater caution. Many will still be found to deny that science can satisfy our imaginative needs to the same extent as art and literature, and the denial does not arise only from conservatism and ignorance. Science, it is said, is impersonal; the highest good must be intimately connected with personality. It is overlooked that the impersonal truth of science is inseparable from its personal meaning. Science, it is said, is mechanical; the accusation at once displays the misunderstanding. A mechanism is certainly something which will produce desired results independent of the attention or volition of a skilled operator, but it is also something which is and must be the individual product of a human mind. A mechanism implies an inventor; it is a means by which one exceptionally endowed man makes his endowments available for the common good; it is something characteristic, not of dead matter, but of the highest spirit of man; it is something that theologians and savages do not understand. If the term is rightly understood science is truly mechanical; for science, like mechanism, is the expression of genius in a form which the dullest can appreciate.

It is curious how even to-day the laity seem unaware of the part played by the genius of great men in the development of science.

They recognise perhaps that the often quoted examples of the greatest achievements of science, the discovery of Neptune or of Hertzian waves, represent something not easily attained by the common mass of mankind; they are willing to admit that Newton or Leverrier, Maxwell or Hertz, must have had some qualities to distinguish them from lesser folk. But they have no knowledge of what these qualities are; they have no idea that their work was an expression of their personality just as surely as the work of Giotto, of Shakespeare, or of Bach. They still tend to contrast the cold-blooded rationalism of the man of science with the passionate dreamings of the artist. But science too has its dreamers, and their dreams come true; they dream, and messages flash across the empty ocean; they dream again, and a new world springs into being and starts upon the course that they have ordained. Nor does the quest of knowledge inspire less passion than the quest of beauty. It is not sickly sentimentality but honest emotion that makes us cry

> Car c'est chose divine
> D'aimer, lorsq'on devine,
> Rêve, invente, imagine
> À peine . . . ,
> Le seul rêve intéresse;
> Vivre sans rêve, qu'est-ce?
> Et moi, j'aime la princesse
> Lointaine.[4]

Nothing could be more absurd than the attempt to distinguish between science and art. Science is the noblest of the arts and men of science the most artistic of all artists. For science, like art, seeks to attain aesthetic satisfaction through the perceptions of the senses; and science, like art, is limited by the impositions of the material world on which it works. The lesser art accepts those limitations; it is content to imitate or to describe Nature and to follow where she leads. The greater refuses to be bound; it imposes itself upon Nature and forces her to submit to its power. The apostle of Art in a previous generation can make no higher claim for the greatest art than this:

> CYRIL. But you don't mean to say that you seriously believe that Life imitates art, that Life in fact is the mirror, and Art the reality?
> VIVIAN. Certainly I do. Paradox though it may seem—and

[4] Rostand, *La Princesse Lointaine*.

paradoxes are always dangerous things—it is none the less true that Life imitates art, far more than Art imitates life. We have all seen in our own day in England how a certain curious and fascinating type of beauty, invented and emphasised by two imaginative painters, has so influenced Life, that, whenever one goes to a private view or to an artistic salon one sees, here the mystic eyes of Rossetti's dream . . . there the sweet maidenhood of the "Golden Stair," the blossom-like mouth and weary loveliness of the "Laus Amoris." . . . And it has always been so. A great artist invents a type, and Life tries to copy it, to reproduce it in a popular form, like an enterprising publisher. . . . Literature always anticipates life. It does not copy it, but moulds it to its purpose. The nineteenth century, as we know it, is largely an invention of Balzac.[5]

But if to lead the way and to bid life follow is the distinctive character of the greatest art, what art can be so great as science? A Newton, a Faraday, or a Maxwell conceives a theory and Life adapts itself for all time to the laws which it predicts; by the force of his imagination he creates no passing fashion, but the permanent structure of the world. He is no puny creature closely bound by the laws of time and sense; he is the creator who lays down those laws; verily the winds and the waves obey him.

Of course such powers are not given to all who pursue science. There are degrees of scientific as of artistic imagination. But the least of us can share in some small measure these achievements. A man need not abandon all pretensions to the proud title of artist because he could not design the Parthenon or write the Fifth Symphony. Most of us who have attempted to advance science have had our all too brief and passing moments of inspiration; we have added a single brick to the mighty structure or finished some corner which the master in his impetuosity has overlooked. And though our tiny efforts rightly pass almost unnoticed by the rest of mankind, they have a value for ourselves beyond what we can tell; one instant we have stood with the great ones of the earth and shared their glory. Even if nothing as yet has stirred in us the creator's joy, we can yet appreciate the success of others. Nobody who has any portion of the scientific spirit can fail to remember times when he has thrilled to a new discovery as if it were his own. He has greeted a new theory

[5] Wilde, *The Decay of Lying.*

with the passionate exclamation, It must be true! He has felt that its eternal value is beyond all reasoning, that it is to be defended, if need be, not by the cold-blooded methods of the laboratory or the soulless processes of formal logic, but, like the honour of a friend, by simple affirmation and eloquent appeal. The mood will and should pass; the impersonal inquiry must be made before the new idea can be admitted to our complete confidence. But in that one moment we have known the real meaning of science, we have experienced its highest value; unless such knowledge and such experience were possible, science would be without meaning and therefore without truth.

Chapter 23
MORITZ SCHLICK
(1882–1936)

INTRODUCTION

MORITZ SCHLICK was born in Berlin on April 14, 1882. Having completed his elementary education and attended the *Realgymnasium* there, he went on to study physics at the Universities of Heidelberg, Lausanne, and Berlin. He received his doctorate of philosophy in 1904 for a thesis on physics written under the direction of Max Planck. By 1911 he had completed a second thesis, entitled "The Nature of Truth According to Modern Logic," with which he qualified as a teacher at Rostock. In 1912 he was appointed to a teaching position in Kiel; ten years later he went to Vienna as professor of the philosophy of inductive science, succeeding Boltzmann and Mach. In Vienna he gathered around him a group of philosophers, among whom were Rudolf Carnap, Otto Neurath, and Friedrich Waismann, and mathematicians and physicists, including Kurt Gödel, Phillip Frank, Karl Menger, and Hans Hahn. This group regularly discussed topics of mutual interest and soon became known as the Vienna Circle. Certain traits characterized the general outlook of the group: an antimetaphysical attitude, a firm belief in the value of radical empiricism, a great faith in the methods of modern logic, and the conviction that the future of philosophy lay in its becoming the logic of science. The group saw themselves as continuing and further developing the positivist tradition of Mach and Boltzmann, but it was certainly also deeply influenced by the ideas of Russell and the earlier thought of Wittgenstein. The group's views were made public by a special manifesto, *Wissenschaftliche Weltauffassung: der Wiener Kreis* (1929); by a series of monographs the first of which appeared in 1929 under the general title *Schriften zur Wissenschaftlichen Weltauffassung;* and later in the journal *Erkenntnis.* After establishing contact with several philosophers of a similar general orientation they started an international movement which was to be known later as logical positivism. Schlick came to the United States in 1929 as a visiting professor at Stanford

University; he also taught at Berkeley in 1931. His death came in Vienna in 1936 when he was murdered on the steps of the university by a former student, apparently deranged.

Schlick concentrated most of his time on philosophical problems connected with natural science. This is not to say, however, that he turned his back on physics in its own right, for after his graduation he continued to maintain close contact with the leading scientists of his time, Planck, Einstein, and Hilbert in particular. In 1917 he published his well-known study on the theory of relativity, *Raum und Zeit in der gegenwartigen Physik,* a book translated into English in 1920 under the title *Space and Time in Contemporary Physics:* the study was one of the first philosophical essays in relativity theory and one of the earliest attempts to introduce the theory to nonphysicists.

Schlick had been schooled in neo-Kantianism, but through his study of modern physical theories, especially relativity, and of the philosophies of Mill and Mach he gradually came to the conclusion that Kant's views on space, time, and causality were untenable, and that the problems connected with these basic scientific concepts can be dealt with only in a radically empirical way.

In 1918 Schlick published his second book, *Allgemeine Erkenntnislehre* ("General Theory of Knowledge"), which dealt with epistemology and a general theory of science. In this book many ideas which would later be central in the philosophy of the Vienna Circle were already defended or at least indicated. A third work on philosophy of science appeared posthumously under the title *Grundzüge der Naturphilosophie* (1948); it was translated into English in 1949 and published as *Philosophy of Nature.*

In addition to these works dealing with epistemology, theory of science, and philosophy of science, Schlick published numerous essays, the most important of which were collected in the posthumous volume *Gesammelte Aufsätze: 1926–1936,* which appeared in 1938, and a study on ethics, *Fragen der Ethik* (1930).

With respect to epistemology and philosophy of science Schlick always belonged to the most moderate wing of the Vienna Circle. He objected, for example, to Carnap's and Neurath's view that all statements allegedly concerning mental events can be translated into statements about physical events. He subscribed to their general view that the meaning of any statement turns entirely on the possibility of its empirical verification, and from this principle drew the conclusion that metaphysics in the traditional sense of the term is impossible and its statements meaningless. However, he did not believe that this made all forms of ethics impossible, nor did he think it impossible on similar grounds to compare statements with reality.

After Wittgenstein published *Tractatus Logico-Philosophicus* in 1921, Schlick was convinced that this conception of philosophy would mark a decisive turning point in its history. From then on he whole-heartedly agreed with Wittgenstein's view that the object of philosophy is the logical clarification of thoughts, and that the result of philosophy is not a number of typically "philosophical" propositions but to make any propositions clear. From this it follows at once that the philosopher's aim is not to construct a new system of philosophy, but to teach a scientific way of philosophizing.

The selection to follow, taken from the posthumously published *Philosophy of Nature*, gives a clear idea of Schlick's final view on philosophy of science. It is not known when he wrote down these lecture notes, but he undoubtedly used them in a series of lectures during the winter term of 1932–1933. The notes are substantially in agreement with the lecture series given in 1922–1923 in Vienna; the latter was published in revised form in Dessoir's *Lehrbuch der Philosophie* (1925). In the later version the influence of Wittgenstein is evident.

SELECTIVE BIBLIOGRAPHY

Schlick, M., *Space and Time in Contemporary Physics* (1917). Oxford, 1920.

———, *Allgemeine Erkenntnislehre*. Berlin, 1918.

———, "Naturphilosophie," in *Lehrbuch der Philosophie*, ed. Dessoir. Berlin, 1925. Vol II.

———, *Gesetz, Kausalität und Wahrscheinlichkeit*. Vienna, 1948.

———, *Philosophy of Nature* (1948). New York, 1949.

Weinberg, J. R., *An Examination of Logical Positivism*. New York, 1936.

Frank, P., *Modern Science and Its Philosophy*. New York, 1961. Pp. 13–61 (*passim*).

Feigel, H., "Moritz Schlick." *Erkenntnis*, 7(1937–1938) 393–419.

Popper, K., *The Logic of Scientific Discovery*. London, 1959 (*passim*).

BASIC ISSUES OF PHILOSOPHY
OF NATURAL SCIENCE

THE TASK OF THE PHILOSOPHY OF NATURE

*T*HE simplest way of defining the essential character of the philosophy of nature is by stating its relation to natural science. From the very beginnings of Western thought up to the age of Newton and even of Kant, no differentiation was made between the philosophy of nature and natural science; but from then on, the speculative method (that of the philosophy of nature) seemed to detach itself from the experimental procedure (that of natural science) until finally, in the middle of the 19th century, it became clear that the speculative method was a deceptive one and led into a blind alley. After that, a period followed during which philosophy was regarded with contempt by the investigators of nature, until, at the beginning of the 20th century, the term "philosophy of nature" regained its esteem and, in consequence of the unprecedented advances in natural science, a general interest in the consideration of its philosophical aspects was revived. As a result of the prevailing attitude towards philosophy in general, the task of a philosophy of nature was at first defined as 1) a synthesis of knowledge for the purpose of obtaining a complete picture of all natural processes, and 2) an epistemological vindication of the foundations of natural science.

This is, however, an unsatisfactory definition, inasmuch as the task, or object, of natural science is the achievement of knowledge concerning all natural events and processes—in other words, it is the statement of the most general propositions, as well as an examination of the truth of the hypotheses.

The consolidation or fusion of the various branches of natural science

This article is from **Philosophy of Nature** by Moritz Schlick, trans. Amethe von Zeppelin (New York, 1949), pp. 1–5 and 17–29. Copyright © 1949 by Philosophical Library, Publishers. Reprinted by permission of the publishers.

—that is to say, the subordination of simple propositions to more and more general ones—can only occur from below, in an upward direction. For in the progress of knowledge in every field, higher and higher levels are reached, and until these levels are attained there is no possibility of achieving the synthesis which is necessary in order to obtain a complete picture. It is likewise impossible for philosophy to achieve this synthesis.—The entire task of natural science consists solely in the persistent and indefatigable examination of the correctness of its propositions which, in consequence, develop into more and more securely established hypotheses. In this way, the assumptions upon which these hypotheses are based, are simultaneously tested within the domain of natural science itself. There is, moreover, no other specifically philosophical vindication of the foundations,—such a vindication would not only be impossible, but superfluous as well—a fact which will be demonstrated in the course of the following considerations.

The task of the philosophy of nature is nevertheless concerned with the hypotheses of natural science—but in quite another sense. Natural knowledge is formulated in propositions; and likewise all the laws of nature are expressed in propositional form. But the knowledge of its meaning is a prerequisite for testing the truth of a proposition. These two concepts are inseparable, and both occur within the domain of natural science. In spite of their inseparability, however, it is possible to distinguish here between two psychological attitudes: the one concerned with testing the truth of hypotheses, and the other with the understanding of their meaning. The typically scientific methods assist in the discovery of truth while the effort of philosophy is directed to the elucidation of meaning. The task of a philosophy of nature is thus to interpret the meaning of the propositions of natural science; and therefore the philosophy of nature is not itself a science, but an activity which is directed to the consideration of the meaning of the laws of nature.

In order to establish our thesis in a complete sense, we should be obliged to define those characteristics of natural science which distinguish it from the Arts and the so-called cultural sciences. We will, however, confine ourselves to the statement that by nature we understand all that is real in so far as it is determined in space and time. All objects or processes which exist or occur in space, exist or occur likewise in time. The converse would appear not to be true; for it would be absurd to attempt to localize feelings and emotions (which are, of course, temporal) as such. They can, however, be attributed to definite individuals (namely to those who possess these feelings and emotions) and in this sense are related to spatial things. Furthermore, since all historical, cultural and linguistic objects are spatio-temporal, they are part of nature and consequently objects of natural science.

Hence, we have the universal quality and all-embracing character of natural science which prevents it from being subordinated to, or ranged alongside, any one of the Arts or cultural sciences. And hence also the unique philosophical significance of natural science: All philosophical progress in the past has arisen out of scientific knowledge and the investigation of scientific problems. It is a very grave mistake—a mistake which was made for the first time during the last hundred years—to believe that the Arts and cultural sciences are in any way equivalent to natural science, or that they are, from the standpoint of philosophy, equally productive.[1]

In addition to its universality, it is the exactitude of natural science which causes it to be, both historically and in actual fact, the most fundamental basis from which to philosophize. Only in the analysis of exact knowledge, is there any hope of achieving true insight. Only here, is there any prospect of attaining definite and final results by means of the elucidation of concepts. The vague, uncertain propositions of the inexact sciences must first be transformed into exact knowledge—that is, they must be translated into, the language of the exact sciences—before their meaning can be fully interpreted. And exact knowledge is knowledge which can be fully and clearly expressed in accordance with the tenets of logic. "Mathematics" is only a name for the method of logically exact formulation. Hence, even Kant, for instance, declared that science only contains as much knowledge as it contains mathematics. In science, more than in any other domain, the stuff or substance of knowledge is derived from an intellectual activity which enables us to arrive at the greatest heights of abstraction. But the higher the level of abstraction attained by a science, the deeper it penetrates the essence of reality.

These are the reasons for the central position occupied by the philosophy of nature.

---

[1] These sentences—as Schlick himself explained in detail in his lectures—are directed against views similar to those of Heinrich Rickert (1863–1936) as expressed in his book: "Die Grenzen der naturwissenschaftlichen Begriffsbildung" (Freiburg 1896). [Tr.]

DESCRIPTION AND EXPLANATION

The first step towards a knowledge of nature consists in the description of nature which is equivalent to the establishment of the facts. And this, in turn, consists in stating, by means of words or symbols, how the facts described are composed of elements, each of which is denoted by the customary symbol (name). For this purpose, certain primitive acts of recognition are always necessary, so that each component can be identified as belonging to a definite class and assigned to a corresponding symbol.[2]

The next step towards a knowledge of nature—explanation—is characterized by the fact that a symbol (concept) which is employed in the description of nature is replaced by a combination of symbols which have already been used in another context. In point of fact, progress in knowledge consists in the discovery that a substitution of this kind is possible. Thus it is a chemical discovery when, instead of the word "water," we can say: "Combination of H and O in proportionate weights of 1:8"; and it is a physical explanation when, instead of speaking of the heat of a body, we can speak of the energy of motion of its smallest particles. And so on. The advantage of this new kind of description lies in the fact that, with its assistance, the ways of behaviour of the things thus designated, can be predicted—inasmuch as this behaviour can be deduced from the behaviour already known, of those things which are denoted by the concepts used in the explanation. If, for example, heat be explained as a form of motion of the smallest particles, we can, as a result, attribute all phenomena of heat to the properties of the invisible motion of a host of the smallest particles, and thereby predict phenomena of heat which had previously been unknown to us. It is obvious that in the progress of knowledge, the number of concepts necessary for a description of nature, will become increasingly reduced; so that what is denoted by the term "world-picture" will become more and more unified. The world will become a "Uni-verse." It is evident from their attempts to reduce the multiplicity of the Universe to a single principle, that even the Greek philosophers of antiquity were conscious in a dim way, of the ultimate goal of knowledge. This idea was at the root of Thales' theory that water is the primal substance of the world; while for Anaximenes and Heraclitus, air and fire respectively, filled this role.

[2] Schlick expressed his views on this subject in detail, in his "Allgemeine Erkenntnislehre" (2nd edition, Berlin 1925). He planned a thorough revision of this work which he considered to be in some respects obsolete. [Tr.]

Explanation means the discovery of like in unlike—of identity in difference. And inasmuch as explanation reduces different species of natural phenomena to the same domain, these different species are included as special cases in the latter. Hence we may say, that explanation is the inclusion of the special in the general. Thus, heat and sound, for example, are both explained in so far as they are regarded as special cases of the motion of the smallest particles.

In the first stages of scientific thought, the discovery of the similar in the dissimilar was interpreted as the discovery of a constant, an invariant—something which remains identical with itself and which, while it is the basis of all variety and change, has no part in them. This constant was called *substance,* and was supposed to occur in a multitude of apparent forms and to be subjected to a variety of processes without its essential nature being altered thereby. This primitive concept of substance, the logical deficiencies of which will become evident later, was even then inadequate. For how this substance came to be differentiated and to undergo such transformations, remained unexplained. Change itself must be rendered intelligible by the discovery in it of the unchanging, or invariable—and for this purpose, the concept of law is necessary.

"General descriptions" constitute a preliminary stage in the procedure (e.g. "a stone which is thrown, falls to the ground"). These general descriptions may even be termed laws; but they still do not constitute an *explanation* of the processes described. Such an explanation can only be achieved when a number of laws of the kind are united in a single law, and when one is recognized as a special case of the other. In that case, *one and the same formula* will describe a number, or indeed, an arbitrary number, of processes. This is the essence of Meyerson's interpretation of the rôle which "identity" plays in the explanation of nature. An explanation is perfect only when this formula is specified with the help of the mathematical concept of "function." For it is only with the help of a formula of this kind that it is possible to obtain a description which is complete in all details.

Galileo was the original creator of this kind of exact natural knowledge. We will first attempt to get a clear idea of the essential character of the natural law which was enunciated by him as the law of falling bodies. We imagine a freely falling body, whose velocity v is measured at many points on its path. We divide these velocities by the time t which the body has taken in order to reach the corresponding points on its path. The quotient will show the same number each time (called g), although the numerator and denominator are constantly changing during the fall. Thus the quotient represents the constant element in the change, or the invariable in the variable. Generally speaking, the formulation of a law concerning any natural process consists in stating the *particular combination* (function) of those variable magnitudes or quantities describing the process, *which remains constant during the whole process*. Galileo "explains" *why* the falling body had traversed a certain distance in a certain time. Newton again, explains Galileo's law, inasmuch as he shows it to be a special case of the law of gravitation. And Einstein explains the law of gravitation inasmuch as he reduces it to a general principle of inertia.

The explanation of nature means a description of nature by means of laws. The function of laws (the meaning of laws) is to *de*-scribe and not to *pre*-scribe. They relate what actually occurs, and not what ought to occur. And when necessity is ascribed to the laws of nature, this means that they are universally valid, and not that they exert force. The laws of a country or State, are forms of compulsion for the citizens of that State. But to speak of compulsion, or force, in the case of the laws of nature, is absurd. One is misled into doing so, because, of the ambiguity of the word "law"—and this, in turn, is due to the half-conscious use of a psychological model.

Psychological models of this kind, in which natural processes are conceived of in accordance with the pattern of psychic events, constitute the basis of the mythical explanation of the world, and of the animistic conception of nature. They are also responsible for a number of metaphysical systems, like that of Schopenhauer, for whom natural processes represented the manifestations of a hidden Will. Bergson's life force (élan vital) plays a similar rôle and likewise represents a primitive psychological model. It is characteristic of both these philosophers that they set up, in opposition to the scientific explanation of nature by means of laws, a philosophical knowledge which they claimed was deeper. This deeper philosophical knowledge does not consist in description, but in a real coalescence with the subject-matter of which knowledge is sought. Only thus, in their opinion, can true understanding be attained. But these philosophers do not realize that description by means of laws achieves all that can possibly be demanded of knowledge; and that psychologically intuitive models only *apparently* advance the understanding of nature—in reality they obstruct it more than does the use of a mechanical model. The word "force" also—the meaning of which we shall analyse later—owes its introduction into science, to a psychological model.[3]

THE CONSTRUCTION OF THEORIES

Theoretical science, as is obvious from its name, consists of theories—that is, of systems of propositions. Propositions constitute a system when they are related to one another through being concerned with the same objects; or even when they can be deduced from one another. The process of formulating a law of nature is, fundamentally, always the same. It consists, in the first place, of recording the observations of a natural process in a table which always contains the relevant measured values of those variable magnitudes which characterize the process. The next step is to discover a function which will represent in a single formula the distribution of values in this table. This formula is then considered to be the law describing the process as long as all new observations are in agreement with it. Inasmuch as the formula always contains more than what is actually observed, and also because it must hold for all processes of a similar kind, the formulation of any law involves a generalization, or a so-called induction. There is no such thing as a logically valid induction going from the particular to the general: the latter can only be conjectured, but never logically inferred. Thus, the universal validity, or truth, of laws must always remain hypothetical. All laws of nature have the character of hypotheses: their truth is never absolutely certain. Hence, natural science consists of a combination of brilliant guesses and exact measurements.

The process of measurement assumed here, raises questions which must be discussed at a later stage.

In the same way as a special law is the result of a series of single observations, a general law is the consequence of the inductive combination of several individual laws, until finally a relatively small number of general propositions which include the totality of natural laws is obtained. Thus to-day, for instance, all chemical laws can, in principle, be reduced to physical laws; and the dividing line between the different domains of physics which used to be externally related to one another (mechanics, acoustics, optics, theory of heat etc.) has long since completely disappeared. At the present time, only me-

---

[3] See "Necessity and Force" in Schlick's "Naturphilosophie," *Lehrbuch der Philosophie*, ed. Dessoir, Vol. II: *Die Philosophie in ihren einzelnen Gebieten* (Berlin 1925, pp. 434–437). [Tr.]

chanics and electro-dynamics are left; and these are in nowise independent of each other, but interpenetrate everywhere. Whether biology will continue to remain a special province, or whether it also will become incorporated in the domain of physics, is a question that will be discussed in due course.[4]

In order to obtain a concrete description of nature (i.e. of nature as it really is), it is not sufficient to formulate laws: the abstract laws must, as it were, be given content. And in addition to these abstract laws, the constellation of reality (at the time of consideration), to which the formulae can be applied, must be stated. Such constellations are called by physicists boundary or initial conditions, and mathematically, they are expressed by the introduction of constants.

Here, we are considering the system of laws in itself, independently of all applications—that is to say, we are only studying general, and not particular, propositions. We can thus select out of this system, a group of the most general propositions from which all the others are derivable. Thus derivation is a purely logical deduction which can be undertaken without knowledge of the meaning of the symbols which occur in the laws. Hence, we will disregard, not only all application to individual cases, but also the meaning of all words and symbols—until the system is reduced to a purely formal structure, or empty framework which does not consist of actual propositions, but only of their forms (in logic, these are known as propositional functions). A system of this kind, which does not represent nature in actuality, but *all the possibilities in nature*, or in other words, its most general form— is known as a hypothetico-deductive system (Pieri). The propositions forming a group at the apex of this system, are called axioms; and the choice as to which propositions shall be taken as axioms is, to a certain extent, arbitrary. We may regard any proposition as an axiom, so long as we fulfil one condition, which is that all the propositions in the system be derivable from the chosen group of axioms. Thus, the quality of being an axiom is not in any sense a natural, intrinsic attribute or characteristic of a law; the only reason for choosing certain propositions as axioms, are those of their expediency or convenience. In the propositions derived from these axioms, further symbols, other than those used in the axioms, are introduced *by definition*. A definition consists of the introduction of new symbols, or signs, for the purpose of abbreviation. The choice as to which of these signs shall

[4] See Schlick, *Philosophy of Nature*, pp. 71–87 [Tr].

be regarded as fundamental symbols and which as derived from the latter by definition, is likewise arbitrary.

Examples:

$$E = \tfrac{1}{2}mv^2 \qquad\qquad M = mv$$

Definition of Energy   *Definition of Momentum*

But instead of mass and velocity, we can also write:

$$\frac{Energy}{Momentum} \quad : \quad v = \frac{2E}{M}$$

Thus, it is immaterial which magnitudes or quantities occur in the axioms.

Hence, the structure of a theory consists of: 1) axioms; 2) derived propositions and 3) definitions. In the symbolic representation of natural science, whether by means of words or of mathematical symbols, the three structural elements cannot be outwardly distinguished from one another.

The symbolic representation of a theory consists of sentences which in their turn are constituted of certain series of spoken or written signs; the theory itself consists primarily of "propositions." The question as to whether a sentence represents a true proposition or only a definition for example,—depends on the interpretations which explain it and give it its meaning. These do not form part of the symbolic representation itself, but are added to it—that is, they are added to a hypothetico-deductive system—from outside as it were, for example, in the form of ostensive definitions. They constitute the rules of the application of the sentences and are conclusive for the philosophical interpretation of the latter. It is, after all necessary to refer to a reality which is described by the system of signs or symbols since, at some time or another, we must break out from this system.[5] Only those sentences which by virtue of their interpretation, represent genuine propositions, can communicate something about nature; the others are merely internal rules for signs and consequently are definitions. We shall discuss later the confusion of true laws of nature with sentences which merely fulfill the function of definitions.[6]

---

[5] Although this sentence is not included in the manuscript, it was dictated by Schlick in his lectures of 1936. [Tr.]

[6] See the observations concerning "Conventionalism" in *Philosophy of Nature*, p. 48ff; and Schlick's essay: "Sind die Naturgesetze Konventionen?", (Gesammelte Aufsätze, Vienna 1938; also reproduced in "Gesetz, Kausalität und Wahrscheinlichkeit" Vienna 1948). [Tr.]

THEORIES AND PICTORIAL MODELS

The connection between theory and reality was formerly always conceived of as though the symbols occurring in the laws of nature represented simple magnitudes, or quantities, which could either be immediately perceived, or could at least be regarded as being of the same nature as such magnitudes or quantities (e.g. a length of 1/100mm). Thus in Newtonian mechanics, the fundamental concepts represented by lines in space, time, and mass, were three terms of which the meaning seemed to be derived immediately from sensory imagery. All three are combined in the concept of motion, which is equivalent to temporal change in the spatial position of a mass. Motion is that process in which the basic requirement for knowledge appears to be fulfilled in a pictorial manner—namely, as the perception of the constant element in change. That which is moved—the mass—fills the rôle of substance, and remains unchanged in sense-perception. And yet, something does change—namely, the position. The whole process seems to be perfectly clear and visually conceivable; and this is the only reason for the predilection for mechanical explanations, and for the desire of earlier physicists to reduce their science to mechanics. Hence also, the peculiar extension in meaning of the word "mechanism."

In the mechanical explanation of nature, it was necessary to assume the existence of invisible motions, in order that observed processes might be reduced to them, and this was successful in both acoustics and the kinetic theory of heat. But in order to explain electromagnetic phenomena and radiation, the hypothesis of the so-called aether of space, to which at first the same properties were ascribed as to the substances accessible to every-day perception, had to be introduced. Thus the aether was conceived of, alternatively, as a gas, fluid or solid body. It was then discovered, however, that in this way, certain self-contradictory properties would have to be attributed to it; and hence, this very crude type of knowledge based on models, was reduced to an absurdity. Actually, the assumption that the aether must possess the same properties as substances that can be weighed, the qualities of which were to be rendered intelligible with its help, were unjustified (Poincaré). The processes which are supposed to take place in the non-perceptible microcosm, need only fulfil the condition that in virtue of their interaction they give rise to those processes which can actually be observed in the domain of the perceptible. We call the laws governing these invisible processes, small-scale, or micro-laws; and those governing perceptible processes, large-scale, or macro-laws. This distinction must occur in every theory; for in every theory, the observed behaviour of things is ascribed to small-scale laws to which the most general hypotheses of natural science are also referred.

Micro-laws and macro-laws could only be identical by a very improbable chance. There is no a priori justification for such an assumption which is essential for the most primitive type of knowledge based on models.

The atomic theories of Democritus, Boscovitsch and Heinrich Hertz, as well as the vortex ring theory of the atom, are all examples of knowledge based on models . . .[7]

Poincaré proved that for every mechanical model discovered, there are always others which can achieve the same results. It is generally true to say that micro-processes can never be unequivocally deduced from observed large-scale processes—a large number of possibilities always exists. In an advanced stage of scientific development, such as that of physics during the first quarter of the present century, the claim to regard micro-processes as analogous to large-scale processes, was, in principle, abandoned. But so long as one retains the assumption, if only in part, that micro-processes are to be construed in perceptual terms, the method of models has not been entirely discarded. Thus, in Bohr's model of the atom, for example, it is still assumed that there is some sense in speaking of an arbitrary physical event with its spatio-temporal magnitudes arbitrarily enlarged or diminished. It is only in the most recent phase of the development of physics that the extension to the realm of the invisibly small, of spatio-temporal conditions prevailing in the regions of the directly measurable, is no longer regarded as permissible. Accordingly, micro-processes conceived in a visualizable manner and the method of representation by models, have been abandoned.

[7] Schlick, "Naturphilosophie," *op. cit.*, pp. 406–422 [Tr.]

Chapter 24
PERCY WILLIAMS BRIDGMAN
(1882–1961)

INTRODUCTION

*P*ERCY WILLIAMS BRIDGMAN was born in Cambridge, Massachusetts, on April 21, 1882. He studied at Harvard, receiving his Ph.D. there in 1908. After graduation he taught mathematics and physics at Harvard, becoming Hollis Professor of Mathematics and Natural Philosophy in 1926, and Higgins University Professor in 1950.

Bridgman spent the largest part of his scientific life involved with investigations in physics; he specialized in the study of materials at high pressures and their thermodynamic behavior. He was awarded the Nobel Prize in physics in 1946 for his extensive work in this area, especially for his investigations of the properties of matter at pressures up to 100,000 atm. These properties included electrical and thermal conductivity, tensile strength, viscosity, and compressibility in a great variety of compounds.

Bridgman also made a substantial contribution to philosophy of science in his propagation of the operationalist point of view. His view was first made public in 1927 in his important study *The Logic of Modern Physics.* In his later publications, *Reflections of a Physicist,* and especially *The Nature of Physical Theory* (1936), and *The Nature of Some of Our Physical Concepts* (1952), Bridgman had considerably changed his original point of view, retaining however the general ideas of operationalism.

In *The Logic of Modern Physics* Bridgman's major concern was the problem of how the basic physical quantities were to be defined since nonambiguity, clarity, and exactness are absolute requirements of empirical science. That this problem had to be posed again was due to the rapid and profound development of modern physics, especially the theory of relativity and quantum mechanics. Bridgman did not approach this problem from a typically logical viewpoint as, for example, Carnap did before him, but rather began with a critical reflection on the methods actually used in physics. In so doing he came to the conclusion that the meanings of the basic physical concepts are to be defined by the measuring operations performed on the objects of physics, and not, as was usually done after Newton, by the a priori determinations of the properties of those objects (*Logic*, p. 4). In other words, whereas classical physics defined most of its basic concepts by means of a procedure quite similar to that used in mathematical sciences, Bridgman demanded that the physicist define his basic concepts using genuine physical experiences, that is to say, operations of measurement. This is why he concludes: "In general, we mean by any concept nothing more than a set of operations; the concept is synonymous with the corresponding set of operations. If the concept is physical, as of length, the operations are actual physical operations, namely, those by which length is measured . . ." (*Ib.* p. 5). Thus instead of a rationalist approach to the problem concerning the meaning of physical concepts, Bridgman pleads for a purely empiricist attitude in physics. The physicist cannot recognize any "*a priori* principles which determine or limit the possibilities of new experience. Experience is determined only by experience" (*Ib.*, p. 3).

Bridgman maintained that if the set of operations equivalent to any physical concept is a unique set of genuine physical operations, our physical concepts then will be clear and distinct. Furthermore, if one consistently follows the view that the proper definition of a physical concept is not in terms of its properties but in terms of actual operations, then there is no danger of having continually to revise our attitude toward nature: if experience is always described in terms of experience, there must always be correspondence between experience and our description. Finally it follows from this point of view that if the operations to which a physical concept is equivalent are actual physical ones, the concept can be defined only within the range of actual experiment and thus is meaningless in regions as yet untouched by experiment (*Ib.*, pp. 6–7).

Bridgman claimed explicitly that in his view Einstein's work in the field of the theory of relativity and quantum mechanics contributed substantially to the change of our attitude toward the meaning of physical concepts (*Ib.*, pp. 3–9), but he admitted on the other hand that this view was prepared by earlier investigations in the realm of philosophy of science. In this context he explicitly mentions Clifford, Stallo, Mach, and Poincaré (*Ib.*, p. *v*). Perhaps Duhem's work could have been mentioned, also.

Finally Bridgman turns his attention to a few very important consequences of accepting an operational point of view in regard to the definition of basic physical concepts. The first consequence of the operational viewpoint is that all our physical knowledge is relative, which means that all physical concepts will be relative to the operations used in the definition of the concepts. Furthermore it means that concepts such as absolute space, absolute time, absolute motion, and the like lose their meaning as physical concepts. There will be only space, time, motion, and the like relative to the point of view of the one who performed the measuring operation. However, Bridgman continues, the "absolute" which was banished according to its original meaning, may return with a different meaning. For one can still maintain that an object has absolute properties, namely, if the numerical magnitude is the same when measured with the same formal procedure by all observers. But this "absolute" is evidently and paradoxically an absolute relative to experiment. Another extremely important consequence of the operationalist point of view is that it gives us an easy criterion for determining whether or not a question is physically meaningful or meaningless. For in this view a physical question is meaningful if and only if it is possible to find operations by which an answer may be given to it (*Ib.*, pp. 25–31). Here in particular it becomes clear that adopting the operational point of view in physics involves much more than a mere restriction of the meaning which we ascribe to the term "physical concept"; it means a far-reaching change in all our habits of thought, in that we shall no longer permit ourselves to use as tools in our thinking concepts of which we cannot give an adequate account in terms of operations (*Ib.*, p. 31).

In *The Logic of Modern Physics* Bridgman analyzes a great variety of physical concepts from the operationalist point of view. The selection gives his view on the concepts of space, time, and causality.

SELECTIVE BIBLIOGRAPHY

Bridgman, P. W., *The Logic of Modern Physics*. New York, 1927.

———, *Dimensional Analysis*, New Haven, Conn., 1931.

———, *The Nature of Physical Theory*. Princeton, N.J., 1936.

———, *The Nature of Thermodynamics*. Cambridge Mass., 1941.

———, *Reflections of a Physicist*. New York, 1950.

———, *The Nature of Some of Our Physical Concepts*. New York, 1952.

Hempel, C. G., *Aspects of Scientific Explanation and Other Essays in the Philosophy of Science.* New York, 1965. Pp. 123–133.

Frank, P., *Modern Science and Its Philosophy,* New York, 1961. Pp. 13–62 (*passim*), and pp. 277–292.

Lindsay, R. B., "A Critique of Operationalism in Physics," in *Philosophy of Science.* Baltimore, 1937. Pp. 456–470.

Bernstein, J., "P. W. Bridgman in Revolt against Formalism," *Synthèse* (1952), 331–341.

Cornelius, B. A., *Operationalism.* Springfield (Ill.), 1955.

DETAILED CONSIDERATION OF
VARIOUS CONCEPTS OF PHYSICS

WE now begin our detailed consideration of the most important concepts of physics. It is entirely beyond the scope of this essay to attempt more than an indication of some of the most important matters. Neither is it to be expected that the parts of this discussion will always have a very close connection with each other; the purpose of the discussion is to aid in acquiring the greatest possible self-consciousness of the whole structure of physics.

THE CONCEPT OF SPACE

A logically satisfying definition of what we understand by the concept of space is doubtless difficult to give, but we shall not be far from the mark if we think of it as the aggregate of all those concepts which have to do with position. Position means position of something. The position of things is determined by some system of measurement; perhaps the simplest is that implied in a Cartesian coördinate system with its three measurements of length. Hence much of the essential discussion of space has already been given in connection with the concept of length. We have seen that measurements of length are made with physical measuring rods applied to some physical object. We cannot measure the distance between two points in empty space, because if space were empty there would be nothing to identify the position of the ends of the measuring rod when we move it from one position to the next. We see, then, from the point of view of operations that the framework of Cartesian geometry, often imagined in an ideal mathematical sense, is really a physical framework, and that what we mean by spatial properties is nothing but the properties of this framework. When we say that space is Euclidean, we mean that the physical space of meter sticks is Euclidean: it is meaningless to ask whether empty space is Euclidean. Geometry, therefore, in so far as its results are expected to apply to the external physical world, and in as far as it is not a logical system built up from postulates, is an experimental science. This view is now well understood and accepted, but there was a time when it was not accepted, but vigorously attacked; the change of attitude toward this question is symptomatic of a change of attitude toward many other similar questions.

We have already emphasized that the space of astronomy is not a physical space of meter sticks, but is a space of light waves. We may, therefore, have different kinds of space, depending on the fundamental operations. The space of meter sticks we have called "tactual space," and the space of light beams "optical space." If we ask whether astronomical space is Euclidean, we mean merely to ask whether those features of optical space which are within the reach of astronomical measurement are Euclidean. The only possible attitude with respect to this question, or such related questions as whether the total volume of space is finite, or whether space has curvature, is that it is entirely for experiment to decide, and that we have no right to form any preconceived notion whatever. It is therefore beyond the scope of this discussion.

It is interesting to notice that the restricted theory of relativity virtually assumes, although often without making the explicit statement, that tactual and optical space are the same. This equivalence results from the properties assumed for light beams. The distance of a mirror may be found equally well by measuring it with meter sticks, or by determining the time required by a light signal to travel there and back. This situation is, however, logically unsatisfying, because it must be assumed that the operations for measuring time are independently defined, and we shall see that they are not. It is a consequence of the assumed equivalence of tactual and optical space that the path of a beam of light is a straight line, as a straight line is determined by operations with meter sticks. When we come to astronomical phenomena, the physical operations with meter sticks can no longer be carried out, and it is meaningless to ascribe to beams of light on an astronomical scale the same geometrical properties that we do on a small scale.

THE CONCEPT OF TIME

According to our viewpoint, the concept of time is determined by the operations by which it is measured. We have to distinguish two sorts of time; the time of events taking place near each other in space, or local time, and the time of events taking place at considerably separated points in space, or extended time. As we now know, the concept of extended time is inextricably mixed up with that of space. This is not primarily a statement about nature at all, and might have been made simply by the observation that the operations by which extended time is measured involve those for measuring space. Of course historically the doctrine of relativity was responsible for the critical attitude which led to an examination of the operations of measuring time, but relativity was not necessary for a realization of the spatial implications of time, any more than the discovery of Planck's quantum unit h was necessary for the invention by Planck of his absolute units of measurement, although historically he was inspired to make this invention by discovering h, and in his own mind seems to have thought of the connection as a necessary one.[1]

The physical operations at the basis of the measurement of time have never been subjected to the critical examination which seems to be required. One method of measurement, for instance, involves the properties of light.

A meter stick is set up with mirrors at the two ends, and a light beam travels back and forth between the two mirrors without absorption. The time required for a single passage back and forth is defined as the unit of time, and time is measured simply by counting these intervals. But such a procedure is unsatisfactory if we are to permit ourselves all those operations which are demanded by even the simplest postulate of relativity, for we must be able to move our clock from place to place, transfer it from one system to another in relative motion, and with it determine the properties of light beams in the stationary or moving system. We recognize in principle that the length of the meter stick may be different when it is in motion, that it may change also during the acceleration incident to moving it from one place to another, and that until proved to the contrary the velocity of light may be a function of velocity or acceleration. The complicated interplay of all these possibilities leaves us in much doubt as to the physical significance of such postulates as, for example, that the velocity of

---

[1] Max Planck, *The Theory of Heat Radiation,* translated by Masius, P. Blakiston's Son & Co., 1914 edition, p. 174.

light is the same in the moving system and the stationary system. In order to ascribe any simple significance to postulates about the velocity of light, it would seem that we must have an instrument for measuring this velocity, and therefore for measuring time, which does not itself involve the properties of light. To do this we might seek to specify the measurement of time in purely mechanical terms, as for instance in terms of the vibration of a tuning fork, or the rotation of a flywheel. But here again we encounter great difficulties, because we recognize that the dimensions of our mechanical clock may change when it is set in motion, and that the mass of its parts may also change. We want to use the clock as a physical instrument in determining the laws of mechanics, which of course are not determined until we can measure time, and we find that the laws of mechanics enter into the operation of the clock.

The dilemma which confronts us here is not an impossible one, and is in fact of the same nature as that which confronted the first physicist who had to discover simultaneously the approximate laws of mechanics and geometry with a string which stretched when he pulled it. We must first guess at what the laws are approximately, then design an experiment so that, in accordance with this guess, the effect of motion on some phenomenon is much greater than the expected effect on the clock, then from measurements with uncorrected clock time find an approximate expression for the effect of motion on mass or length, with which we correct the clock, and so on ad infinitum. However, so far as I know, the possibility of such a procedure has not been analyzed, and until the analysis is given, our complacency is troubled by a real disquietude, the intensity of which depends on the natural skepticism of our temperament.

In practice, the difficulties of such a logical treatment are so great that the matter has been entirely glossed over. It is convenient to postulate a clock, of unknown construction, but such that the velocity of light, when measured in terms of it, has certain properties. Such, for example, is the point of view in Birkhoff's recent book.[2] The difficulty with this method is that the resulting edifice is as divorced from physical reality as is the logical geometry of postulates. We cannot be at all sure that the properties of light as measured with our physical clocks are the same as the theoretical properties. The difficulty is particularly important and fundamental in the general theory of relativity; the basis of the whole theory is the infinitesimal interval ds,

[2] G. D. Birkhoff. *Relativity and Modern Physics*, Harvard University Press, 1923.

which is supposed to be given. Once given, the mathematics follows. But in a physical world, *ds* is *not* given, but must be found by physical operations, and these operations involve measurements of length and of time with clocks whose construction is not specified. In any actual physical application the question must be answered whether the physical instrument used in measuring the temporal part of *ds* is really a clock or not. There is at present no criterion by which this question can be answered. If the vibrating atom is a clock, then the light of the sun is shifted toward the infra-red, but how do we know that the atom is a clock (some say yes, others no)? If we find the displacement physically have we thereby proved that general relativity is physically true, or have we proved that the atom is a clock, or have we merely proved that there is a particular kind of connection between the atom and the rest of nature, leaving the possibility open that neither is the atom a clock nor general relativity true? In practice, of course, we shall adopt the solution which is simplest and most satisfying aesthetically, and doubtless shall say that the atom is a clock and relativity true. But if we adopt this simple view, we must also cultivate the abiding consciousness that at some time in the future troubles may have their origin here.

It seems to me that the logical position of general relativity theory is merely this: Given any physical system, then it is possible to assign values to *ds* such that relations mathematically deduced by the principle of relativity correspond to relations between measurable quantities in the physical system; but that the things that we physically call *ds* are anything more than approximately connected with the *ds*'s required to give the mathematical relations, is at present no more than a pious faith.

To return to the concept of time, we have already stated that there are two main problems, that of measuring time at a single point of space, and that of spreading a time system over all space. The second aspect of the problem is that to which attention has been directed by relativity theory; the following detailed examination shows how the operations of relativity for setting and synchronizing clocks at distant places involve the measurement of space. It is a fundamental postulate that the adjustment of the clocks is to be accomplished by light signals. The synchronization of the clocks is now simple enough. We merely demand that light signals sent from the master clock at intervals of one second arrive at any distant clock at intervals of one second as measured by it, and we change the rate of the distant clock until it measures these intervals as one second. After its rate has been adjusted, the distant clock is to be so *set* that when a light signal is despatched from the master clock at its indicated zero of time the time of arrival recorded at the distant clock shall be such that the distance of the clock from the master clock divided by the time of arrival shall give the velocity of light, assumed already known. This operation involves a measurement of the distance of the distant clock, so that in spreading the time coördinates over space the measurement of space is involved by definition, and the measurement of time is, therefore, not a self-contained thing. This is the physical basis for the treatment of space and time as a four dimensional manifold. Although mathematically the numbers measuring space and time enter the formulas symmetrically, nevertheless the physical operations by which these numbers are obtained are entirely distinct and never fuse, and I believe it can lead only to confusion to see in the possibility of a four dimensional treatment anything more than a purely formal matter.

The notion of extended time, therefore, involves the measurement of space. It is an interesting question whether the notion of local time also involves the measurement of space. A rigorous answer to this question involves giving the specifications for the construction of a clock, which we have seen has not yet been done. It seems to me probable, however, that the construction of even a single local clock involves in some way the *measurement* of space. If, for example, we use a vibrating tuning fork, we must find how the time of vibration depends on the amplitude of vibration, and this involves space measurement, or if we use a rotating flywheel, we have to correct for the change of moment of inertia due to the change of dimensions when it is set into motion or brought into a gravitational field, and all this involves space measurement. However, these considerations are not certain, and perhaps the question is not important.

There is now the further consideration that actually in practice the concept of local time is not entirely divorced from that of extended time, for two bodies cannot occupy the same space at the same time, and the time of any event is actually measured on an instrument at some distance, communication being maintained by light or elastic signals. But experience convinces us that in the limit, as the phenomenon to be measured gets closer to the clock, there is no measurable difference, whether communication with the clock is maintained by light, or acoustical or tactual signals, so that we have come in physical practice to accept measurement of the time of events in the immediate neighborhood of the clock (local time) as one of the ultimately simple things behind which we do not attempt to go.

Local time is, therefore, a concept treated by the physicist even now as simple and unanalyzable. This concept is what most people have in mind when they think of time. Time, according to this concept, is something with the properties of local time; it was something of this kind that Newton must have meant by his absolute time, and it is the tacit retention of this sort of concept that is responsible for the difficulty so often found in grasping the idea of the relativity of simultaneity, which is of course entirely foreign to our experience of simultaneity in local time. An examination of the operations involved in extending time has shown how the concept of extended time is different from that of simple local time; this difference leads to appreciably different numerical relations when we are dealing with high velocities or great distances. Local time is proved by experience not to be a satisfactory concept for dealing with events separated by great distances in space or with phenomena involving high velocities. For instance, we must not talk about the age of a beam of light, although the concept of age is one of the simplest derivatives of the concept of local time. Neither must we allow ourselves to think of events taking place in Arcturus *now* with all the connotations attached to events taking place *here* now. It is difficult to inhibit this habit of thought, but we must learn to do it. The naïve feeling is very strong that it does *mean* something to talk about the entire present state of the universe independent of the process by which news of the condition of distant parts is determined by us. I believe that an examination of this feeling will show that it is psychological in character; what we mean by the totality of the present is merely the entire present content of our consciousness. This is apparently a simple direct thing; we do not appreciate until we make further analysis that our present consciousness of the existence of the moon or a star is due to light signals, and that therefore the apparently simple immediate consciousness of events distant in space involves complicated physical operations.

Similarly, if we continue to use local time, we get into trouble, when we go to high velocities, with our simple concept of velocity, which may be defined in terms of a combination of space and time concepts. The concept of local time thus loses its value and becomes merely a blunted tool when we try to carry it out of its original range. But the concept of extended time, with which we have to replace local time, is a complicated thing, to which we have not yet got ourselves accustomed; it may perhaps prove to be so complicated as never to be a very useful intuitive tool of thought.

All these considerations about time have been concerned only with intervals of such an order of magnitude that they are readily experienced by any individual. If we have to deal with intervals either very long or very short, it is obvious that our entire procedure changes, and consequently the concept changes. In extending the time concept to eras remote in the past, for example, we try as always, to choose the new operations so as to piece on continuously with those of ordinary experience. A precise analysis of the change in the concept of time when applied to the remote past does not seem to be of great significance for our present physical purpose, and will not be attempted here. It is perhaps worth while to point out, however, that all our other concepts, as well as that of time, must be modified when applied to the remote past; an example is the concept of truth. It is amusing to try to discover what is the precise meaning in terms of operations of a statement like this: "It is true that Darius the Mede arose at 6:30 on the morning of his thirtieth birthday."

Of more concern for our physical purposes is the modification which the time concept undergoes when applied to very short intervals. What is the meaning, for example, in saying that an electron when colliding with a certain atom is brought to rest in 10^{-18} seconds? Here I believe the situation is very similar to that with regard to short lengths. The nature of the physical operations changes entirely, and as before, comes to contain operations of an electrical and optical character. The immediate significance of 10^{-18} is that of a number, which when substituted into the equations of optics, produces agreement with observed facts. Thus short intervals of time acquire meaning only in connection with the equations of electrodynamics, whose validity is doubtful and which can be tested only in terms of the space and time coördinates which enter them. Here is the same vicious circle that we found before. Once again we find that concepts fuse together on the limit of the experimentally attainable.

This discussion of the concept of time will doubtless be felt by some to be superficial in that it makes no mention of the *properties* of the physical time to which the concept is designed to apply. For instance, we do not discuss the one dimensional flow of time, or the irrevocability of the past. Such a discussion, however, is beyond our present purpose, and would take us deeper than I feel competent to go, and perhaps beyond the verge of meaning itself. Our discussion here is from the point of view of operations: we assume the operations to be given, and do not attempt to ask why precisely these operations were chosen, or whether others might not be more suitable. Such properties of time as its irrevocability are implicitly contained in the operations themselves, and the physical essence of time is buried in that long physical experience that taught us what operations are adapted to describing and correlating nature. We may digress, however, to consider one question. It is quite common to talk about a reversal of the direction of flow of time. Particularly, for example, in discussing the equations of mechanics, it is shown that if the direction of flow of time is reversed, the whole history of the system is retraced. The statement is sometimes added that such a reversal is actually impossible, because it is one of the properties of physical time to flow always forward. If this last statement is subjected to an operational analysis, I believe that it will be found not to be a statement about nature at all, but merely a statement about operations. It is *meaningless* to talk about time moving backward: by definition, *forward* is the direction in which time flows.

THE CAUSALITY CONCEPT

The causality concept is unquestionably one of the most fundamental, perhaps as fundamental as that of space and time, and therefore at least equally entitled to a first place in the discussion. But as ordinarily understood, there are certain spatial and temporal implications in the causality concept, so that it can best be discussed in this order after our examination of space and time.

There is an aspect of the causality concept that in many respects is closely related to the question of "explanation," for to find the causes of an event usually involves at the same time finding its explanation. But there are nevertheless sufficient differences to warrant a separate discussion.

It seems fairly evident that there was originally in the causality concept an animistic element much like that in the concept of force to be discussed later. The physical essence of the concept as we now have it, freed as much as possible from the animistic element, seems to be somewhat as follows. We assume in the first place an isolated system on which we can perform unlimited identical experiments, that is, the system may be started over again from a definite initial condition as often as desired.[3] We assume further that when so

___

[3] We must include in general in the concept of "initial condition" the past history of the system. In order not to make this condition so broad as to defeat itself, we have to add the observation that actually identity of past history is necessary

started, the system always runs through exactly the same sequence of events in all its parts. This contains the assumption that the course of events runs independent of the absolute time at which they occur—there is no change with time of the properties of the universe.[4] It is a result of experience that systems with these properties actually exist. An alternative way of stating our fundamental hypothesis is that two or more isolated similar systems started from the same initial condition run through the same future course of events. Upon the system given in this way, which by itself runs a definite course of events, we assume that we can superpose from the outside certain changes, which have no connection with the previous history of the system, and are completely arbitrary. Now of course in nature, as we observe it, there is no such thing as an arbitrary change, without connection with past history, so that strictly our assumption is a pure fiction. It is here that the animistic element still seems to persist, although perhaps not necessarily. We regard our acts as not determined by the external world, so that changes produced in the external world by acts of our wills are, to a certain degree of approximation, arbitrary. The system, then, on which we are experimenting, is one capable of isolation from us in that we may regard ourselves as outside the system, and having no connection with it. The system, furthermore, is capable of isolation from the rest of the physical universe, in that events taking place outside the system have no connection with those taking place inside.[5] Experience gives the justification for assuming that physical isolation of this sort is possible. Actually, of course,

---

over only a comparatively short interval of time. Logical precision seems unattainable here—the physical concepts themselves have not the necessary precision.

[4] As so often in physics, we appear to be doing two things at once here. It is doubtful whether we can give a meaning to "definite initial condition" apart from the future behavior of the system, so that we have no real right to infer from uniform future behavior both a constancy of the laws of nature, independent of time, and a constancy of initial condition. I very much question whether a thoroughgoing operational analysis would show that there are really two independent concepts here, and whether the use of two formally quite different concepts is anything more than a convenience in expression. It seems to me that it may be just as meaningless to ask whether the laws of nature are independent of time as it was to ask Clifford whether the absolute scale of magnitude may not be changing as the solar system travels through space.

[5] Here again, the concept of "isolation" or "connection" is defined only in terms of the behavior of the system, and it is not clear whether this is really an operationally independent concept or not.

isolation is never complete, but only partial, up to presumably any desired degree of approximation.

The statement that two exactly similar isolated systems, starting from the same initial conditions (including past history in the general idea of initial condition) will run through the same future course of events involves as a corollary that if differences develop in the behavior of two such apparently similar systems these differences are evidence of other previous differences. The thesis that this corresponds to experience may be called the thesis of essential connectivity and is perhaps the broadest we have: it is the thesis that differences between the behavior of systems do not occur isolated but are associated with other differences. It is essentially the same thesis as that already mentioned in connection with "explanation," namely that it is possible to correlate any of the phenomena of nature with other phenomena.

If now the connectivity or correlation between phenomena is of a special kind, we have a causal connection; namely, if whenever we arbitrarily impress event A on a system we find that event B always occurs, whereas if we had not impressed A, B would not have occurred, then we say that A is the cause of B, and B the effect of A. By suitably choosing the event A, we may find the effect of any event of which the system is susceptible.

The relation between A and B is an unsymmetrical one, by the very nature of the definition, the cause being the arbitrary variable element, and the effect that which accompanies it. Furthermore, A may obviously be the cause of more than one event B, and may cause a whole train of events.

The causal concept analyzed in this way is not simple by any means. We do not have a simple event A causally connected with a simple event B, but the whole background of the system in which the events occur is included in the concept, and is a vital part of it. If the system, including its past history, were different, the nature of the relation between A and B might change entirely. The causality concept is therefore a relative one, in that it involves the whole system in which the events take place.

In practice we now take an exceedingly pregnant step and seek to extend the concept, and rid ourselves as much as possible of its relativity. It is a matter of experience that there are often a great number of systems in which A is the cause of B. In many cases the causal relation persists through such a very wide range of systems that we lose sight entirely of the system, and come to assume that we have an *absolute* causal connection between A and B. For instance, when

I strike a bell, and hear the sound, the causal connection persists through such a great number of different kinds of system that I might think that here is an absolute causal connection. Such an absolute causal connection would mean that always under all circumstances, the striking of the bell is accompanied by a sound. But *all* conditions means only *all* those conditions covered by experiment. Thus in the case of the bell, all our experiments were made in the presence of the atmosphere. The causal connection between the striking of the bell and the sound should have been always recognized in principle as relative to the presence of the atmosphere. Indeed, later experiments in the absence of the atmosphere show that the atmosphere does play an essential part. Now as a matter of fact, the atmosphere is so comparatively easy to remove that we very readily include the atmosphere in the chain of causal connection. But if the atmosphere had been impossible to remove, like the old ether of space, our idea of the causal connections between the striking of the bell and its sound might have been quite different. In actual physical applications of the causality concept, the constant background which is maintained during all the variations by which the causal connection is established usually has to be inferred from the context.

It is a matter of perhaps universal experience that the event A is accompanied by not only one event, which is the effect of A by definition, but A entails a whole causal train of events. It seems to be a generalization from experience that the causally connected train of events started by A is a never ending train, provided the system is large enough. This is perhaps not necessary in the general case, but if the event A involves imparting external energy to the system, or the action of external force (momentum change), there can be no question.

That there is a causal train started by A is particularly evident if A and B are separated in space. Thus in the case of the bell, the impulse given to the air by the vibration of the bell is propagated through the air as an elastic wave, which thus constitutes the causal train of events. The phenomenon of propagation is characteristic of causal connections of a mechanical character, and is the justification for the introduction of the time concept in connection with the causality concept, where it now appears for the first time. It is evident that when a disturbance is propagated to a distant point, the effect *follows* the cause in time, as time is usually measured.

We extend this result, and usually think that the effect *necessarily* follows the cause. We now examine whether this is a necessary result of the causality concept. If we are to talk about the time of events at different places, we must have some way of setting clocks all over space. If this is done arbitrarily, there is no necessary connection between the local clock times of a cause and its effect, but nevertheless the causality concept involves a certain temporal relation even in this most general case. Suppose that event A takes place at point 1 and its effect, event B, at point 2. We station a confederate at 2 who sends a light signal (or any other sort of signal) to 1, as soon as the event B occurs at 2. Then it is a consequence of the nature of the causality concept that the signal cannot arrive at 1 before event A occurs. For if it did arrive before A, we should merely omit to perform A, which by hypothesis is arbitrary, and entirely in our control, and then our assumption would be violated that the system is such that the event B occurs only when A also occurs. The same argument shows *a fortiori* that if the effect B occurs at the same place as its cause A, it cannot precede it in time. I ·cannot see that the nature of the causality concept imposes any further restriction on the time of B. The restricted principle of relativity, however, in postulating that no signal can be propagated faster than a light signal, virtually makes a further assumption about the temporal connection of causally connected events, namely, that the event B at 2 cannot occur before the arrival at 2 of a light signal which started from 1 at the instant that A occurred at 1. For if B did occur earlier, we could use events A and B as a signaling code, thus violating our hypothesis.

There is thus a closest connection in time, when time is extended over space as the theory of relativity directs, between cause and effect, depending on their separation in space; from this arises the relativity concept of the causal cone, which in the four dimensional manifold of space-time divides the aggregate of all those events which may be causally related from the aggregate of those which are separated by such a small interval in time and such a large interval in space that communication by light signals and therefore a causal connection is not possible. Given now two events A and B which are related as cause to effect in one system of reference, then they must be causally related also in every other system of reference. For if they were not, we could by definition of causality suppress the event A in one of the systems in which the causal relation does not hold, and this, because of the nature of the concept of event, involves suppressing A in all the systems, thus violating our hypothesis of a causal connection in the original system. The concept of event involved in this argument will be examined later. It appears then, that the fundamental postulate of relativity (that the form of natural laws is the same in all reference systems) demands that the temporal order of events causally connected be the same in all reference systems.

The whole universe at this present moment is often supposed to be causally connected with all succeeding states. This means that if we could repeat experience, starting from the same initial conditions, the future course of events would always be found to be the same. The truth of this conviction can never be tested by direct experiment, but it is something at which we arrive by the usual physical process of successive approximation. It is difficult to formulate precisely what we mean by "present" state of the universe, and there is every reason to think that such a formulation is not unique, but the concept contains the necessary implication that none of the events constituting the "present" can be causally connected. The events in distant places which constitute the present must be separated by an interval of time less than time required by light to travel between the two places.

The conviction, arising from experience, that the future is determined by the present and correspondingly the present by the past, is often phrased differently by saying that the present causally determines the future. This is in a certain sense a generalization of the causality concept. It is one of the principal jobs of physics to analyze this complex causal connection into components, representing as far as possible the future state of the system as the sum of independent trains of events started by each individual event of the present. How far such an analysis is possible must be decided by experiment. It is certainly possible to a very large extent in most cases, but there seems to be no reason to expect that a complete analysis is possible. So far as the system is describable in terms of linear differential equations, the causal trains started by different events propagate themselves in space and time without interference and with simple addition of effects, and conversely the present may be analyzed back into the simple sum of elementary events in the past, but if the equations governing the motion of the system are not linear, effects are not additive, and such a causal analysis into elements is not possible. No emphasis is to be laid here on the *differential* aspect of the equations: it is quite possible that finite difference equations may have the same property of additivity. Although there can be no question that linear equations enormously preponderate, neither can there be any doubt that some phenomena cannot be described in terms of linear equations (*e.g.*, ferro-magnetism), so that there seems no reason to think that a causal analysis is always possible. I believe, however, that the assumption that such an analysis into small scale elements is possible is tacitly made in the thought of many physicists. If the analysis is not possible, we may expect to find results following the coöperation of several events which cannot be built up from the results of the events occurring individually.

When a causal analysis is possible, finding the simplest events which act as the origin of independent causal trains is equivalent to finding the ultimate elements in a scheme of explanation, so that here we merge with the concept of explanation, as already mentioned. As was true of the explanatory sequence, so here there can be no *formal* end of the causal sequence, because we can always ask for the cause of the last member. But it may be physically meaningless to extend the causal sequence beyond a certain point. We have seen from the point of view of operations that the causal concept demands the possibility of variation in the system. It is therefore meaningless to say that A is the cause of B unless we can experience systems in which A does not occur. Now if in extending the causal sequence, we eventually arrive at a condition so broad that physically no further variation can be made, our causal sequence has to stop.

Corresponding to this property of the causality concept, the causal sequence may be terminated either formally, by postulate, or naturally, by the intrinsic physical nature of the elements of the sequence. Thus if we say that light gets from point to point because it is propagated by a medium of unalterable properties, which fills all space, which is always present and can never be eliminated physically, we have by the postulated properties of the medium brought the possibility of further inquiry to a close, because to take the next step and ask the cause of the properties of the ether, demands that we be able to perform experiments with the ether altered or absent. Such an ending of the sequence is evidently pure formalism, without physical significance. But other considerations may give physical significance. Thus if there are other sorts of experiment that can be explained by assuming a universal medium of the same properties, the concept proves not only to be useful, but to have a certain degree of physical significance. An example of an inevitable termination of the causal sequence is afforded by the possibility, already mentioned, that the value of the gravitational constant may be determined by the total quantity of matter in the universe. Without further qualification, this is an entirely sterile statement, but if it can be shown that there is a simple numerical connection, the matter takes on interest, and we may seek further for a correlation between the numerical relation and other things.

This analysis of the causality concept does not pretend to be complete and leaves many interesting questions untouched. Perhaps one of the most interesting of these questions is whether we can separate into cause and effect two phenomena which *always* accompany each other, and whether therefore the classification of phenomena into causally connected groups is an exhaustive classification. But the discussion is broad enough for our purpose here; the most important points of view to acquire are that the causality concept is relative to the whole background of the system which contains the causally connected events, and that we must assume the possibility of an unlimited number of identical experiments, so that the causality concept applies only to sub-groups of events separated out from the aggregate of all events.

INDEXES

Name Index

Index of Subjects

A

Absolute elasticity (of small particles), 163–64
Absolute space, 247–248
  alternative suggestions, 248, 256
  criticism of, 182–183, 248, 279–280
  reason for introducing this concept, 248; *see also* space
Absolute time, 248–249
  criticism of, 182–183; *see also* time
Abstraction, 299–306
  and classification, 349–350
  and formation of concepts, 349–350
Analogy (between nature and our understanding of it), 342–343, 344–345; *see also* model
Analytic geometry, 114, 120–125
Appropriateness (of theories), 225
  of Hertz's mechanical theory, 239–241
Atom
  as economic shorthand conception of science, 200–202
  not real, 200
Atomism, 149, 340–341, 345, and *passim*
Axioms
  and definitions, 62
  axioms and definitions taken together express our fundamental ideas incompletely, 63
  combination of axioms must take place according to the rules of deductive logic, 60–61
  express the necessary conditions which flow from the funda-

mental ideas of each science, 59–60, 62–63
  of geometry
    aprioric character of, 111–12, 111n
    are not propositions of the pure doctrine of space, 130–131
    contain a not necessary presupposition, 114
    do not depend upon the native form of our perceptive faculty, 130
    Euclid's method of constructive intuition, 113–114
    founded in experience, 109
    have real import only if they are connected with mechanical principles, 132
    imply mechanical considerations, 130–132
    meaning of, 112–113
    origin of, 125–132
    significance of, 111–132

C

Categories
  deduction of, 23n
Causal cone, 477
Causality
  and explanation, 473, 479
  aprioric character of, 179
  operational definition of, 473–480
  principle of, 334–336, 346–347; *see also* cause
Causation, law of, 94–97, 144
  and induction, 97
  and probability, 190
  evidence of this law, 98–103

Causation, law of (*cont.*)
the law itself an instance of induction, 99–100
Cause
as the group of positive and negative conditions which with more or less probability precede an event, 135
axioms related to the idea of cause, 70–74
contains an element of necessity which cannot be given by experience, 69
definition of, 36–37, 96–97, 139–141; and *passim*
fallacious use of the term, 137–139
idea a priori, 66
its notion as a condition of all phenomena, 100
measured by their effects, 71–74
not derived from experience, 66–67
not merely a constant succession of events, 68–69
not the invariable and necessary condition of an event, 135
relation of cause and effect, 136–137
some abstract quality or power by which change is produced, 69–70
Chemistry
not a science but a systematic art, 18, 20; and *passim*
Classification
and identification, 350
does not presuppose abstraction, but precedes it, 349–350
Cogredience, of events, 418–419; *see* event
Colligation of facts, 74–75
Combinational logic, 133–134
Concealed mass, 227–229, 240
Concealed motion, 227–229
Concept (of science)
abstract symbol, 300, 356–357
and facts, 360
founded on naïve common sense as modified by philosophy, 414
functional concept, 350
functional concepts in natural science, 352–364
not formed by abstraction, 349–350

operationalist definition of, 462–463
physical concepts of series, 362–364
scientific concepts are functional concepts, 350
serial construction of concepts, 360–362
their meaning depends upon the theory in which they appear, 300–301
see also fundamental concepts *and* fundamental ideas
Conceptions (of science)
do not refer to real things, 192–193
economical shorthand expressions which enable us to classify, describe, and predict phenomena, 192–193, 196–197, 203–205
transcend the realm of perceptions, 191–192
Concreteness (of an element of nature), 418
Congruence, principle of
and geometrical measurement, 126–130
Connectivity
the thesis of essential connectivity, 475–476; *see also* causality
Continuity, 194–195, 202–203
Conventionalism, 171, 264–265
Correctness (of theories), 225
and Hertz's theory of mechanics, 237–239
only one theory can be correct, 241–242
Creative imagination (in science), 440–443, 445–446
Critique of science
as trend in philosophy of science, 206, 328

D

Deduction
a form of apparent reasoning, 81
and induction, 61
in geometry and empirical science, 59–63
its meaning, 82
Definitions (in demonstrative sciences), 60–63
depend upon the fundamental ideas of each science, 62–63

Description, and explanation, 453–
456
Determinism, criticism of, 206–208,
218–222, 267–273
Discontinuity
and perception, 199
conceptual discontinuity, 199–202
Dogmatism, criticism of, 215–217
Duration (as ultimate fact of experience), 416
Dynamics
as part of the metaphysics of nature, 25
head of all natural sciences, 42
its limits are determined by that of
pure mathematics only, 42

E

Economy of thought (in science),
171, 174–187, 192–193, 204–
205, 338
in general, 171–177
in mathematics, 177–178
in physics, 178–184
in psychology, 184–187
not the object of science, 428, 437–
440
Energetism, 315–316, 318–327
Energy
as ultimate substance, 322
definition of, 322
law of the conservation of, 322–
323
problematic character of this concept, 226, 227, 228
Empirical Psychology
mathematics is inapplicable here,
20–21
not a natural science proper, 20–
21
not a systematic art of analysis or
experimental doctrine, 20–21
only possible as historical doctrine
of the internal sense, 21
Empiricism, 189, 203–205; and *passim*
criticism of, 360; and *passim*
Essence, 17n
and nature, 17, 17n
Ether, 12–15, 158–160
and action at a distance, 15
a theoretical conception, not a
real thing, 202–203, 203n; and
passim

Evidence, definition of, 99
Experience, 56–59
and experiment, 74–75
and observation, 57
source of all knowledge, 31–35
unable to prove a proposition to be
necessarily and universally
true, 57–59
Experiment, 32
as active observation, 32
Experimental physics, subject matter
of, 229
Explanation
and description, 453–456
and hypotheses, 152
and theories, 432
definition of, 152, 352–354
Extension (of events), 416–418
Event
and cogredience, 418–419
and extension, 416–418
and object, 419–420
and sense-object, 420–421
definition of, 404
occupied and unoccupied event,
422
part of a duration, 416
primary type of physical facts, 416
Event-particle, 417

F

Fact
and concept, 360
and theory, 356–357
Force
as characteristic element in mechanical laws, 210–212, 217
genesis of the idea, 36–37
its importance in natural science,
37
problematic character of this concept, 224, 225–226, 227, 228
Formalization (in science), 299–306
Foundation of science, task of, 421
Fundamental concepts (of mechanics), 224–226, 227–229
Fundamental ideas (of science), 52–
66
exposition and discussion of these
ideas is the task of philosophy
of science, 66
not derived from experience, 63–65
their aprioric character, 64–65